Water-in-Plants Bibliography
volume 10 1984

References no. 12912–14765 / ABD–ZUR

Editors J. Pospíšilová and J. Solárová

1986 **DR W. JUNK PUBLISHERS**
a member of the KLUWER ACADEMIC PUBLISHERS GROUP
DORDRECHT / BOSTON / LANCASTER

Contributors

J. Solárová
J. Pospíšilová
J. Čatský
Z. Šesták
I. Tichá
D. Hodáňová

ISBN-13:978-90-6193-576-6   e-ISBN-13:978-94-009-4816-7
DOI:10.1007/978-94-009-4816-7

PREFACE

The tenth volume of Water-in-Plants Bibliography includes papers in all fields of plant water relations research which appeared during the year 1984 - from theoretical considerations about the state of water in cells and its membrane transport to drought resistance of plants or physiological significance of irrigation. In addition to papers devoted entirely to plant water relations, papers on other topics are included if they contain data on plant hydration level, water vapour efflux, rate of water uptake or water transport, etc., or if they contain valuable methodological information (measurement of selected microclimatic factors, soil moisture etc.).

We have tried to cover fully the relevant papers which have been published in important scientific periodicals and books. Articles appeared in local journals, mimeographed booklets, abstracts of thesis and of symposia contributions, etc., were chosen mostly from reprints received directly from authors. The courtesy of those is highly appreciated. The manuscript is usually prepared in May and June of the year following the year which it covers. Unfortunately some reprints come later and thus the respective references appear in the following volume, with one year delay.

To maximize the value of the bibliography the references are arranged alphabetically according to the authors' names, and each volume is provided with three indexes. The authors' index contains all names of authors, co-authors and editors. Plant genera used as experimental material are indexed according to their Latin names. The subject index covers primary items chosen according to the interest of water relations researchers. Its preparation was based not only on the titles, key words and abstracts but also on the whole content of the article. By combining two or more items a more detailed information may be obtained.

This volume is accompanied with cumulative indexes to volumes 6 - 10.

Since more than 1500 relevant papers dealing with plant water relations and relative topics are published every year and included in this bibliography, and since all citations have been checked with the originals, collecting and preparing for publication such a large amount of material would have been impossible without the collaboration of our colleagues from the Department of Physiology of Photosynthesis and Water Relations of the Institute of Experimental Botany of the Czechoslovak Academy of Sciences in Prague. We have also acknowledge with thanks the cooperation of Mrs. Ludmila Hávová and Mrs. Jindřiška Srbová who helped in typing card material and the librarians of our Institute Mrs. Zora Zawoyská and Mrs. Taťana Borská who helped us with checking the references.

Dr. Jana Pospíšilová and Ing. Jarmila Solárová

Institute of Experimental Botany
Czechoslovak Academy of Sciences

Flemingovo nám. 2
160 00 PRAHA 6
Czechoslovakia

Praha, 30 October 1985

INSTRUCTIONS FOR USE

All references are arranged alphabetically according to the authors' names. They are numbered and these numbers are used in the indexes. An asterisk preceding the number denotes the reference published in the preceding period (1975 - 1981).

Authors' names are presented in the spelling used in the original paper. If this spelling does not correspond to the spelling usually used by the author (e.g. Russian papers of English authors), one spelling is referred to the other in the Authors' index. Like the transcriptions they are alphabetically arranged mostly according to the authors' own references. Nevertheless, the editors apologize for some misinterpretations which are partly corrected by the cross-indexing in the Authors' index.

The references contain the original unshortened title of the paper (book). English, French, and German titles are cited in the preceding language. Titles in other languages are supplemented with a translation in English (using the title of the respective English abstract, if it is presented). Titles of Japanese, Chinese etc. papers are given in English translation only. In both these cases the abbreviations of the original language and the language of the abstract are given in brackets at the end of the reference. The following abbreviations are used most frequently:

Belorussian	**Japan**ese
Bulgarian	**Latv**ian
Chinene	**Lithua**nian
Croatian	**Norweg**ian
Danish	**Pol**ish
English	**Russ**ian
Esthonian	**Rouma**nian
French	**Slov**ak
German	**Span**ish
Georgian	**Swed**ish
Hungarian	**Ukra**inian
Italian	**Uzb**eg

The transliteration of Cyrillic characters is in accordance with the BSI-ASA/SC-Z39 draft table, i.e.:

a	а		p	п
b	б		r	р
ch	ч		s	с
d	д		sh	ш
e	е		shch	щ
ê	э		t	т
f	ф		ts	ц
g	г		u	у
i	и		v	в
ï	й		y	ы
k	к		ya	я
kh	х		yu	ю
l	л		z	з
m	м		zh	ж
n	н		"	ъ
o	о		'	ь

Several exceptions apply for Ukrainian and Belorussian:

Ukrainian:	y	и
	i	i
	ï	ï
Belorussian:	ŭ	ў

The journals' names are abbreviated mainly according to the Style Manual for Biological Journals (Second Edition, American Institute of Biological Sciences, Washington, D.C. 1964), e.g.:

Abhandlungen
Abstract
Abteilung
Academy
Acker
Acta
Advances
Africa (-ican)
agricultural
Agriculture
Agrobiology (-ogiya)
Agrobotanica
Agrokémia
Agronomy
agropecuaria
Akademie (-emiya)
Algology
allgemeine
Amélioration
America
American
Anais (-alele)
Analysis
analytical
Anatomy
angewante
animal
Annales (-als)
annual
anorganic (-anisch)
applied
aquatic
Arbeit
Archiv
Argentina
Association
Atmosphere
atmospheric
atomic
Australia (-ralian)
Azerbaĭdzhanskaya
Bacteriology
Beiheft
Beiträge
Belgique
Belorusskaya
Berichte
biochemical
Biochemie
Biochemistry
biochimica
biokhimicheskiĭ
Biokhimiya
Bioklimatologie
Biologia (-ogy)
biological (-ogisk)
biophysical
Biophysics
Bodenkunde
Boletin (-ettino)
Bolgarskiĭ
botanica (-anicorum)
botanical (-anisca)
Botanika (-any)

Brasileira
Brazil
Breeding
British
Bulletin (-etins)
Byulleten
California
Canada (-adian)
cellular (-ulaire)
Center
central
Centralblatt
Československý
chemical
Chemistry
chimicus
Chinese
Chromatography
Chronicle
Ciencia
cientificas
College
Commision
Communication
comparative
Comptes Rendus
Conference
Congress
Conservation
Contamination
Contribution
Control
Croatica
cultural
Culture
current
Cytobiology
Cytochemistry
Cytology
Czechoslovak
Danske
dendrological
Dendrology
Department
Deutsche (-schland)
Development
Disease
Dissertation
Division
Doklady
Dopovidi
Drainage
ecological
Ecology
Economy
Edafology
Education
Ékologiya
éksperimental'nyĭ
Embryology
Encyclopedia
Engineering
ɛnology
ɛntomology

environmental
Enzymology
Estonskaya
European
Experiment
experimental
Faculty
Federation
Fizika
Fiziologiya
Flurbereinigung
forestiere
Forestry
Forschung
Foundation
France
Gazette
general
genetical
geneticheskiĭ
Genetics (-ika)
Geobotany
Geofizika
Geophysics
Gesellschaft
Giornale
gosudarstvennyĭ
Government
Grassland
Gruzinskaya
Helveticus
Histochemistry
Histoire (-ory)
Histology
horticultural
Horticulture
Hungaricae
Hungaricus
Husbandry
Hydrobiology
Hydrology
Indian
Industry
inorganic
Institute
Institutului
international
Investigation
Irrigation
Isotopes
issledovatel'skiĭ
Italian (-y)
Izvestiya
Jahrbuch
Japan (-anese)
Journal
Khimiya
Klasse
Kongelige
Közlemenyek
kul'turnykh
Laboratory
Landbauforschung
Landwirtschaft

lesní (-ího)
Letters
Limnology
Linnean
Litovskoĭ
Lucrarile
Magazin
Management
marina (-ine)
Material
Mathematics
Mededelingen
mediterranean
Meldinger
Meteorology
Microbiology
Midland
Mitteilungen
Modeling
modern
molecular
Monographiae (-aphy)
Moskovskiĭ (-ovskogo)
Mycology
national
natural
Naturalist
naturelle
naturkundliche
Naturforschung
nauchnye (-nyĭ)
Neerlandica
Netherland
New Zealand
Norges
Norwegian
Notiser
nuclear
Nutrition
obshcheĭ (-iĭ)
Oceanography
Oecologia
Ökologie
Optics
opytnaya (-yĭ)
organic
original
ornamental
Otdelenie
Paleobotany
Palynology
Pathology
pedagogicheskiĭ
Pesquisa
Pesticide
Pflanzen-
Pflanzenernährung
Pflanzenphysiology
Pflanzenzüchtung
Philosophy
Photogrammetric
Phycology
physical
Physics

physiological	rolniczych	SSSR	Ukrains'kaya
Physiology	Rostlin (-lina)	Stantsii (-ntsiya)	Universidad (-ersity
Phytologist	rostlinná	Statio	US. USA
Phytopathology	Roumaine	stiintifice	USSR
Phytotaxonomy	royal	subtropical (-icale)	Uzbekskiĭ (-ekskaya)
Plantarum	Russian	summary	vědecké (-ecký)
Polonica (-ska)	Russkiĭ	Supplement	vegetable
Pollution	Sborník	Survey	végétale
Práce	Scandinavica	Swedish	Verhandlungen
Practice	Scandinavicus	Symposium	Veröffentlichungen
prikladnoĭ	School	System	Vestnik
Proceedings	Science	Tagungsberichte	Videnskabernes
Progress	scientific	technical (-nische)	Virology
Publishers	Section	Technology	Virusforschungen
Quality	Selskabs	theoretical	Volume
quantitative	Sel'skokhozyaĭstvo	thermal	Voprosy
Quarterly	Series (-iya)	Tidsskrift	vostochnyĭ
Radiation	Service	Tijdschrift	vsesoyuznyĭ
Radiobiology	Shkola (-oly)	Toxicology	vyssheĭ (-iĭ)
Rasteniĭ	Sibirskiĭ (-skogo)	Transactions	výzkumný (-umného)
Rastenievodstvo	Skrifter	Travail (-aux)	Weekblatt
Recherche (-erches)	Slovak (-enská)	tropical (-icale)	Wetenschappen
Report	Society	Trudy	Wissenschaft
Research	Soobshcheniya	Turkmenskaya	Zapiski
Resources	Sovetskiĭ (-iet)	uchenye	Zeitschrift
Review (-ista, -ue)	sovremennyĭ	Ugeskrift	Zeitung
Rivista	special	United Kingdom	Zentralblatt
Roczniky	sperimentale	Ukrainian	Zhurnal

The numbers at the end of each reference of a journal article denote: volume (issue) : first page - last page, year of publication. The number of issue is given only in journals where each issue is paginated separately.

Book titles are cited according to the title page, not to the book jacket or cover. The publishing house, place and year of publication are included.

Printers' errors in the original papers are marked by underlining the respective words (letters).

12912 - **ABDEL-RAHMAN, A.M., ABDEL-HADI, A.H.:** Possibilities to reduce adverse effects of salinity by indole-3-acetic acid. - Biol. Plant. 26: 81-87, 1984.

12913 - **ABDUL-JABBAR, A.S., LUGG, D.G., SAMMIS, T.W., GAY, L.W.:** A field study of plant resistance to water flow in alfalfa. - Agron. J. 76: 765-769, 1984.

12914 - **ACEVEDO, H.E., MASSARDO, V.C.:** Efecto del déficit hídrico en dos etapas del desarrollo en el cultivo del tomate (*Lycopersicon esculentum* Mill.). [The effect of water deficit in two developmental stages on the growth of the tomato (*Lycopersicon esculentum* Mill.).] - Fyton 44: 151-166, 1984. [In Span, ab: E.]

*12915 - **ACHARYA, C.L., CRUIZIAT, P., DAUDET, F.A.:** Comparison of psychrometric and pressure chamber techniques for measurement of leaf water potential. - J. Ind. Soc. Soil Sci. 28: 286-289, 1980.

12916 - **ACOCK, M.C., GARNER, J.O., Jr.:** Effect of fertilizer and watering methods on growth and yields of pot-grown sweet potato genotypes. - HortScience 19: 687-689, 1984.

12917 - **ADAMEC, L.:** The effect of plasmolysis and deplasmolysis on the permeability of plant membranes. - Biol. Plant. 26: 128-131, 1984.

12918 - **ADEDEJI, F.O.:** The effect of temperature, soil-water potential, irradiance, and their interactions on CO_2 exchange rates of two sub-dominant tropical weeds. - J. exp. Bot. 35: 1252-1259, 1984.

12919 - **ADJAHOSSOU, D.F., LOUGUET, P., VIEIRA da SILVA, J.B.:** Corrélations entre les résistances stomatiques de divers croisements de palmier à huile (*Elaeis guineensis* Jacq) et la tolérance à la sécheresse. - Acta Oecol. - Oecol. Plant. 5: 163-178, 1984.

*12920 - **AGGARWAL, P.K., SINHA, S.K.:** Water stress and water-use efficiency in field grown wheat: a comparison of its efficiency with that of C_4 plants. - Agr. Meteorol. 29: 159-167, 1983.

12921 - **AGGARWAL, P.K., SINHA, S.K.:** Effect of water stress on grain growth and assimilate partitioning in two cultivars of wheat contrasting in their yield stability in a drought-environment. - Ann. Bot. 53: 329-340, 1984.

12922 - **AGGARWAL, P.K., SINHA, S.K.:** Differences in water relations and physiological characteristics in leaves of wheat associated with leaf position on the plant. - Plant Physiol. 74: 1041-1045, 1984.

12923 - **AGRAWAL, P.K., FOCK, H.P.:** Carbon dioxide exchange in attached flag leaves and ears of hexaploid triticale in relation to seed shrivelling. - Field Crops Res. 8: 315-321, 1984.

12924 - **AHMAD, I., HELLEBUST, J.A.:** Osmoregulation in the extremely euryhaline marine microalga *Chlorella autotrophica*. - Plant Physiol. 74: 1010-1015, 1984.

*12925 - **AHMAD, R., ABDULLAH, Z.:** Biomass production of food and fiber crops using highly saline water under desert conditions. - In: SAN PIETRO, A. (ed.): Biosaline Research: A Look to the Future. Pp. 149-163. Plenum Publishing Corporation, New York 1982.

12926 - **AHMADI, N., PUARD, M.:** Comportements hydriques de deux types variétaux de riz a l'égard des potentiels décroissants de l'eau dans le milieu nutritif. - Agron. trop. 39: 236-242, 1984.

12927- **AIKEN, S.G., LEFKOVITCH, L.P.:** The taxonomic value of using epidermal characteristics in the Canadian rough fescue complex (*Festuca altaica, F. cam-
pestris, F. hallii, "F. scabrella"*). - Can. J. Bot. 62: 1864-1870, 1984.

12928 - **ALBERDA, T.:** Production and Water Use of Several Food and Fodder Crops
under Irrigation in the Desert Area of Southwestern Peru. (Agricultural
Research Report 928). - Pudoc, Wageningen 1984.

#12929 - **ALBINEŢ, E.:** Comportarea în cultura neirigată şi irigată a unor hibrizi de
porumb pentru boabe în zona nord-estică a Cîmpiei Moldovei. [Behaviour of
some corn-hybrids for grain in irrigated and nonirrigated crop in the
north-eastern region of Moldavia Plain.] - Lucrări Ştiinţ. Ser. Agron. 25:
41-43, 1981. [In Roum, ab: E.]

12930 - **ALBINEŢ, E.:** Influenţa soiului a epocii de recoltare şi a irigaţiei asupra
producţiei si calităţii tehnologice la sfecla de zahăr în condiţiile
Cîmpiei Moldovei. [Influence of variety, harvesting period and irrigation
upon production and technological quality of sugar-beet in the conditions
of Moldavia Plain.] - Lucrări Ştiinţ. Ser. Agron. 25: 71-74, 1981. [In
Roum, ab: E.]

#12931 - **ALBINEŢ, E.:** Rezultatele cercetărilor privind consumul de apă la sfecla de
zahăr irigată în condiţiile cîmpiei colinare din nord-estul Moldovei.
[Results of the research concerning water consumption at irrigated sugar
-beet in the hill-plain conditions of north-eastern Moldavia.] - Lucrări
Ştiinţ. Ser. Agron. 25: 75-78, 1981. [In Roum, ab: E.]

#12932 - **ALBINEŢ, E.:** Corelatia dintre producţia de boabe şi fertilizarea cu azot pe
fond constant de fosfor porumbul neirigat şi irigat în incinta îndiguită
Trifeşti-Sculeni. [The correlation between seed production and fertilization with N on a constant background with P at non-irrigated and irrigated
maize from the embanked zone Trifeşti-Sculeni.] - Lucrări Ştiinţ. Ser.
Agron. 26: 47-52, 1982. [In Roum, ab: E.]

#12933 - **ALBINEŢ, E., DĂNILĂ, R., BEJENARU, A., ENOIU, E., PECINGINE, N., TOMA, C.:**
Studii privind consumul de apă prin metoda de calcul Thornthwaite la princi-
palele culturi de cîmp irigate în zona Huşi. [Researches concerning water
consumption, using the Thornthwaite calculation method at the principal
irrigated plain crops in Huşi zone.] - Lucrări Ştiinţ. Ser. Agron. 25:
11-14, 1981. [In Roum, ab: E.]

#12934 - **ALBINEŢ, E., PURICE, C.:** Consumul de apă la cîteva culturi irigate predo-
minante în condiţiile cîmpiei superioare a Jijiei şi posibilităţile de
economisire a apei de irigat. [Water consumption at several prevalent
irrigated cultures within the conditions of Superior Jijia plain and the
possibilities to save water.] - Lucrări Ştiinţ. Ser. Agron. 26: 9-14,
1982. [In Roum, ab: E.]

12935 - **ALBRECHT, S.L., BENNETT, J.M., BOOTE, K.J.:** Relationship of nitrogenase
activity to plant water stress in field-grown soybeans. - Field Crops Res.
8: 61-71, 1984.

12936 - **ALDERMAN, S.C., LACY, M.L.:** Influence of temperature and water potential
on growth of *Botrytis allii.* - Can. J. Bot. 62: 1567-1570, 1984.

12937 - **ALDERMAN, S.C., LACY, M.L.:** Influence of temperature and moisture on growth
and sporulation of *Botrytis squamosa.* - Can. J. Bot. 62: 2793-2797, 1984.

12938 - **ALEKSANDROV, A.Yu., USPENSKAYA, N.Ya., NOVAKOVA, A.A., KUZMIN, R.N., RUBIN, A.B.:** Mössbauer spectroscopy of the Fe(III) $\rightleftarrows$ Fe(II) transitions in the active centers of membrane-bound iron-sulphur proteins of *Rhodopseudomonas sphaeroides*. - Studia biophys. 102: 129-134, 1984.

12939 - **ALESHIN, E.P., VOROB'EV, N.V.:** Obrazovanie vschodov risa pri vozdeĭstvii ponizhennykh temperatur i razlichnom urovne uvlazhneniya pochvy. [Development of rice seedlings under low temperatures and various soil moisture levels.] -Sel'skokhoz. Biol. 1984(7): 52-54, 1984. [In R, ab: E.]

*12940 - **ALEXANDRI, A.A., ELIADE, E., NÄGLER, M.:** Cîteva aspecte ale influenţei unor fungicide asupra plantelor de tomate. [Aspects of the influence of some fungicides on tomato plants.] - Anal. Inst. Cercet. Protecţ. Plant. 16: 331-350, 1981. [In Roum, ab: E, R.]

*12941 - **ALI, S.M., NAIDU, A.P.:** Screening for drought tolerance in maize. - Ind. J. Genet. 42: 381-388, 1982.

12942 - **ALINA, B.A., BAĬMUKHASHEVA, B.G., KLYSHEV, L.K.:** Deĭstvie khloridnogo zasoleniya na sostoyanie ribosom khloroplastov gorokha. [Effect of chloride salinity on the state of pea chloroplast ribosomes.] - Fiziol. Rast. 31: 1077-1082, 1984. [In R, ab: E.]

12943 - **ALLAWAY, W.G., PITMAN, M.G., STOREY, R., TYERMAN, S., ASHFORD, A.E.:** Water relations of coral cay vegetation on the Great Barrier Reef: water potentials and osmotic content. - Aust. J. Bot. 32: 449-464, 1984.

12944 - **ALLEN, S., RAVEN, J.A.:** NH_4^+ as N source caused increased water uptake by *Ricinus*. - In: CRAM, W.J., JANÁČEK, K., RYBOVÁ, R., SIGLER, K. (ed.): Membrane Transport in Plants. Pp. 105-106. Academia, Praha 1984.

12945 - **ALLEY, W.M.:** On the treatment of evapotranspiration, soil moisture accounting, and aquifer recharge in monthly water balance models. - Water Resour. Res. 20: 1137-1149, 1984.

12946 - **ALMÁDI, L.:** Vízháztartási vizsgálatok II. [Study of water regime II.] - Bot. Közlem. 71: 33-50, 1984. [In Hung, ab: G.]

12947 - **ALONI, B., ROSENSHTEIN, G.:** Proline accumulation: A parameter for evaluation of sensitivity of tomato varieties to drought stress? - Physiol. Plant. 61: 231-235, 1984.

12948 - **ALPERT, P., OECHEL, W.C.:** Microdistribution and water loss resistances of selected bryophytes in an Alaskan *Eriophorum* tussock tundra. - Holarctic Ecol. 7: 111-118, 1984.

12949 - **AMAKI, W., SHINOHARA, Y., HAYATA, Y., SANO, H., SUZUKI, Y.:** Effects of bulb desiccation and storage on the *in vitro* propagation of hyacinth. - Scientia Hort. 23: 353-360, 1984.

12950 - **AMBROS, Z.:** Modelling of the moisture régime of forest ecosystems. - Ekológia (Bratislava) 3: 129-138, 1984.

12951 - **AMIRDZHANOV, A.G.:** Radiatsionnye faktory i programmirovanie urozhaev vinograda. [Radiation factors and grape-vine yield programming.] - Sel'skokhoz. Biol. 1984(9): 15-22, 1984. [In R, ab: E.]

12952 - **AMIRO, B.D., GILLESPIE, T.J., THURTELL, G.W.:** An energy balance model for estimating leaf cuvette environments. - Environ. exp. Bot. 24: 321-330, 1984.

*12953 - **AMORIM NETO, M. da S., VILLA NOVA, N.A.**: Novo sistema de medidas de evapo-
ração para o tangue classe "A". [A new evaporation measurement system for
the class "A" pan.] - Pesq. agrop. Bras. 18: 695-702, 1983. [In Port, ab: E.]

*12954 - **ANDERSEN, P.C., BUCHANAN, D.W., ALBRIGO, L.G.**: Antitranspirant effects on
water relations and fruit growth of rabbiteye blueberry. - J. Amer. Soc.
hort. Sci. 104: 378-383, 1979.

*12955 - **ANDERSEN, P.C., BUCHANAN, D.W., ALBRIGO, L.G.**: Water relations and yield of
three rabbiteye blueberry cultivars with and without drip irrigation. -
J. Amer. Soc. hort. Sci. 104: 731-736, 1979.

12956 - **ANDERSEN, P.C., LOMBARD, P.B., WESTWOOD, M.N.**: Effect of root anaerobiosis
on the water relations of several *Pyrus* species. - Physiol. Plant. 62:
245-252, 1984.

12957 - **ANDERSEN, P.C., LOMBARD, P.B., WESTWOOD, M.N.**: Leaf conductance, growth, and
survival of willow and deciduous fruit tree species under flooded soil
conditions. - J. Amer. Soc. hort. Sci. 109: 132-138, 1984.

12958 - **ANDERSEN, P.C., ODNEAL, M.B.**: Effect of soil profile modification and drip
irrigation on the establishment of 'Vidal Blanc' and 'Chancellor' grapes. -
HortScience 19: 562-565, 1984.

12959 - **ANDERSEN, P.C., PROEBSTING, W.M.**: Water and ion fluxes of abscisic acid
-treated root systems of pear, *Pyrus communis*. - Physiol. Plant. 60:
143-148, 1984.

12960 - **ANDERSON, C.A.**: Development of leaf water deficits in detached green and
lime-chlorotic leaves of seedlings from populations of *Eucalyptus obliqua*
L'Hérit. - Plant Soil 77: 171-181, 1984.

12961 - **ANDERSON, W.K., GELLERMAN, J.L., SCHLENK, H.**: Effect of drought on phytyl
wax esters in *Phaseolus* leaves. - Phytochemistry 23: 2695-2696, 1984.

12962 - **ANDONOV, D.**: Vliyanie na polivniya rezhim i na ravnishcheto na torene
v"rkhu nyakoi ot pokazatelite za otsenka na tekhnologichnite kachestva
na zakharnoto tsveklo. [Effect of irrigation regime and fertilizer level
on some criteria for the evaluation of sugar beet technological qualities.]
- Rasteniev. Nauki 21(3): 49-56, 1984. [In Bulg, ab: F, R.]

12963 - **ANGUS, D.E., WATTS, P.J.**: Evapotranspiration - how good is the Bowen ratio
method? - Agr. Water Manage. 8: 133-150, 1984.

12964 - **ANTONOV, I.**: Izmenchivost i nasledyavane na broya i tegloto na lista pri
zakharnoto tsveklo. [Variability and inheritance of leaf number and weight
in sugar beets.] - Genet. Selekt. 17(1): 3-9, 1984. [In Bulg, ab: E, R.]

*12965 - **APARICIO-TEJO, P.M., BOYER, J.S.**: Significance of accelerated leaf senescence at low water potentials for water loss and grain yield in maize. -
Crop Sci. 23: 1198-1202, 1983.

12966 - **APEL, P., BAUME, H., OHLE, H.**: Hybrids between *Brassica alboglabra* and
Moricandia arvensis and their photosynthetic properties. - Biochem. Physiol.
Pflanz. 179: 793-797, 1984.

12967 - **APPEL, D.N., STIPES, R.J.**: Canker expansion on water-stressed pin oaks
colonized by *Endothia gyrosa*. - Plant Dis. 68: 851-853, 1984.

*12968 - **ARES, J.O., VILLA, A., MANDADORI, G.**: Air pollutant uptake by xerophytic
vegetation: fluoride. - Environ. exp. Bot. 20: 259-269, 1980.

*12969 - **ARKLEY, R.J.:** Transpiration and productivity. - In: RECHCIGL, M., Jr. (ed.): CRC Handbook of Agricultural Productivity. Volume 1. Plant Productivity. Pp. 209-211. CRC Press, Inc., Boca Raton 1982.

12970 - **ARMITAGE, A.M., TU, Z.P., VINES, H.M.:** The influence of chlormequat and daminozide on net photosynthesis, transpiration, and photorespiration of hybrid geranium. - HortScience 19: 705-707, 1984.

12971 - **ASHWORTH, E.N., ABELES, F.B.:** Freezing behavior of water in small pores and the possible role in the freezing of plant tissues. - Plant Physiol. 76: 201-204, 1984.

12972 - **ASLAM, M., HUFFAKER, R.C., RAINS, D.W.:** Early effects of salinity on nitrate assimilation in barley seedlings. - Plant Physiol. 76: 321-325, 1984.

12973 - **ASPINALL, D.:** Water deficit and wheat. - In: PEARSON, C.J. (ed.): Control of Crop Productivity. Pp. 91-107. Academic Press, Harcourt Brace Jovanovich Publishers, Sydney - Orlando - San Diego - Petamula - New York - London - Toronto - Montreal - Tokyo 1984.

12974 - **ASRAR, G., HIPPS, L.E., KANEMASU, E.T.:** Assessing solar energy and water use efficiencies in winter wheat: a case study. - Agr. Forest Meteorol. 31: 47-58, 1984.

*12975 - **ASSAF, R., LEVIN, I., BRAVDO, B.:** Apple fruit growth as a measure of irrigation control. - HortScience 17: 59-61, 1982.

12976 - **ASSAF, R., LEVIN, I., BRAVDO, B.:** Effect of drip irrigation on the yield and quality of Golden Delicious and Jonathan apples. - J. hort. Sci. 59: 493-499, 1984.

12977 - **ASTON, A.R.:** Evaporation from eucalypts growing in a weighing lysimeter: a test of the combination equations. - Agr. Forest Meteorol. 31: 241-249, 1984.

12978 - **ASTON, A.R.:** The effect of doubling atmospheric CO_2 on streamflow: a simulation. - J. Hydrol. 67: 273-280, 1984.

12979 - **ATTIWILL, P.M., CLAYTON-GREENE, K.A.:** Studies of gas exchange and development in a subhumid woodland. - J. Ecol. 72: 285-294, 1984.

12980 - **AUGUSTIN, B.J., SNYDER, G.H.:** Moisture sensor-controlled irrigation for maintaining bermudagrass turf. - Agron. J. 76: 848-850, 1984.

12981 - **AUSSENAC, G., GRANIER, A., IBRAHIM, M.:** Influence du dessèchement du sol sur le fonctionnement hydrique et la croissance du douglas (*Pseudotsuga menziesii* (Mirb.) Franco). - Acta Oecol. - Oecol. Plant. 5: 241-253, 1984.

12982 - **AWASTHI, D.K., KUMAR, V., RAWAT, R.:** Stomatal studies in Amaryllidaceae with special reference to stomatal abnormalities. - Proc. Ind. Acad. Sci. (Plant Sci.) 93: 629-633, 1984.

*12983 - **AYAZLOO, M., BELL, J.N.B., GARSED, S.G.:** Modification of chronic sulphur dioxide injury to *Lolium perenne* L. by different sulphur and nitrogen nutrient treatments. - Environ. Pollut., Ser. A 22: 295-307, 1980.

12984 - **AYERBE, L., TENORIO, J.L., VENTAS, P., FUNES, E., MELLADO, L.:** *Euphorbia lathyris* as an energy crop - Part 1. Vegetative matter and seed productivity. - Biomass 4: 283-293, 1984.

12985 - **AZAM-ALI, S.N., GREGORY, P.J., MONTEITH, J.L.**: Effects of planting density
on water use and productivity of pearl millet (*Pennisetum typhoides*) grown
on stored water. II. Water use, light interception and dry matter production.
- Exp. Agr. 20: 215-224, 1984.

*12986 - **AZCÓN-BIETO, J.**: Inhibition of photosynthesis by carbohydrates in wheat
leaves. - Plant Physiol. 73: 681-686, 1983.

12987 - **BAAS, P.**: Vegetative anatomy and the taxonomic status of *Ilex collina* and
Nemophanthus (Aquifoliaceae). - J. Arnold Arboretum 65: 243-250, 1984.

12988 - **BAAS, P., CHENGLEE, L., XINYING, Z., KEMING, G., YUEFEN, D.**: Some effects of
dwarf growth on wood structure. - IAWA Bull. 5: 45-63, 1984.

*12989 - **BAČIĆ, G., BOŽOVIĆ, B., RATKOVIĆ, S.**: Proton magnetic relaxation in plant
cells and tissues: effect of external ion concentration on spin-lattice
relaxation time (T_1) and water content in primary roots of *Zea mays*. -
Studia biophys. 70: 31-43, 1978.

*12990 - **BAČIĆ, G., RATKOVIĆ, S.**: A near-infrared study of H_2O-D_2O exchange in maize
root. - Period. biol. 84: 111-114, 1982.

12991 - **BAČIĆ, G., RATKOVIĆ, S.**: Water exchange in plant tissue studied by proton
NMR in the presence of paramagnetic centers. - Biophys. J. 45: 767-776,
1984.

12992 - **BADHAM, E.R.**: Modeling growth, development, transpiration and translocation
in the mushroom *Psilocybe cubensis*. - Bull. Torrey bot. Club 111: 159-164,
1984.

12993 - **RAGAUTDINOVA, R.I., FILIPPOVA, T.L., FEDOSEEVA, G.P.**: Transpiratsionnye
koêffitsienty i produktivnost' rasteniĭ kartofelya. [Transpiration coefficients in potato plants and their productivity.] - Sel'skokhoz. Biol. 1984
(3): 60-62, 1984. [In R, ab: E.]

*12994 - **BAILEY, S.W., AITKEN, R.L., HUGHES, J.D.**: The effect of soil water status
on critical phosphorus concentrations in *Stylosanthes hamata* cv. Verano. -
Plant Soil 74: 467-472, 1983.

12995 - **BAKER, D.N., ENOCH, H.Z.**: Plant growth and development. - In: LEMON, E.R.
(ed.): CO_2 and Plants. The Response of Plants to Rising Levels of Atmospheric
Carbon Dioxide. Pp. 107-130. American Association of Advanced Science,
Washington 1983.

12996 - **BAKRADZE, N.G., BALLA, Yu.I., METREVELI, I.M., SHARIMANOV, Yu.G.**: Model'noe
opisanie protsessa vykhoda vody iz kletok rasteniĭ pri ikh okhlazhdenii.
[Model description of water loss from plant cells at cooling.] - Biofizika
29: 105-108, 1984. [In R, ab: E.]

12997 - **BAKRADZE, N.G., KESHELASHVILI, L.V., MOISTSRAPISHVILI, K.M., NAMCHEVADZE, È.N.**:
Temperaturnaya zavisimost' osmoticheskogo davleniya vnutrikletochnoĭ i mezh-
kletochnoĭ sredy v rastitel'nykh tkanyakh. [Temperature relationship of
osmotic pressures of intra- and intercellular solutions of plant tissues.] -
Biofizika 29: 143-146, 1984. [In R, ab: E.]

*12998 - **BALASIMHA, D.**: Seasonal changes in nitrate reductase activity and other
indicators of plant water stress in field cacao (*Theobroma cacao*) plants. -
Plant Physiol. Biochem. 9: 74-79, 1982.

*12999 - **BALASIMHA, D.**: Proline accumulation in water stressed potatoes. - J. Ind.
Potato Assoc. 10: 56-59, 1983.

*13000 - BALASIMHA, D.: Water relations and physiological responses to water stress in cacao. - Plant Physiol. Biochem. 10: 65-71, 1983.

*13001 - BALASIMHA, D.: Effect of abscisic acid and kinetin on growth and proline accumulation in cacao seedlings under water stress. - Ind. J. Plant Physiol. 26: 139-142, 1983.

*13002 - BALASIMHA, D., VIRK, M.S.: Effect of water stress on tuber yield and metabolism in potato. - J. Ind. Potato Assoc. 5: 104-107, 1978.

*13003 - BALASUBRAMANIAN, V., CHARI, A.V.: Effect of irrigation scheduling on grain yield and nitrogen use efficiency of irrigated wheat at Kadawa and Bakura, northern Nigeria. - Fertilizer Res. 4: 201-210, 1983.

13004 - BALDY, C., MADJIDIEH, H.: Propriétés optiques de feuilles de *Quercus ilex* L. provenant d'un arbre âgé d'une protégée du Var. - C.R. Acad. Sci. Paris, Sér. III 299: 617-620, 1984.

13005 - BALL, M.C., FARQUHAR, G.D.: Photosynthetic and stomatal responses of two mangrove species, *Aegiceras corniculatum* and *Avicennia marina*, to long term salinity and humidity conditions. - Plant Physiol. 74: 1-6, 1984.

13006 - BALL, M.C., FARQUHAR, G.D.: Photosynthetic and stomatal responses of the grey mangrove, *Avicennia marina*, to transient salinity conditions. - Plant Physiol. 74: 7-11, 1984.

13007 - BALL, M.C., TAYLOR, S.E., TERRY, N.: Properties of thylakoid membranes of the mangroves, *Avicennia germinans* and *Avicennia marina*, and the sugar beet, *Beta vulgaris*, grown under different salinity conditions. - Plant Physiol. 76: 531-535, 1984.

13008 - BALLA, Yu.I., BAKRADZE, N.G., SHARIMANOV, Yu.G.: Effektivnaya vodoproni-tsaemost' kompleksa membrannykh struktur rastitel'nykh kletok pri subnulevykh temperaturakh. [Efficient water permeability in plant cellular membranes at subzero temperatures.] - Biofizika 29: 864-867, 1984. [In R, ab: E.]

13009 - BARCIKOWSKI, W., NOBEL, P.S.: Water relations of cacti during desiccation: distribution of water in tissues. - Bot. Gaz. 145: 110-115, 1984.

*13010 - BARLOW, E.W.R.: Water relations of the mature leaf. - In: DALE, J.E., MILTHORPE, F.L. (ed.): The Growth and Functioning of Leaves. Pp. 315-345. Cambridge University Press, Cambridge - London - New York - New Rochelle - Melbourne - Sydney 1983.

*13011 - BARREIRO NETO, M., SOUZA, J.G., de, BRAGA SOBRINHO, R., VIEIRA, R.de M.: Arquitetura da planta e queda de botões, flores e frutos, em algodoeiro herbáceo tolerante à seca. [Observations on the architecture and shedding of upland cotton selected for drought tolerance.] - Pesq. agrop. Bras. 18: 1085-1088, 1983. [In Port, ab: E.]

13012 - BARTA, A.L.: Ethanol synthesis and loss from flooded roots of *Medicago sativa* L. and *Lotus corniculatus* L. - Plant Cell Environ. 7: 187-191, 1984.

*13013 - BARUAH, K.K.: Effect of flooding on the growth and yield of barley (*Hordeum vulgare* L.). - Ind. J. Plant Physiol. 25: 432-436, 1982.

*13014 - BASER, D.K., JAGGI, I.K., SINHA, S.B.: Root and shoot growth and transpiration from maize plant as influenced by various depths of water table. - Z. Acker- Pflanzenbau 150: 390-399, 1981.

13015 - **BASHE, D., MASCARENHAS, J.P.**: Changes in potassium ion concentrations during pollen dehydration and germination in relation to protein synthesis. - Plant Sci. Lett. 35: 55-60, 1984.

13016 - **BATCHELOR, C.H.**: The accuracy of evapotranspiration estimated with the FAO modified Penman equation. - Irrig. Sci. 5: 223-233, 1984.

13017 - **BATCHELOR, J.T., SCOTT, H.D., SOJKA, R.E.**: Influence of irrigation and growth stage on element concentrations of soybean plant parts. - Commun. Soil Sci. Plant Anal. 15: 1083-1109, 1984.

13018 - **BAUER, A., FRANK, A.B., BLACK, A.L.**: Estimation of spring wheat leaf growth rates and anthesis from air temperature. - Agron. J. 76: 829-835, 1984.

13019 - **BAUER, H.**: Net photosynthetic CO_2 compensation concentrations of some lichens. - Z. Pflanzenphysiol. 114: 45-50, 1984.

*13020 - **BAUER, U.**: Zur Leistungsdauer des Saatgraslandes. 2. Mitteilung: Saatgrasland zur Mähnutzung. - Arch. Acker- Pflanzenbau Bodenk. 26: 779-787, 1982.

13021 - **BAZZAZ, F.A., CARLSON, R.W.**: The response of plants to elevated CO_2 I. Competition among an assemblage of annuals at two levels of soil moisture. - Oecologia 62: 196-198, 1984.

13022 - **BECANA, M., APARICIO-TEJO, P.M., SÁNCHEZ-DÍAZ, M.**: Effects of water stress on enzymes of ammonia assimilation in root nodules of alfalfa (*Medicago sativa*). - Physiol. Plant. 61: 653-657, 1984.

13023 - **BECK, E., SCHLÜTTER, I., SCHEIBE, R., SCHULZE, E.-D.**: Growth rates and population rejuvenation of East African giant groundsels (*Dendrosenecio keniodendron*). - Flora 175: 243-248, 1984.

13024 - **BECK, E., SCHULZE, E.-D., SENSER, M., SCHEIBE, R.**: Equilibrium freezing of leaf water and extracellular ice formation in Afroalpine 'giant rosette' plants. - Planta 162: 276-282, 1984.

*13025 - **BEHAIRY, A.K.A., EL-SAYED, M.M.**: Biochemical composition of some marine brown algae from Jeddah Coast, Saudi Arabia. - Ind. J. mar. Sci. 12: 200-201, 1983.

13026 - **BEHL, R., HARTUNG, W.**: Transport and compartmentation of abscisic acid in roots of *Hordeum distichon* under osmotic stress. - J. exp. Bot. 35: 1433-1440, 1984.

13027 - **BELCHEVA, L., NEDYALKOV, I.**: Mikroflora na echemicheno z"rno s povishena vlazhnost, s"khranyavano bez dost"p na v"zdukh. [Microflora of barley seeds with higher moisture content stored without access of air.] - Rasteniev. Nauki 21(3): 97-103, 1984. [In Bulg, ab: E, R.]

13028 - **BELESKY, D.P., FISHMAN, M.L., WILKINSON, S.R.**: Effect of nitrogen fertilization and mild water stress on the distribution of nitrogen in tall fescue. - Plant Soil 77: 295-303, 1984.

13029 - **BELESKY, D.P., WILKINSON, S.R., EVANS, J.J.**: Amino acid composition of fractions of 'Kentucky-31' tall fescue as affected by N fertilization and mild water stress. - Plant Soil 81: 257-267, 1984.

13030 - **BELESKY, D.P., WILKINSON, S.R., McHAN, F.**: Yield, composition and quality of tall fescue as influenced by N fertility and soil water availability. - Commun. Soil Sci. Plant Anal. 15: 945-968, 1984.

*13031 - BELFORD, R.K., THOMSON, R.J.: Effects of waterlogging on the growth and
          yield of winter barley. - J. Sci. Food Agr. 32: 410-411, 1981.

*13032 - BELMANS, C., WESSELING, J.G., FEDDES, R.A.: Simulation model of the water
          balance of a cropped soil: SWATRE. - J. Hydrol. 63: 271-286, 1983.

*13033 - BELOT, Y., GUENOT, J., CAPUT, C., BOURDEAU, F.: Incorporation of tritium
          into organic matter of terrestrial plants exposed to tritiated-water
          releases of short duration. - Health Phys. 44: 666-668, 1983.

 13034 - BENEŠOVÁ, H., ZIMA, J., POSPÍŠILOVÁ, J., ŠESTÁK, Z.: Photochemical activities
          of chloroplasts immobilized in Ca^{2+}-alginate gel as affected by age and water
          potential of leaves. - In: Biogenez, Struktury i Funktsii Fotosinteticheskogo
          Apparata. P. 5. UEB ČSAV, Praha 1984.

 13035 - BENNERT, H.W., SCHMIDT, B.: On the osmoregulation in *Atriplex hymenelytra*
          (Torr.) Wats.(*Chenopodiaceae*). - Oecologia 62: 80-84, 1984.

 13036 - BENNETT, J.H., LEE, E.H., HEGGESTAD, H.E.: Biochemical aspects of plant
          tolerance to ozone and oxyradicals: superoxide dismutase. - In: KOZIOŁ, M.J.,
          WHATLEY, F.R. (ed.): Gaseous Air Pollutants and Plant Metabolism. Pp. 413-
          424. Butterworths, London - Boston - Durban - Singapore - Sydney - Toronto -
          Wellington 1984.

 13037 - BENNETT, J.M., ALBRECHT, S.L.: Drought and flooding effects on N$_2$ fixation,
          water relations, and diffusive resistance of soybean. - Agron. J. 76: 735-
          740, 1984.

 13038 - BENNETT, J.M., BOOTE, K.J., HAMMOND, L.C.: Relationships among water poten-
          tial components, relative water content, and stomatal resistance of field
          -grown peanut leaves. - Peanut Sci. 1984(11): 31-35, 1984.

*13039 - BENNETT, J.M., HAMMOND, L.C.: Assessment of maize growth and development in
          response to irrigation management under subtropical conditions. - In:
          METZNER, H. (ed.): Photosynthesis and Plant Productivity. Pp. 74-77. Wissen-
          schaftliche Verlagsgesellschaft, Stuttgart 1983.

 13040 - BENNETT, M.A., WATERS, L., Jr.: Influence of seed moisture on lima bean
          stand establishment and growth. - J. Amer. Soc. hort. Sci. 109: 623-626,
          1984.

 13041 - BENOIT, G.R., GRANT, W.J., ISMAIL, A.A., YARBOROUGH, D.E.: Effect of soil
          moisture and fertilizer on the potential and actual yield of lowbush blue-
          berries. - Can. J. Plant Sci. 64: 683-689, 1984.

 13042 - BENSON, D.M.: Influence of pine bark, matric potential, and pH on sporangium
          production by *Phytophthora cinnamomi*. - Phytopathology 74: 1359-1363, 1984.

 13043 - BENZ, L.C., DOERING, E.J., REICHMAN, G.A.: Water-table contribution to
          alfalfa evapotranspiration and yields in sandy soils. - Trans. ASAE 27:
          1307-1312, 1984.

*13044 - BÉRCZI, A., OLÁH, Z., ERDEI, L.: Nutrition of winter wheat during the life
          cycle. I. Yield and accumulation of dry matter and minerals. - Physiol.
          Plant. 58: 124-130, 1983.

 13045 - BERG, W.A., SIMPS, P.L.: Herbage yields and water-use efficiency on a loamy
          site as affected by tillage, mulch, and seeding treatments. - J. Range
          Manage. 37: 180-184, 1984.

13046 - BERGMANN, H., ECKERT, H.: Einfluss von Glycinbetain auf die Wasserausnut-
         zung von Winterweizen *Triticum aestivum* L. - Biol. Plant. 26: 384-387,
         1984.

13047 - BERGMANN, H., ECKERT, H., SEMBDNER, G.: Einfluss von Gibberellin A_3 und
         synthetischen Wirkstoffen auf die Wasserausnutzung in der biologischen
         Stoffproduktion von Gerste (*Hordeum vulgare* L.). - Biochem. Physiol.
         Pflanz. 179: 573-584, 1984.

13048 - BESL, H., FISCHER, M.A., HAGEMANN, F., HÖLL, W.: Botanik (Studienhilfe
         zu Strasburger: Lehrbuch der Botanik, 32. Auflage). 3. Auflage. - Gustav
         Fischer Verlag, Stuttgart - New York 1984.

13049 - BEUKES, D.J.: Transpiration of apple trees as related to different meteoro-
         logical, plant and soil factors. - J. hort. Sci. 59: 151-159, 1984.

13050 - BEUKES, D.J.: Apple root distribution as affected by irrigation at diffe-
         rent soil water levels on two soil types. - J. Amer. Soc. hort. Sci. 109:
         723-728, 1984.

13051 - BHADORIA, P.B.S.: Movement of chloride towards root as influenced by soil
         moisture and transpiration. - Z. Acker- Pflanzenbau 153: 257-259, 1984.

*13052 - BHATIA, R.C.: Foliar epidermal and stomatal studies of *Glinus lotoides* L. -
         Folia geobot. phytotaxon. 18: 13-16, 1983.

13053 - BHATIA, V.S., BHARGAVA, S.C., SINHA, S.K.: Effect of irrigation on repro-
         ductive efficiency of bunch and spreading types of groundnut (*Arachis
         hypogaea* L.). - J. agr. Sci. 102: 505-508, 1984.

13054 - BHELLA, H.S., KWOLEK, W.F.: The effects of trickle irrigation and plastic
         mulch on zucchini. - HortScience 19: 410-411, 1984.

13055 - BHIVARE, V.N., NIMBALKAR, J.D.: Salt stress effects on growth and mineral
         nutrition of French beans. - Plant Soil 80: 91-98, 1984.

13056 - BIERHUIZEN, J.F., BIERHUIZEN, J.M., MARTAKIS, G.F.P.: The effect of light
         and CO_2 on photosynthesis of various pot plants. - Gartenbauwissenschaft
         49: 251-257, 1984.

*13057 - BIL', K.Ya., LYUBIMOV, V.Yu., DEMIDOVA, R.N., GEDEMOV, T.: Fenomen koope-
         rativnogo funktsionirovaniya v liste CAM- i C_4-fotosinteza. [Phenomenon
         of cooperated functioning in CAM- and C_4-photosynthesis of leaf.] - In:
         Fotosinteticheskiĭ Metabolizm Ugleroda. Pp. 47-57. Ural'skiĭ Gossudarstven-
         nyĭ Universitet Imeni A.M. Gor'kogo, Sverdlovsk 1983. [In R, ab: E.]

13058 - BINGHAM, F.T., GLAUBIG, B.A., SHADE, E.: Water, salinity, and nitrate
         relations of a citrus watershed under drip, furrow, and sprinkler irriga-
         tion. - Soil Sci. 138: 306-313, 1984.

13059 - BIRKENHEAD, K., WILLMER, C.M.: Carbon dioxide fixation by guard cell proto-
         plasts of *Commelina communis*. - J. exp. Bot. 35: 1260-1264, 1984.

13060 - BISSON, M.A., BARTHOLOMEW, D.: Osmoregulation or turgor regulation in
         Chara? - Plant Physiol. 74: 252-255, 1984.

13061 - BISWAS, A.K., CHOUDHURI, M.A.: Effect of water stress at different deve-
         lopmental stages of field-grown rice. - Biol. Plant. 26: 263-266, 1984.

13062 - BJÖRKMAN, O., POWLES, S.B.: Inhibition of photosynthetic reactions under
         water stress: interaction with light level. - Planta 161: 490-504, 1984.

11   13063 - 13078 / BLA - BOI

13063 - **BLACK, R.A.:** Water relations of *Quercus palustris*: field measurements on an experimentally flooded stand. - Oecologia 64: 14-20, 1984.

13064 - **BLACK, V.J.:** The effect of air pollutants on apparent respiration. - In: KOZIOŁ, M.J., WHATLEY, F.R. (ed.): Gaseous Air Pollutants and Plant Metabolism. Pp. 231-248. Butterworths, London - Boston - Durban - Singapore - Sydney - Toronto - Wellington 1984.

13065 - **BLACKMAN, P.G., DAVIES, W.J.:** Age-related changes in stomatal response to cytokinins and abscisic acid. - Ann. Bot. 54: 121-125, 1984.

13066 - **BLACKMAN, P.G., DAVIES, W.J.:** Modification of the CO_2 responses of maize stomata by abscisic acid and by naturally-occurring and synthetic cytokinins. - J. exp. Bot. 35: 174-179, 1984.

13067 - **BLAKE, T.J., TSCHAPLINSKI, T.J., EASTHAM, A.:** Stomatal control of water use efficiency in poplar clones and hybrids. - Can. J. Bot. 62: 1344-1351, 1984.

13068 - **BLATT, C.R.:** Irrigation, mulch, and double row planting related to fruit size and yield of 'Bounty' strawberry. - HortScience 19: 826-827, 1984.

13069 - **BLEKHMAN, G.I.:** Ochistka i izuchenie fermenta s 2´:3´-nukleotidaznoĭ aktivnost'yu v list'yakh pshenitsy Diamant i Bezenchukskaya 98 pri obezvozhivanii. [Purification and study of the enzyme possessing 2´: 3´-nucleotidase activity from wheat cultivars Diamant and Bezenchukskaya 98 subjected to dehydration.] - Fiziol. Rast. 31: 1083-1091, 1984. [In R, ab: E.]

13070 - **BLICKSTAD, E.:** The effect of water activity on growth and end-product formation of two *Lactobacillus* spp. and *Brochothrix thermosphacta* ATCC 11509ᴵ. - Appl. Microbiol. Biotechnol. 19: 13-17, 1984.

✶13071 - **BLOEMEN, G.W.:** Calculation of hydraulic conductivities and steady state capillary rise in peat soils from bulk density and solid matter volume. - Z. Pflanzenernähr. Bodenk. 146: 460-473, 1983.

13072 - **BLOOM, A., EPSTEIN, E.:** Varietal differences in salt-induced respiration in barley. - Plant Sci. Lett. 35: 1-3, 1984.

13073 - **BOAG, S., PORTIS, A.R.,Jr.:** Inhibited light activation of fructose and sedoheptulose bisphosphatase in spinach chloroplasts exposed to osmotic stress. - Planta 160: 33-40, 1984.

13074 - **BOCZ, E., PEPÓ, P.:** Az őszi búza fajták öntözési reakciójának vizsgálata csernozjom talajon. [Irrigation response of winter wheat varieties on chernozem soil.] - Növénytermelés 33: 337-349, 1984. [In Hung, ab: E.]

13075 - **BOCZ, E. PEPÓ, P.:** A műtrágyázás és öntözés hatása az őszi búzafajták minőségére. [Effect of fertilization and irrigation on quality in varieties of winter wheat.] - Növénytermelés 33: 407-416, 1984. [In Hung, ab: E.]

13076 - **BODE, J., WILD, A.:** The influence of (2-chloroethyl)trimethylammoniumchloride (CCC) on growth and photosynthetic metabolism of young wheat plants (*Triticum aestivum* L.). - J. Plant Physiol. 116: 435-446, 1984.

13077 - **BOIS, J.F., COUCHAT, P., MOUTONNET, P.:** Etude de la reponse à un stress hydrique de quelques varietés de riz pluvial et de riz irrigué. I. Incidence sur la transpiration. - Plant Soil 80: 227-236, 1984.

13078 - **BOIS, J.F., COUCHAT, P., MOUTONNET, P.:** Etude de la reponse à un stress hydrique de quelques varietés de riz pluvial et de riz irrigué. II. Incidence sur les echanges de CO_2 et l'efficience de l'eau. - Plant Soil 80: 237-246, 1984.

*13079 - **BOIS, J.F., ORSTOM, A., COUCHAT, P.:** Effets d'un stress hydrique sur le comportement racinaire et aerien du riz pluvial. - In: Isotope and Radiation Techniques in Soil Physics and Irrigation Studies 1983. Pp. 551-560. International Atomic Energy Agency, Vienna 1983.

13080 - **BOLAÑOS, J.A., LONGSTRETH, D.J.:** Salinity effects on water potential components and bulk elastic modulus of *Alternanthera philoxeroides* (Mart.) Griseb. - Plant Physiol. 75: 281-284, 1984.

13081 - **BONETTI, R., MONTANHEIRO, M.N.S., SAITO, S.M.T.:** The effects of phosphate and soil moisture on the nodulation and growth of *Phaseolus vulgaris*. - J. agr. Sci. 103: 95-102, 1984.

13082 - **BOOBATHI BABU, D., SINGH, S.P.:** Studies on transpiration suppressants on spring sorghum in north-western India in relation to soil moisture regimes. I. Effect on yield and water use efficiency. - Exp. Agr. 20: 151-159, 1984.

13083 - **BOOBATHI BABU, D., SINGH, S.P.:** Studies on transpiration suppressants on spring sorghum in north-western India in relation to soil moisture regimes. II. Effect on growth and nutrient uptake. - Exp. Agr. 20: 161-170, 1984.

13084 - **BOUSLAMA, M., SCHAPAUGH, W.T.,Jr.:** Stress tolerance in soybeans. I. Evaluation of three screening techniques for heat and drought tolerance. - Crop Sci. 24: 933-937, 1984.

13085 - **ROWLING, D.J.F., EDWARDS, A.:** pH gradients in the stomatal complex of *Tradescantia virginiana.* - J. exp. Bot. 35: 1641-1645, 1984.

13086 - **BOX, J.E.,Jr., LANGDALE, G.W.:** The effects of in-row subsoil tillage and soil water on corn yields in the Southeastern Coastal Plain of the United States. - Soil Tillage Res. 4: 67-78, 1984.

13087 - **BOYER, J.S., WESTGATE, M.E.:** Water transport for cell enlargement. - In: CRAM, W.J., JANÁČEK, K., RYBOVÁ, R., SIGLER, K. (ed.): Membrane Transport in Plants. Pp. 96-102, Academia, Praha 1984.

13088 - **BRADY, C.J., GIBSON, T.S., BARLOW, E.W.R., SPEIRS, J., WYN JONES, R.G.:** Salt-tolerance in plants. I. Ions, compatible organic solutes and the stability of plant ribosomes. - Plant Cell Environ. 7: 571-578, 1984.

*13089 - **BRAUE, C.A., WAMPLE, R.L., KOLATTUKUDY, P.E., DEAN, B.B.:** Relationship of potato tuber periderm resistance to plant water status. - Amer. Potato J. 60: 827-837, 1983.

13090 - **BRAUN, H.J.:** The significance of the accessory tissues of the hydrosystem for osmotic water shifting as the second principle of water ascent, with some thoughts concerning the evolution of trees. - IAWA Bull. 5: 275-294, 1984.

13091 - **BRENNAN, R.F,, GARTRELL, J.W., ROBSON, A,D.:** Reactions of copper with soil affecting its availability to plants. III. Effect of incubation temperature, - Aust, J. Soil Res. 22: 165-172, 1984.

13092 - **RRIGGS, S.P., SCHEFFER, R.P., HAUG, A.R.:** Osmotic conditions affect sensitivity of oat tissues to toxin from *Helminthosporium victoriae*. - Physiol. Plant Pathol. 25: 103-110, 1984,

13093 - **BRINCKMANN, E., TURNER, N.C., SHACKEL, K.A., GOLLAN, T., SCHULZE, E.-D.:** Effects of atmospheric and soil drought on leaf water status and stomatal response. - In: CRAM, W.J., JANÁČEK, K., RYBOVÁ, R., SIGLER, K. (ed.): Membrane Transport in Plants, Pp, 135-140. Academia, Praha 1984.

13094 - BRINCKMANN, E., TYERMAN, S.D., STEUDLE, E., SCHULZE, E.-D.: The effect of different growing conditions on water relations parameters of leaf epidermal cells of *Tradescantia virginiana* L. - Oecologia 62: 110-117, 1984.

*13095 - BRITTON, C.M., CLARK, R.G., SNEVA, F.A.: Effects of soil moisture on burned and clipped Idaho fescue. - J. Range Manage. 36: 708-710, 1983.

13096 - BRLANSKY, R.H., TIMMER, L.W., LEE, R.F., GRAHAM, J.H.: Relationship of xylem plugging to reduced water uptake and symptom development in citrus trees with blight and blightlike declines. - Phytopathology 74: 1325-1328, 1984.

*13097 - BROSHCHILOV, K., BROSHCHILOVA, M.: Intenzivnost na fotosintezata i produktivnost na trigodishni iglolistni fidanki v zavisimost ot vlazhnostta na pochvata. [Photosynthetic rate and productivity of three year old nursery trees in dependence on soil moisture.] - In: Fiziologiya na Rasteniyata. Vol. 6. Pp. 584-587. BAN, Sofia 1982. [In Bulg.]

*13098 - BROSHCHILOVA, M.: Vliyanie na SO_2 v'rkhu rastezha i intenzivnostta na fotosintezata i transpiratsiyata na fidanki ot *Populus simonii* var. *fastigiata* C.S., *Salix madtsudana* Koidz. f. *tortuosa* Rehd., *Tilia argentea* Desf. [Effect of SO_2 on growth and photosynthetic and transpiration rates of the nursery poplar, willow and linden trees.] - Gorskostop. Nauka 20(1): 20-32, 1983. [In Bulg, ab: E, R.]

13099 - BROWN, D., KERSHAW, K.A.: Photosynthetic capacity changes in *Peltigera* 2. Contrasting seasonal patterns of net photosynthesis in two populations of *Peltigera rufescens*. - New Phytol. 96: 447-457, 1984.

*13100 - BROWN, R.H., BOUTON, J.H.: Photosynthetic characteristics of *Panicum* species in the *Laxa* group. - In: RANDALL, D.D., BLEVINS, D.G., LARSON, R.L. (ed.): Current Topics in Plant Biochemistry and Physiology. Volume 1. Pp. 78-89, University of Missouri, Columbia 1983.

*13101 - BROWN, R.H., WILSON, J.R.: Nitrogen response of *Panicum* species differing in CO_2 fixation pathways. II. CO_2 exchange characteristics. - Crop Sci. 23: 1154-1159, 1983.

13102 - BRUCKLER, L.: Utilisation des micropsychromètres pour la mesure du potentiel hydrique du sol en laboratoire et *in situ*. - Agronomie 4: 171-182, 1984.

13103 - BRULFERT, J., GUERRIER, D., QUEIROZ, O.: Rôle de la photopériode dans l'adaptation à la sécheresse: cas d'une plante à métabolisme crassulacéen, l'*Opuntia ficus-indica* Mill. - Bull. Soc. Bot. France, Actual. bot. 131: 69-77, 1984.

13104 - BRUN, L.J., DEIBERT, E.J., FRENCH, E.W., HOAG, B.K.: Technology impact on evapotranspiration - yield relationships for spring wheat in North Dakota. - North Dakota Farm Res. 42(2): 8-9, 25, 1984.

13105 - BUCHANAN-BOLLIG, I.C., SMITH, J.A.C.: Circadian rhythms in crassulacean acid metabolism: phase relationships between gas exchange, leaf water relations and malate metabolism in *Kalanchoë daigremontiana*. - Planta 161: 314-319, 1984.

13106 - BÜCHNER, K.-H., BENZ, R., WENDLER, S., ZIMMERMANN, U.: The role of mobile charges in plant membranes for the turgor pressure sensing. - In: CRAM, W.J., JANAČEK, K., RYBOVÁ, R., SIGLER, K. (ed.): Membrane Transport in Plants. Pp. 107-108. Academia, Praha 1984.

13107 - BUNCE, J.A.: Effects of humidity on photosynthesis. - J. exp. Bot. 35: 1245-1251, 1984.

13108 - **BUNCE, J.A.**: Identifying soybean lines differing in gas exchange sensitivity to humidity. - Ann. appl. Biol. 105: 313-318, 1984.

13109 - **BURCHETT, M.D., FIELD, C.D., PULKOWNIK, A.**: Salinity, growth and root respiration in the grey mangrove, *Avicennia marina*. - Physiol. Plant. 60: 113-118, 1984.

13110 - **BURGASS, R.W., POWELL, A.A.**: Evidence for repair processes in the invigoration of seeds by hydration. - Ann. Bot. 53: 753-757, 1984.

13111 - **BURIOL, G.A., MENOUX, Y., PARCEVAUX, S., de**: Determination de la masse d'eau et des propriétés optiques d'une feuille à partir de modifications de son bilan énergétique. II. - Applications en conditions artificielles et naturelles. - Agronomie 4: 501-506, 1984.

13112 - **BURIOL, G.A., SANTIBAÑEZ, F., MENOUX, Y., PARCEVAUX, S.,de, BERTOLINI, J.-M.**: Détermination de la masse d'eau et des propriétés optiques d'une feuille à partir de modifications de son bilan énergétique. I. - Bases théoretiques de la méthode et technique de mesure. - Agronomie 4: 493-500, 1084.

13113 - **BURKINA, Z.S., GUSEĬNOVA, G.M.**: Postuplenie vody v korni kukuruzy v zavisimosti ot temperatury. [Water uptake by maize roots in relation to temperature.] - Fiziol. Rast. 31: 1107-1112, 1984. [In R, ab: E.]

13114 - **BURROWS, F.J.**: Trees and forest restoration. - In: PEARSON, C.J. (ed.): Control of Crop Productivity. Pp. 269-286. Academic Press, Sydney - Orlando - San Diego - Petaluma - New York - London - Toronto - Montreal - Tokyo 1984.

13115 - **BURTSEVA, E.I., KONONOV, K.E.**: Statisticheskiĭ analiz rastitel'nosti solonchakovatykh lugov poĭmy reki Leny. IV. Kompozitsionnaya ordinatsiya po faktoru uvlazhneniya i vidovaya klassifikatsiya rastitel'nosti v zavisimosti ot faktorov zasoleniya i uvlazhneniya. [Statistical analysis of saline meadow flora of Lena river low land. IV. Compositional ordination on moisture factor and flora species classification in dependence on salinization and moisture factors.] - Biol. Nauki 1984 (7): 71-76, 1984. [In R, ab: E.]

13116 - **BUSBY, C.H., GUNNING, B.E.S.**: Microtubules and morphogenesis in stomata of the water fern *Azolla*: an unusual mode of guard cell and pore development. - Protoplasma 122: 108-119, 1984.

*13117 - **BUTTERFASS, T.**: A nucleotypic control of chloroplast reproduction. - Protoplasma 118: 71-74, 1983.

13118 - **CAEMMERER, S.,von, FARQUHAR, G.D.**: Effects of partial defoliation, changes of irradiance during growth, short-term water stress and growth at enhanced p(CO_2) on the photosynthetic capacity of leaves of *Phaseolus vulgaris* L. - Planta 160: 320-329, 1984.

13119 - **CALKIN, H.W., PEARCY, R.W.**: Leaf conductance and transpiration, and water relations of evergreen and deciduous perennials co-occuring in a moist chaparral site. - Plant Cell Environ. 7: 339-346, 1984.

13120 - **CALKIN, H.W., PEARCY, R.W.**: Seasonal progressions of tissue and cell water relations parameters in evergreen and deciduous perennials. - Plant Cell Environ. 7: 347-352, 1984.

*13121 - **CALLANDER, B.A., WOODHEAD, T.**: Canopy conductance of estate tea in Kenya. - Agr. Meteorol. 23: 151-167, 1981.

13122 - **CALLOW, J.A.:** Cellular and molecular recognition between higher plants and fungal pathogens. - In: LINSKENS, H.F., HESLOP-HARRISON, J. (ed.): Cellular Interactions. Pp. 212-237. Springer-Verlag, Berlin - Heidelberg - New York - Tokyo 1984.

*13123 - **CALMÉS, J., VIALA, G., CAVALIÉ, G.:** Photorespiratory metabolism in soybean plant; effects of shading or water stress. - In: METZNER, H. (ed.): Photosynthesis and Plant Productivity. Pp. 167-171. Wissenschaftliche Verlagsgesellschaft, Stuttgart 1983.

*13124 - **CAMILLERI, J.C.:** Leaf thickness of mangroves (*Rhizophora mangle*) growing in different salinities. - Biotropica 15: 139-141, 1983.

*13125 - **CAMILLO, P., SCHMUGGE, T.J.:** Estimating soil moisture storage in the root zone from surface measurements. - Soil Sci. 135: 245-264, 1983.

13126 - **CAMILLO, P.J., GURNEY, R.J.:** A sensitivity analysis of a numerical model for estimating evapotranspiration. - Water Resour. Res. 20: 105-112, 1984.

*13127 - **CAMILLO, P.J., GURNEY, R.J., SCHMUGGE, T.J.:** A soil and atmospheric boundary layer model for evapotranspiration and soil moisture studies. - Water Resour. Res. 19: 371-380, 1983.

*13128 - **CAMPBELL, G.S.:** Watering - critique II. - In: TIBBITTS, T.W., KOZLOWSKI, T.T. (ed.): Controlled Environment Guidelines for Plant Research. Pp. 301-321. Academic Press, New York - London - Toronto - Sydney - San Francisco 1979.

*13129 - **CAMPBELL, G.S., CAMPBELL, M.D.:** Irrigation scheduling using soil moisture measurements: theory and practice. - Adv. Irrig. 1: 25-42, 1982.

13130 - **CARBONELL, J.V., MADARRO, A., PIÑAGA, F., PEÑA, J.L.:** Deshidratación de frutas y hortalizas con aire ambiente. IV. Cinética de adsorción y desorción de agua en zanahorias. [Dehydration of fruits and vegetables with ambient air. IV. Kinetics of water adsorption and desorption in carrots.] - Rev. Agroquím. Tecnol. Alimentos 24: 94-104, 1984. [In Span, ab: E.]

13131 - **CARBONNEAU, A.:** Action de la concentration de la solution nutritive sur quelques caractéristiques physiologiques et technologiques chez *Vitis vinifera* L. cv. Cabernet-Sauvignon. III. - Régime hydrique, photosynthèse brute et corrélations intra-plante. - Agronomie 4: 535-541, 1984.

13132 - **CARGILL, S.M., JEFFERIES, R.L.:** Nutrient limitation of primary production in a sub-arctic salt marsh. - J. appl. Ecol. 21: 657-668, 1984.

13133 - **CARLSON, D.J., CARLSON, M.L.:** Reassessment of exudation by fucoid macroalgae. - Limnol. Oceanogr. 29: 1077-1087, 1984.

*13134 - **CARRIJO, O.A., OLITTA, A.F.L., MINAMI, K., MENEZES SOBRINHO, J.A.,de :** Efeito de diferentes quantidades de água sobre a produção de duas cultivares de alho. [Effect of different water levels on the production of two garlic cultivars under drip irrigation.] - Pesq. agropec. Bras. 17: 783-790, 1982. [In Port, ab: E.]

*13135 - **CARTER, D.L.:** Salinity and plant productivity. - In: RECHCIGL, M.,Jr. (ed.): CRC Handbook of Agricultural Productivity. Volume 1. Plant Productivity. Pp. 117-133. CRC Press, Inc., Boca Raton 1982.

*13136 - **CASSELL, D., HERING, T.F.:** The effect of water potential on soil-borne diseases of wheat seedlings. - Ann. appl. Biol. 101: 367-375, 1982.

13137 - **CASTALDO-COBIANCHI, R., GIORDANO, S.:** An adaptive pattern for water conduction in the ectohydric moss *Zygodon viridissimus* var. *rupestris* Hartm. - J. Bryol. 13: 235-239, 1984.

*13138 - **CASTLE, W.S.:** Antitranspirant and root and canopy pruning effects on mechanically transplanted eight-year-old 'Murcott' citrus trees. - J. Amer. Soc. hort. Sci. 108: 981-985, 1983.

*13139 - **CAVALIERI, A.J., BOYER, J.S.:** Water potentials induced by growth in soybean hypocotyls. - Plant Physiol. 69: 492-496, 1982.

13140 - **ČERMÁK, J., JENÍK, J., KUČERA, J., ŽÍDEK, V.:** Xylem water flow in a crack willow tree (*Salix fragilis* L.) in relation to diurnal changes of environment. - Oecologia 64: 145-151, 1984.

13141 - **CEULEMANS, R., IMPENS, I., STEENACKERS, V.:** Stomatal and anatomical leaf characteristics of 10 *Populus* clones. - Can. J. Bot. 62: 513-518, 1984.

*13142 - **CHAKRABARTI, S., SAHA, S.:** Comparative study of some morphological and anatomical characteristics of rice cultivars in relation to yield. - Ind. Agr. 27: 101-110, 1983.

*13143 - **CHAKRAVARTY, N.V.K., SASTRY, P.S.N.:** Biomass production in mung bean (*Vigna radiata* L. Wilczek) in relation to pan evaporation and ambient temperature. - Agr. Meteorol. 29: 57-62, 1983.

13144 - **CHANG, H., SIEGEL, B.Z., SIEGEL, S.M.:** Salinity-induced changes in isoperoxidases in taro, *Colocasia esculenta*. - Phytochemistry 23: 233-235, 1984.

13145 - **CHARBENEAU, R.J.:** Kinematic models for soil moisture and solute transport. - Water Resour. Res. 20: 699-706, 1984.

13146 - **CHEESEMAN, J.M.:** The interrelationships between transpiration, sodium and potassium transport in rooted cuttings of *Aster simplex*. - In: CRAM, W.J., JANÁČEK, K., RYBOVÁ, R., SIGLER, K. (ed.): Membrane Transport in Plants. Pp. 151-152. Academia, Praha 1984.

*13147 - **CHEN, C.-Y., CHANG, H.-S., HUANG, C.-L.:** [Characteristics of water metabolism of corn. I. Water absorption and stomatal aperture.] - J. Sci. Eng. 18: 1-14, 1981. [In Chin, ab: E.]

13148 - **CHEN, J.:** Uncoupled multi-layer model for the transfer of sensible and latent heat flux densities from vegetation. - Boundary-Layer Meteorol. 28: 213-225, 1984.

*13149 - **CHEN, S.-S., BLACK, C.C.,Jr.:** Diurnal changes in volume and species tissue weight of Crassulacean acid metabolism plants. - Plant Physiol. 71: 373-378, 1983.

13150 - **CHENGA REDDY, V., BHASKAR, C.V.S., RAGHAVENDRA, A.S., DAS, V.S.R.:** Photosynthesis of wheat cultivars in relation to photosynthetic unit and stomatal conductance. - Photosynthetica 18: 226-230, 1984.

*13151 - **CHERNYSHEVA, S.V., UDOVENKO, G.V.:** Razvitie vo vremeni reaktsii fotosinteticheskogo apparata pshenits na vozdeľstvie stressovykh faktorov sredy. [Time course of responses of wheat photosynthetic apparatus to environmental stress factors.] - Fiziol. Rast. 30: 787-791, 1983. [In R.]

*13152 - CHEW, W.Y., WILLIAMS, C.N., RAMLI, K.: Studies on the availability to
         plants of soil nitrogen in Malaysian tropical oligotrophic peat. III -
         Effects of initial soil air-drying and frequency of watering. - Trop. Agr.
         57: 11-19, 1980.

13153 - CHONAN, N., KAWAHARA, H., MATSUDA, T.: [Ultrastructure of vascular bundles
        and fundamental parenchyma in relation to movement of photosynthate in
        leaf sheath of rice.] - Jap. J. Crop Sci. 53: 435-444, 1984. [In Jap, ab: E.]

*13154 - CHOUDHURI, G.N., SINHA, P.: Water relations and life history strategies of
         two sympatric weeds of Indo-Gangetic Plains. - Acta bot. Ind. 11: 173-179,
         1983.

13155 - CHOUDHURY, R.J., FEDERER, C.A.: Some sensitivity results for corn canopy
        temperature and its spatial variation induced by soil hydraulic hetero-
        geneity. - Agr. Forest Meteorol. 31: 297-317, 1984.

13156 - CHOW, K.H.: Alcohol dehydrogenase synthesis and waterlogging tolerance in
        maize. - Trop. Agr. 61: 302-304, 1984.

13157 - CHRISTIANSEN, M.N., ST.JOHN, J.B.: Chemical modification of plant response
        to temperature extremes. - In: ORY, R.L., RITTIG, F.R. (ed.): Bioregulators.
        Chemistry and Uses. Pp. 235-243. American Chemical Society, Washington 1984.

13158 - CHRISTIE, E.K.: Natural grasslands. - In: PEARSON, C.J. (ed.): Control of
        Crop Productivity. Pp. 199-215. Academic press, Sydney - Orlando - San Diego
        - Petaluma - New York - London - Toronto - Montreal - Tokyo 1984.

*13159 - CLARKE, J.M.: Time of physiological maturity and post-physiological maturity
         drying rates in wheat. - Crop Sci. 23: 1203-1205, 1983.

13160 - CLARKE, J.M.: Drying rates of standing compared to windrowed wheat. - Agron.
        J. 76: 803-806, 1984.

13161 - CLARKE, J.M., TOWNLEY-SMITH, T.F.: Drying rates of spring triticale compared
        to wheat. - Agron. J. 76: 454-456, 1984.

13162 - CLARKE, J.M., TOWNLEY-SMITH, T.F., McCAIG, T.N., GREEN, D.G.: Growth ana-
        lysis of spring wheat cultivars of varying drought resistance. - Crop Sci.
        24: 537-541, 1984.

13163 - CLARKSON, D.T.: Calcium transport between tissues and its distribution in
        the plant. - Plant Cell Environ. 7: 449-456, 1984.

13164 - CLEMENTI, F., ROSSI, J.: Effect of drying and storage conditions on survival
        of Leuconostoc oenos. - Amer. J. Enol. Viticult. 35: 183-186, 1984.

13165 - CLINT, G.M., MACROBBIE, E.A.C.: Effects of fusicoccin in 'isolated' guard
        cells of Commelina communis L. - J. exp. Bot. 35: 180-192, 1984.

13166 - CLINT, G.M., MACROBBIE, E.A.C.: The effects of fusicoccin on ion fluxes in
        stomatal guard cells. - In: CRAM, W.J., JANÁČEK, K., RYBOVÁ, R., SIGLER, K.
        (ed.): Membrane Transport in Plants. Pp. 157-158. Academia, Praha 1984.

13167 - CLOUGH, B.F.: Growth and salt balance of the mangroves Avicennia marina
        (Forsk.) Vierh. and Rhizophora stylosa Griff. in relation to salinity. -
        Aust. J. Plant Physiol. 11: 419-430, 1984.

13168 - **CLOUGH, B.F.:** Mangroves. - In: PEARSON, C.J. (ed.): Control of Crop Productivity. Pp. 253-265. Academic Press, Sydney - Orlando - San Diego -
Petaluma - New York - London - Toronto - Montreal - Tokyo 1984.

*13169 - **CLOUGH, J.M., TEERI, J.A., TONSOR, S.J.:** Photosynthetic adaptation of
Solanum dulcamara L. to sun and shade environments. IV. Comparison of
North American and European genotypes. - Oecologia 60: 348-352, 1983.

13170 - **CLOUTIER, Y., ANDREWS, C.J.:** Efficiency of cold hardiness induction by
desiccation stress in four winter cereals. - Plant Physiol. 76: 595-598,
1984.

13171 - **COALE, F.J., EVANGELOU, V.P., GROVE, J.H.:** Effects of saline-sodic soil
chemistry on soybean mineral composition and stomatal resistance. - J.
environ. Qual. 13: 635-639, 1984.

*13172 - **COHEN, M., PELOSI, R.R., BRLANSKY, R.H.:** Nature and location of xylem
blockage structures in trees with citrus blight. - Phytopathology 73:
1125-1130, 1982.

*13173 - **COHEN, S., COHEN, Y.:** Field studies of leaf conductance response to
environmental variables in citrus. - J. appl. Ecol. 20: 561-570, 1983.

*13174 - **COHEN, S.S., GALE, J., SHMIDA, A., POLJAKOFF-MAYBER, A., SURAQUI, S.:**
Xeromorphism and potential rate of transpiration on Mount Hermon, an east
Mediterranean mountain. - J. Ecol. 69: 391-403, 1981.

*13175 - **COLLICARD, J.J., CARLIER, G.:** Paramètres relatifs à l'eau dans les feuilles
de Châtaigniers (*Castanea sativa* Mill.) de taillis du Bas-Dauphine. - Bull.
Écol. 14: 201-212, 1983.

13176 - **COLLIER, G.F., TIBBITTS, T.W.:** Effects of relative humidity and root temperature on calcium concentration and tipburn development in lettuce. - J.
Amer. Soc. hort. Sci. 109: 128-131, 1984.

13177 - **COLLINS, G., STIBBE, E., KROESBERGEN, B.:** Influence of soil moisture stress
and soil bulk density on the imbibition of corn seeds in a sandy soil. -
Soil Tillage Res. 4: 361-370, 1984.

13178 - **COLQUHOUN, I.J., RIDGE, R.W., BELL, D.T., LONERAGAN, W.A., KUO, J.:**
Comparative studies in selected species of *Eucalyptus* used in rehabilitation of the northern jarrah forest, Western Australia. I. Patterns of
xylem pressure potential and diffusive resistance of leaves. - Aust. J.
Bot. 32: 367-373, 1984.

13179 - **COMES, R.D., MARQUIS, L.Y., KELLEY, A.D.:** Response of concord grapes
(*Vitis labrusca*) to 2,4-D in irrigation water. - Weed Sci. 32: 455-459,
1984.

13180 - **COMSTOCK, J., EHLERINGER, J.:** Photosynthetic responses to slowly decreasing
leaf water potentials in *Encelia frutescens*. - Oecologia 61: 241-248, 1984.

13181 - **CONNER, L.N., CONNER, A.J.:** Comparative water loss from leaves of *Solanum
laciniatum* plants cultured *in vitro* and *in vivo*. - Plant Sci. Lett. 36:
241-246, 1984.

13182 - **CONOVER, D.G., GEIGER, D.R.:** Germination of Australian channel millet
(*Echinochloa turnerana* (Domin) J.M. Black) seeds. I. Dormancy in relation
to light and water. - Aust. J. Plant Physiol. 11: 395-408, 1984.

13183 - **CONOVER, D.G., GEIGER, D.R.:** Germination of Australian channel millet
(*Echinochloa turnerana* (Domin) J.M.Black) seeds. II. Effects of anaerobic
conditions, continuous flooding and low water potential. - Aust. J. Plant
Physiol. 11: 409-417, 1984.

13184 - **COOK, B.G., MULDER, J.C.:** Response of nine tropical grasses to nitrogen
fertilizer under rain-grown conditions in south-eastern Queensland. 1.
Seasonal dry matter productivity. - Aust. J. exp. Agr. anim. Husb. 24:
410-414, 1984.

*13185 - **COOMBS, J., HALL, D.O., CHARTIER, P. (ed.):** Plants as Solar Collectors. -
D. Reidel Publishing Company, Dordrecht - Boston - Lancaster 1983.

13186 - **CORNISH, K., ZEEVAART, J.A.D.:** Abscisic acid metabolism in relation to
water stress and leaf age in *Xanthium strumarium*. - Plant Physiol. 76:
1029-1035, 1984.

13187 - **CORNISH, P.S., McWILLIAM, J.R., SO, H.B.:** Root morphology, water uptake,
growth and survival of seedlings of ryegrass and phalaris. - Aust. J. agr.
Res. 35: 479-492, 1984.

13188 - **CORNISH, P.S., SO, H.B., McWILLIAM, J.R.:** Effects of soil bulk density and
water regimen on root growth and uptake of phosphorus by ryegrass. - Aust.
J. agr. Res. 35: 631-644, 1984.

13189 - **COSBY, B.J., HORNBERGER, G.M., CLAPP, R.B., GINN, T.R.:** A statistical
exploration of the relationships of soil moisture characteristics to the
physical properties of soils. - Water Resour. Res. 20: 682-690, 1984.

13190 - **COSGROVE, D.:** Hydraulic aspects of plant growth. - What's New Plant Physiol.
15: 5-8, 1984.

*13191 - **COSGROVE, D.J.:** Photocontrol of extension growth: a biophysical approach. -
Phil. Trans. roy. Soc. London B 303: 453-465, 1983.

13192 - **COSGROVE, D.J., Van VOLKENBURGH, E., CLELAND, R.E.:** Stress relaxation of
cell walls and the yield threshold for growth. Demonstration and measurement by micro-pressure probe and psychrometer techniques. - Planta 162:
46-54, 1984.

*13193 - **COSTIGAN, P.A., GREENWOOD, D.J., McBURNEY, T.:** Variation in yield between
two similar sandy loam soils. I. Description of soils and measurement of
yield differences. - J. Soil Sci. 34: 621-637, 1983.

13194 - **COUDRET, A., CAUDERON, Y.:** Effets de modifications du génome D et d'un
choc osmotique sur les paramètres hydriques et sur le comportement stomatique de *Triticum aestivum* cv. "Chinese Spring". - Agronomie 4: 37-46,
1984.

13195 - **COUOT-GASTELIER, J., LAFFRAY, D., LOUGUET, P.:** Étude comparée de l'ultrastructure des stomates ouverts et fermés chez le *Tradescantia virginiana*. -
Can. J. Bot. 62: 1505-1512, 1984.

13196 - **COWAN, I.:** Optimization of productivity: carbon and water economy in
higher plants. - In: PEARSON, C.J. (ed.): Control of Crop Productivity.
Pp. 13-31. Academic Press, Sydney - Orlando - San Diego - Petaluma - New
York - London - Toronto - Montreal - Tokyo 1984.

*13197 - **COX, C.S., McEVOY, P.B.:** Effect of summer moisture stress on the capacity of
tansy ragwort (*Senecio jacobaea*) to compensate for defoliation by cinnabar
moth (*Tyria jacobaeae*). - J. appl. Ecol 20: 225-234, 1983.

13198 - **COX, E.F.:** The influence of water supply on the establishment of trans-planted lettuce seedlings raised in paperpots. - J. hort. Sci. 59: 95-99, 1984.

13199 - **COX, E.F.:** The effects of irrigation on the establishment and yield of lettuce and leek transplants raised in peat blocks. - J. hort. Sci. 59: 431-437, 1984.

*13200 - **COXSON, D.S., KERSHAW, K.A.:** The ecology of *Rhizocarpon superficiale*. II. The seasonal response of net photosynthesis and respiration to temperature, moisture, and light. - Can. J. Bot. 61: 3019-3030, 1983.

13201 - **COXSON, D.S., KERSHAW, K.A.:** Low-temperature acclimation of net photo-synthesis in the crustaceous lichen *Caloplaca trachyphylla*. - Can. J. Bot. 62: 86-95, 1984.

13202 - **COYNE, P.I., BRADFORD, J.A.:** Leaf gas exchange in 'Caucasian' bluestem in relation to light, temperature, humidity, and CO_2. - Agron. J. 76: 107-113, 1984.

13203 - **CRAFTS-BRANDNER, S.J., BELOW, F.E., WITTENBACH, V.A., HARPER, J.E., HAGEMAN, R.H.:** Differential senescence of maize hybrids following ear removal. II. Selected leaf. - Plant Physiol. 74: 368-373, 1984.

13204 - **CRAM, W.J.:** Mannitol transport and suitability as an osmoticum in root cells. - Physiol. Plant. 61: 396-404, 1984.

13205 - **CRAM, W.J.:** Mannitol as an osmoticum. - In: CRAM, W.J., JANÁČEK, K., RYBOVÁ, R., SIGLER, K. (ed.): Membrane Transport in Plants. Pp. 483-484. Academia, Praha 1984.

13206 - **CRAM, W.J., JANÁČEK, K., RYBOVÁ, R., SIGLER, K. (ed.):** Membrane Transport in Plants. - Academia, Praha 1984.

*13207 - **CRANE, J.L.,Jr., DICKMANN, D.I., FLORE, J.A.:** Photosynthesis and transpi-ration by young *Larix kaempferi* trees: C_3 responses to light and tempera-ture. - Physiol. Plant. 59: 635-640, 1983.

13208 - **CRAWFORD, R.M.M.:** Anaerobic respiration and flood tolerance in higher plants. - In: PALMER, J.M. (ed.): The Physiology and Biochemistry of Plant Respira-tion. Pp. 67-75. Cambridge University Press, Cambridge - London - New York - New Rochelle - Melbourne - Sydney 1984.

13209 - **CREBER, G.T., CHALONER, W.G.:** Influence of environmental factors on the wood structure of living and fossil trees. - Bot. Rev. 50: 357-448, 1984.

13210 - **CRUIZIAT, P.:** Some aspects of the water relations between different organs of the same plant. - Wiss. Z. Humboldt-Univ., math. naturwiss. Reihe 33: 356-359, 1984.

13211 - **CRUIZIAT, P., BODET, C.:** Use of a big pressure chamber to obtain pressure -volume curves of large samples of plants: method, first results, questions. - In: CRAM, W.J., JANÁČEK, K., RYBOVÁ, R., SIGLER, K. (ed.): Membrane Transport in Plants. Pp. 109, Academia, Praha 1984.

13212 - **CRUZ, R.T., O'TOOLE, J.C.:** Dryland rice response to an irrigation gradient at flowering stage. - Agron. J. 76: 178-183, 1984.

*13213 - **CURTAIN, C.C., LOONEY, F.D., REGAN, D.L., IVANCIC, N.M.:** Changes in the ordering of lipids in the membrane of *Dunaliella* in response to osmotic -pressure changes. An e.s.r. study. - Biochem. J. 213: 131-136, 1983.

13214 - **CURTIS, W.F.:** Photosynthetic potential of sun and shade *Viola* species. - Can. J. Bot. 62: 1273-1278, 1984.

13215 - **CURTIS, W.F., KINCAID, D.T.:** Leaf conductance responses of *Viola* species from sun and shade habitats. - Can. J. Bot. 62: 1268-1272, 1984.

13216 - **DACEY, J.W.H., HOWES, B.L.:** Water uptake by roots controls water table movement and sediment oxidation in short *Spartina* marsh. - Science 224: 487-489, 1984.

13217 - **DAIE, J., WYSE, R., HEIN, M., BRENNER, M.L.:** Abscisic acid metabolism by source and sink tissues of sugar beet. - Plant Physiol. 74: 810-814, 1984.

*13218 - **DAINES, R.H., BRENNAN, E., LEONE, I.A.:** Air pollution, plant response, and productivity. - In: RECHCIGL, M.,Jr. (ed.): CRC Handbook of Agricultural Productivity. Volume 1. Plant Productivity. Pp. 375-399. CRC Press, Inc., Boca Raton 1982.

*13219 - **DAIYA, K.S., SHARMA, H.K., CHAWAN, D.D.:** Effect of different moisture regimes on the growth of *Cassia* species. - Folia geobot. phytotaxon. 18: 189-193, 1983.

*13220 - **DALE, J.E.:** The Growth of Leaves. - Edward Arnold, London 1982.

13221 - **DALEY, P.F., CLOUTIER, C.F., McNEIL, J.N.:** A canopy porometer for photosynthesis studies in field crops. - Can. J. Bot. 62: 290-295, 1984.

13222 - **DALZIEL, J., LAWRENCE, D.K.:** Biochemical and biological effects of kaurene oxidase inhibitors, such as paclobutrazol. - In: MENHENETT, R., LAWRENCE, D.K. (ed.): Biochemical Aspects of Synthetic and Naturally Occurring Plant Growth Regulators. Pp. 43-57. British Plant Growth Regulator Group, Wantage 1984.

13223 - **DASGUPTA, J., BEWLEY, J.D.:** Variations in protein synthesis in different regions of greening leaves of barley seedlings and effects of imposed water stress. - J, exp. Bot. 35: 1450-1459, 1984.

13224 - **DAVID, H., JARLET, E., DAVID, A.:** Effects of nitrogen source, calcium concentration and osmotic stress on protoplasts and protoplast-derived cell cultures of *Pinus pinaster* cotyledons. - Physiol. Plant. 61: 477-482, 1984.

13225 - **DAVIDSON, H.R., CAMPBELL, C.A.:** Growth rates, harvest index and moisture use of Manitou spring wheat as influenced by nitrogen, temperature and moisture. - Can. J. Plant Sci. 64: 825-839, 1984.

13226 - **DAVIES, F.S., WILCOX, D.:** Waterlogging of containerized rabbiteye blueberries in Florida. - J. Amer. Soc. hort. Sci. 109: 520-524, 1984.

13227 - **DAVIES, M.S.:** The response of contrasting populations of *Erica cinerea* and *Erica tetralix* to soil type and waterlogging. - J. Ecol. 72: 197-208, 1984.

13228 - **DAVY de VIRVILLE, J., CHAUVEAU, M., PERSON-DEDRYVER, F.:** Modifications de la croissance et de l'intensité respiratoire de racines de blé infectées par *Heterodera avenae* Woll. - Agronomie 4: 813-818, 1984.

13229 - **DAWSON, P., WESTE, G.:** Impact of root infection by *Phytophthora cinnamomi* on the water relations of two *Eucalyptus* species that differ in susceptibility. - Phytopathology 74: 486-490, 1984.

*13230 - **DEAN, C.A., SANDS, R.:** Stomatal response to evaporative demand and soil water status in families of radiata pine. - Aust. Forest Res. 13: 179-182, 1983.

13231 - **DeRELL, D.S., HOOK, D.D., McKEE, W.H.,Jr., ASKEW, J.L.:** Growth and physiology of loblolly pine roots under various water table level and phosphorus treatments. - Forest Sci. 30: 705-714, 1984.

13232 - **De ROER, A.H.:** Hydrostatic pressure applied to roots: its effect on respiration and trans-root electrical potential. - In: CRAM, W.J., JANÁČEK, K., RYBOVÁ, R., SIGLER, K. (ed.): Membrane Transport in Plants. Pp. 155-156. Academia, Praha 1984.

*13233 - **De BOER, A.H., PRINS, H.B.A.:** A study of the electrophysiological organization in roots of two *Plantago* species: direct measurement of ion transport to the xylem using excised roots. - In: SARIĆ, M.R., LOUGHMAN, B.C. (ed.): Genetic Aspects of Plant Nutrition. Pp. 93-100. Martinus Nijhoff / Dr W. Junk Publishers, The Hague - Boston - Lancaster 1983.

13234 - **De ROER, A.H., PRINS, H.B.A.:** Trans-root electrical potential in roots of *Plantago media* L. as affected by hydrostatic pressure: the induction of an O_2 deficien root core. - Plant Cell Physiol. 25: 643-655, 1984.

*13235 - **De BOER, A.H., PRINS, H.B.A., ZANSTRA, P.E.:** Bi-phasic composition of trans-root electric potential in roots of *Plantago* species: involvement of spatially separated electrogenic pumps. - Planta 157: 259-266, 1983.

13236 - **DeJONG, T.M., DOYLE, J.F.:** Leaf gas exchange and growth responses of mature 'Fantasia' nectarine trees to paclobutrazol. - J. Amer. Soc. hort. Sci. 109: 878-882, 1984.

13237 - **DeJONG, T.M., TOMBESI, A., RYUGO, K.:** Photosynthetic efficiency of kiwi (*Actinidia chinensis* Planch.) in response to nitrogen deficiency. - Photosynthetica 18: 139-145, 1984.

*13238 - **DE KONING,A., HURD, R.G.:** A comparison of winter-sown tomato plants grown with restricted and unlimited water supply. - J. hort. Sci. 58: 575-581, 1983.

13239 - **DELUCIA, E.H., BERLYN, G.P.:** The effect of increasing elevation on leaf cuticle thickness and cuticular transpiration in balsam fir. - Can. J. Bot. 62: 2423-2431, 1984.

13240 - **DEMARTY, M., MORVAN, C., THELLIER, M.:** Calcium and the cell wall. - Plant Cell Environ. 7: 441-448, 1984.

13241 - **DENMEAD, O.T.:** Plant physiological methods for studying evapotranspiration: problems of telling the forest from the trees. - Agr. Water Manage. 8: 167-189, 1984. Also in: SHARMA, M.L. (ed.): Evapotranspiration from Plant Communities. Pp, 167-189. Elsevier Science Publishers, Amsterdam - Oxford - New York - Tokyo 1984.

*13242 - **DENNIS, C., HÖCKER, J.:** Effect of relative humidity on chilling sensitivity of sporangiospores of *Rhizopus* species. - Trans. Brit. mycol. Soc. 77: 179-183, 1981.

*13243 - **DENNIS, W.D., WOLEDGE, J.:** The effect of shade during leaf expansion on photosynthesis by white clover leaves. - Ann. Bot. 51: 111-118, 1983.

*13244 - **DERCO, M.**: Effect of vegetation factors optimization on field crop yields
under irrigation conditions in semi-humid climate. - Ved. Práce výskum.
Úst. závlahového Hospodárstva Bratislave 15: 235-301, 1981.

13245 - **DERCO, M., BARTA, V.**: Vplyv závlahy, organizácie porastu a poveternosti
na tvorbu úrody zrna odrôd jarného jačmeňa. [The effect of irrigation,
stand organization and weather on grain yield formation in cultivars of
spring barley.] - Rost. Výroba (Praha) 30: 607-616, 1984. [In Slov, ab: E, R.]

*13246 - **DESHMUKH, P.S., SRIVASTAVA, G.C.**: Variation in proline accumulation in
sunflower genotypes under moisture stress conditions. - Ind. J. Plant
Physiol. 25: 396-399, 1982.

13247 - **DESJARDINS, R.L., BUCKLEY, D.J., AMOUR, G.S.**: Eddy flux measurements of CO_2
above corn using a microcomputer system. - Agr. Forest Meteorol. 32: 257-
265, 1984.

13248 - **DE SOUZA, F.**: Modelo matemático da irrigação por sulcos. [Mathematical
model of furrow irrigation.] - Pesq. agropec. Bras. 19: 1135-1143, 1984.
[In Port, ab: E.]

13249 - **De STIGTER, H.C.M., BROEKHUYSEN, A.G.M.**: Weight-change patterns in cut and
intact rose shoots. II. Floral and foliar contributions as related to competition for water. - J. Plant Physiol. 115: 319-329, 1984.

13250 - **De VISSER, R., POORTER, H.**: Growth and root nodule nitrogenase activity of
Pisum sativum as influenced by transpiration. - Physiol. Plant. 61: 637-
642, 1984.

13251 - **DEVITT, D., JARRELL, W.M., JURY, W.A., LUNT, O.R., STOLZY, L.H.**: Wheat
response to sodium uptake under zonal saline-sodic conditions. - Soil Sci.
Soc. Amer. J. 48: 86-92, 1984.

13252 - **DEVITT, D., STOLZY, L.H., JARRELL, W.M.**: Response of sorghum and wheat to
different K^+/Na^+ ratios at varying osmotic potentials. - Agron. J. 76:
681-688, 1984.

*13253 - **DE WIT, C.T., PENNING de VRIES, F.W.T.**: Crop growth models without hormones.
- Neth. J. agr. Sci. 31: 313-323, 1983.

13254 - **DIAMANTOGLOU, S., KULL, U.**: Kohlenhydratgehalte und osmotische Verhältnisse
bei Blättern und Rinden von *Arbutus unedo* L. und *Arbutus andrachne* L. im
Jahresgang. - Ber. Deut. bot. Ges. 97: 433-441, 1984.

13255 - **DICKSON, M.H., BOETTGER, M.A.**: Emergence, growth, and blossoming of bean
(*Phaseolus vulgaris*) at suboptimal temperatures. - J. Amer. Soc. hort. Sci.
109: 257-260, 1984.

13256 - **DIX, N.J.**: Minimum water potentials for growth of some litter-decomposing
agarics and other basidiomycetes. - Trans. Brit. mycol. Soc. 83: 152-153,
1984.

13257 - **DIX, N.J.**: Moisture content and water potential of abscissed leaves in
relation to decay. - Soil Biol. Biochem. 16: 367-370, 1984.

13258 - **DIXON, M., GRACE, J.**: Effect of wind on the transpiration of young trees. -
Ann. Bot. 53: 811-819, 1984.

13259 - **DIXON, M.A., GRACE, J., TYREE, M.T.**: Concurrent measurements of stem density,
leaf and stem water potential, stomatal conductance and cavitation on a
sapling of *Thuja occidentalis* L. - Plant Cell Environ. 7: 615-618, 1984.

13260 - **DIXON, M.A., TYREE, M.T.:** A new stem hygrometer, corrected for temperature gradients and calibrated against the pressure bomb. - Plant Cell Environ. 7: 693-697, 1984.

*13261 - **DOBROLYUBSKIĬ, O.K., FEDORENKO, I.V., TANURKOV, G.R.:** Vliyanie mikroèlementov na soderzhanie form vody v rastenii vinograda. [Effects of microelements on free and bound water content in grapes.] - Agrokhimiya 1981 (12): 86-89, 1981. [In R.]

13262 - **DOLEY, D.:** Experimental analysis of fluoride susceptibility of grape vine (*Vitis vinifera* L.): foliar fluoride accumulation in relation to ambient concentration and wind speed. - New Phytol. 96: 337-351, 1984.

*13263 - **DOLGOV, S.I., ZHITKOVA, A.A., VINOGRADOVA, G.B.:** Produktivnost' ispol'zovaniya rasteniyami pochvennoĭ vlagi pri razlichnoĭ vlazhnosti pochvy. [Effective utilization of soil moisture by plants under varying conditions of soil moisture content.] - Pochvovedenie 1979(11): 88-94, 1979. [In R, ab: E.]

13264 - **DONE, A.A., MYERS, R.J.K., FOALE, M.A.:** Responses of grain sorghum to varying irrigation frequency in the Ord Irrigation Area. I. Growth, development and yield. - Aust. J. agr. Res. 35: 17-29, 1984.

13265 - **DONNELLY, D.J., VIDAVER, W.E.:** Leaf anatomy of red raspberry transferred from culture to soil. - J. Amer. Soc. hort. Sci. 109: 172-176, 1984.

13266 - **DONNELLY, D.J., VIDAVER, W.E.:** Pigment content and gas exchange of red raspberry *in vitro* and *ex vitro*. - J. Amer. Soc. hort. Sci. 109: 177-181, 1984.

13267 - **DOW, E.W., DAYNARD, T.B., MULDOON, J.F., MAJOR, D.J., THURTELL, G.W.:** Resistance to drought and density stress in Canadian and European maize (*Zea mays* L.) hybrids. - Can. J. Plant Sci. 64: 575-585, 1984.

13268 - **DOWDELL, R.J., WEBSTER, C.P.:** Effect of drought and irrigation on the fate of nitrogen applied to cut permanent grass swards in lysimeters: experimental design and crop uptake. - J. Sci. Food. Agr. 35: 1092-1104, 1984.

13269 - **DOWNTON, W.J.S.:** Salt tolerance of food crops: prospectives for improvements. - CRC critical Rev. Plant Sci. 1: 183-201, 1984.

13270 - **DRAKE, B.G.:** Light response characteristics of net CO_2 exchange in brackish wetland plant communities. - Oecologia 63: 263-270, 1984.

13271 - **DRAKE, B.G., GALLAGHER, J.L,:** Osmotic potential and turgor maintenance in *Spartina alterniflora* Loisel. - Oecologia 62: 368-375, 1984.

13272 - **DROZDOV, S.N., KURETS, V.K., TITOV, A.F.:** Termorezistentnost' Aktivno Vegetiruyushchikh Rasteniĭ. [Thermoresistance of Actively Growing Plants.] - Nauka, Leningradskoe Otdelenie, Leningrad 1984. [In R.]

*13273 - **DU, Z.-C., YANG, Z.-G..:** [Studies on characteristics of photosynthetic ecology in *Leymus chinensis*.] - Acta bot. Sin. 25: 370-379, 1983. [In Chin, ab: E.]

13274 - **DUNDON, C.G., SMART, R.E.:** Effects of water relations on the potassium status of Shiraz vines. - Amer. J. Enol. Viticult. 35: 40-45, 1984.

13275 - **DUNIN, F.X., ASTON, A.R.:** The development and proving of models of large scale evapotranspiration: an Australian study. - Agr. Water Manage. 8: 305-323, 1984. Also in: SHARMA, M.L. (ed.): Evapotranspiration from Plant

Communities. Pp. 305-323. Elsevier Science Publishers, Amsterdam - Oxford - New York - Tokyo 1984.

13276 - DUNLEAVY, P.J., COBB, A.H.: Bentazone-induced stomatal movement in epidermal peels from *Chenopodium album* L. I. Preliminary studies on the effect of light and carbon dioxide. - New Phytol. 97: 115-120, 1984.

13277 - DUNLEAVY, P.J., COBB, A.H.: Bentazone-induced stomatal movement in epidermal peels from *Chenopodium album* L. II. Characterization of the response to different concentrations of KCl and CO_2. - New Phytol. 97: 121-128, 1984.

13278 - DURAND, B.J., ORCAN, L., YANKO, U., ZAUBERMAN, G., FUCHS, Y.: Effects of waxing on moisture loss and ripening of 'Fuerte' avocado fruit. - Hort-Science 19: 421-422, 1984.

13279 - DÜRING, H.: Evidence for osmotic adjustment to drought in grapevines (*Vitis vinifera* L.). - Vitis 23: 1-10, 1984.

13280 - DURU, J.O.: Blaney-Morin-Nigeria evapotranspiration model. - J. Hydrol. 70: 71-83, 1984.

*13281 - DURZAN, D.J.: Metabolism of tritiated water during imbibition and germination of Jack pine seeds. - Can. J. Forest Res. 13: 1204-1218, 1983.

13282 - DUMAYRI, M.: Comparison of wheat cultivars grown in the field under different levels of moisture. - Cereal Res. Commun. 12: 27-34, 1984.

13283 - DVOŘÁK, I., DAINTY, J., JANÁČEK, K.: Time hierarchy in plant cell volume changes, - In: CRAM, W.J., JANÁČEK, K., RYBOVÁ, R., SIGLER, K. (ed.): Membrane Transport in Plants, Pp. 45-46. Academia, Praha 1984.

13284 - DWYER, L.M., STEWART, D.W.: Indicators of water stress in corn (*Zea mays* L.). - Can. J. Plant Sci. 64: 537-546, 1984.

13285 - DYBING, C.D., YARROW, G.L.: Morphactin effects on soybean leaf anatomy and chlorophyll content. - J. Plant Growth Regulation 3: 9-21, 1984.

*13286 - EAGLES, C.F., WILSON, D.: Photosynthetic efficiency and plant productivity. - In: RECHCIGL, M.,Jr. (ed.): CRC Handbook of Agricultural Productivity. Volume I. Plant Productivity. Pp. 213-247. CRC Press, Inc., Boca Raton 1982.

13287 - EAMUS, D., JENNINGS, D.H.: Determination of water, solute and turgor potentials of mycelium of various basidiomycete fungi causing wood decay. - J. exp. Bot. 35: 1782-1786, 1984.

13288 - EAMUS, D., WILSON, J.M.: A model for the interaction of low temperature, ABA, IAA, and CO_2 in the control of stomatal behaviour. - J. exp. Bot. 35: 91-98, 1984.

13289 - EAMUS, D., WILSON, J.M.: The effect of chilling temperatures on the water relation of leaf epidermal cells of *Rhoeo discolor*. - Plant Sci. Lett. 37: 101-104, 1984.

13290 - EASTHAM, J., OOSTERHUIS, D.M., WALKER, S.: Leaf water and turgor potential threshold values for leaf growth of wheat. - Agron. J. 76: 841-847, 1984.

13291 - ECK, H.V.: Irrigated corn yield response to nitrogen and water. - Agron. J. 76: 421-428, 1984.

*13292 - EDGLEY, M., BROWN, A.D.: Yeast water relations: physiological changes induced by solute stress in *Saccharomyces cerevisiae* and *Saccharomyces rouxii*. - J. gen. Microbiol. 129: 3453-3463, 1983.

13293 - EDWARDS, A., BOWLING, D.J.F.: An electrophysiological study of the stomatal complex of *Tradescantia virginiana*. - J. exp. Bot. 35: 562-567, 1984.

13294 - EDWARDS, W.R.N., BOOKER, R.E.: Radial variation in the axial conductivity of *Populus* and its significance in heat pulse velocity measurement. - J. exp. Bot. 35: 551-561, 1984.

13295 - EDWARDS, W.R.N., WARWICK, N.W.M.: Transpiration from a kiwifruit vine as estimated by the heat pulse technique and the Penman-Monteith equation. - N. Zeal. J. agr. Res. 27: 537-543, 1984.

13296 - EGGLI, U.: Stomatal types of *Cactaceae*. - Plant Syst. Evol. 146: 197-214, 1984.

13297 - EHLERINGER, J.R.: Intraspecific competitive effects on water relations, growth and reproduction in *Encelia farinosa*. - Oecologia 63: 153-158, 1984.

13298 - EHLERINGER, J.R., COOK, C.S.: Photosynthesis in *Encelia farinosa* Gray in response to decreasing leaf water potential. - Plant Physiol. 75: 688-693, 1984.

*13299 - EHRET, D.L., JOLLIFFE, P.A.: A semi-open system for measurements of plant gas exchange rates. - Photosynthetica 17: 523-531, 1983.

13300 - EHRLER, W.L., NAKAYAMA, F.S.: Water stress status in guayule as measured by relative leaf water content. - Crop Sci. 24: 61-66, 1984.

13301 - EHWALD, R.: Physikalische Grenzen der Nährstoffaufnahme durch die Wurzel. - Wiss. Z. Humboldt-Univ., math. naturwiss. Reihe 33: 328-329, 1984.

13302 - EHWALD, R., RICHTER, E., SCHLANGSTEDT, M.: Solute leakage from the isolated parenchyma of *Allium cepa* and *Kalanchoë daigremontiana*. - J. exp. Bot. 35: 1095-1103, 1984.

13303 - EIBACH, R., ALLEWELDT, G.: Einfluss der Wasserversorgung auf Wachstum, Gaswechsel und Substanzproduktion traubentragender Reben. II. Der Gaswechsel. - Vitis 23: 11-20, 1984.

13304 - EICH, D., KÖPPEN, D., SCHWÄRZEL, H., FICHTNER, E.: Einfluss von Beregnung und Jahreswitterung auf den Ertrag von Zuckerrüben. - Arch. Acker- Pflanzenbau Bodenk. 28: 471-476, 1984.

13305 - ELIÁŠ, P.: Adaptations of understorey species to exist in temperate deciduous forests. - In: MARGARIS, N.S., ARIANOUSTOU-FARRAGITAKI, M., OECHEL, W.C. (ed.): Being Alive on Land. Pp. 157-165. Dr W. Junk Publishers, The Hague 1984.

*13306 - EL-LAKANY, M.H., LUARD, E.J.: Comparative salt tolerance of selected *Casuarina* species. - Aust. Forest Res. 13: 11-20, 1982.

*13307 - ELLENSON, J.L., AMUNDSON, R.G.: Delayed light imaging for the early detection of plant stress. - Science 215: 1104-1106, 1982.

*13308 - ELLIOT, M.C.: The regulation of plant growth. - In: THOMAS, T.H. (ed.): Plant Growth Regulator Potential and Practice, Pp. 57-98. British Plant Growth Regulator Group, The British Crop Protection Council, Croydon 1982.

*13309 - **EL-MELEIGI, A., CLAFLIN, L.E., RANEY, R.J.:** Effect of seedborne *Fusarium moniliforme* and irrigation scheduling on colonization of root and stalk tissue, stalk rot incidence, and grain yields. - Crop Sci. 23: 1025-1028, 1983.

*13310 - **EL-SHAFEI, Y.Z., DARWISH, A.:** Effect of n-fertilization on wheat yield as affectd by soil moisture stress and rainfall under improved irrigation practice. - Agrochimica 24: 69-77, 1980.

13311 - **EL-SHARKAWY, M.A., COCK, J.H.:** Water use efficiency of cassava. I. Effects of air humidity and water stress on stomatal conductance and gas exchange. - Crop Sci. 24: 497-502, 1984.

13312 - **EL-SHARKAWY, M.A., COCK, J.H., De CADENA, G.:** Stomatal characteristics among cassava cultivars and their relation to gas exchange. - Exp. Agr. 20: 67-76, 1984.

13313 - **EL-SHARKAWY, M.A., COCK, J.H., De CADENA, G.:** Influence of differences in leaf anatomy on net photosynthetic rates of some cultivars of cassava. - Photosynthesis Res. 5: 235-242, 1984.

13314 - **EL-SHARKAWY, M.A., COCK, J.H.: HELD, K., A.A.:** Water use efficiency of cassava. II. Differing sensitivity of stomata to air humidity in cassava and other warm-climate species. - Crop Sci. 24: 503-507, 1984.

13315 - **EMEL'YANOV, L.G., BEL'KOVICH, T.M., ANISIMOV, A.V.:** Vodoobmen list'ev yachmenya v razlichnykh usloviyakh vodoobespecheniya. [Water exchange of barley leaves under various conditions of water supply.] - Fiziol. Biokhim. kul't. Rast. 16: 72-78, 1984. [In R, ab: E.]

13316 - **EMEL'YANOV, L.G., LOSEVA, N.L., GORDON, L.Kh., PETROV, V.E.:** Energetika prorostkov yachmenya v zavisimosti ot uslovii vlagoobespecheniya i teplovogo rezhima. [Energetics of barley seedlings depending on water supply and heat regime.] - Fiziol. Biokhim. kul't. Rast. 16: 335-341, 1984. [In R, ab: E.]

*13317 - **EMMERT, F.H.:** A system for concurrent measurement of centripetal passage of radiophosphorus and of water flow across roots to the xylem. - J. Plant Nutr. 6: 1017-1024, 1983.

13318 - **EMMERT, F.H.:** Concurrent measurement of centripetal passage of radiophosphorus and of water flow across intact roots exposed to various water stress agents. - J. Plant Nutr. 7: 1019-1026, 1984.

*13319 - **ENGENHART, M., BURIAN, K.:** Die Wirkung von Bleiionen auf die Durchflusskapazität von intakten *Phaseolus*-Wurzel-Systemen. - Flora 173: 415-427, 1983.

*13320 - **ENOCH, H.Z., COHEN, Y.:** Sensitivity of an infrared gas analyzer used in the differential mode, to partial gas pressures of carbon dioxide and water vapor in the bulk air. - Agr. Meteorol. 24: 131-138, 1981.

13321 - **ERICSON, M.C., ALFINITO, S.H.:** Proteins produced during salt stress in tobacco cell culture. - Plant Physiol. 74: 506-509, 1984.

13322 - **ERWEE, M.G., GOODWIN, P.B.:** Characterization of the *Egeria densa* leaf symplast: Response to plasmolysis, deplasmolysis and to aromatic amino acids. - Protoplasma 122: 162-168, 1984.

13323 - **ESIKOV, A.D., IVLEV, A.A., ONIPCHENKO, V.G., SHNOL', S.È.:** O fraktsioniro-
vanii izotopov pri dvizhenii vody v ksileme vysshikh rasteniĭ. [On isotope
fractionation at water movement in the xylem of higher plants.] - Biofizika
29: 1041-1045, 1084. [In R, ab: E.]

*13324 - **ESPINOZA, W.:** Efeito da densidade de plantio sobre a evapotranspiração do
milho irrigado na época da seca, em cerrado do distrito federal. [Effect
of plant population on corn evapotranspiration in an oxisol of the "cerrados".] - Pesq. agropec. Bras. 14: 343-350, 1979. [In Port, ab; E.]

*13325 - **ESPINOZA, W.:** Extração de água pelo milho em latossolo da região dos cerrados. [Rates and patterns of water uptake from an oxisol of the "cerrados"
of Brazil by corn crop.] - Pesq. agropec. Bras. 15: 69-78, 1980. [In Port,
ab: E.]

*13326 - **ESPINOZA, W., De AZEVEDO, J.A., Dos REIS, A.E.G.:** Variação do regime hídrico
em dois solos sob cerrados LVE e LVA em funcao da cobertura vegetal.
[Influence of soil cover on the field moisture regimes of two "cerrado"
oxisols: LVE and LVA.] - Pesq. agropec. Bras. 15: 283-295, 1980. [In Port,
ab: E.]

*13327 - **ESSIAMAH, S.K.:** Spring sap of trees. - Ber. Deut. bot. Ges. 93: 257-267,
1980.

13328 - **ETHERINGTON, J.R.:** Relationship between morphological adaptation to grazing,
carbon balance and waterlogging tolerance in clones of Dactylis glomerata L.
- New Phytol. 98: 647-658, 1984.

13329 - **ETHERINGTON, J.R.:** Comparative studies of plant growth and distribution in
relation to waterlogging. X. Differential formation of adventitious roots
and their experimental excision in Epilobium hirsutum and Chamerion angustifolium. - J. Ecol. 72: 389-404, 1984.

13330 - **EVANS, D.W., PEADEN, R.N.:** Seasonal forage growth rate and solar energy
conversion of irrigated vernal alfalfa. - Crop Sci. 24: 981-984, 1984.

13331 - **EVANS, L.S.:** Botanical aspects of acidic precipitation. - Bot. Rev. 50:
449-490, 1984.

13332 - **EVANS, L.S., THOMPSON, K.H.:** Comparison of experimental designs used to
detect changes in yields of crops exposed to acidic precipitation. - Agron.
J. 76: 81-84, 1984.

13333 - **EVELING, D.W., BATAILLÉ, A.:** The effect of deposits of small particles on
the resistance of leaves and petals to water loss. - Environ. Pollut. Ser.
A. 36: 229-238, 1984.

13334 - **EWERS, F.W., ZIMMERMANN, M.H.:** The hydraulic architecture of balsam fir
(Abies balsamea). - Physiol. Plant. 60: 453-458, 1984.

13335 - **EWERS, F.W., ZIMMERMANN, M.H.:** The hydraulic architecture of eastern hemlock
(Tsuga canadensis). - Can. J. Bot. 62: 940-946, 1984.

*13336 - **FAENSEN-THIEBES, A.:** Veränderungen im Gaswechsel, Chlorophyllgehalt und
Zuwachs von Nicotiana tabacum L. und Phaseolus vulgaris L. durch Ozon und
deren Beziehung zur Ausbildung von Blattnekrosen. - Angew. Bot. 57: 181-191,
1983.

13337 - **FAGERSTEDT, K.:** Comparisons of flooding tolerance in four barley cultivars
(Hordeum vulgare) grown in Finland, and in Carex rostrata. - Ann. bot. Fenn.
21: 293-298, 1984.

13338 - **FAHEY, T.J., YOUNG, D.R.:** Soil and xylem water potential and soil water content in contrasting *Pinus contorta* ecosystems, southeastern Wyoming, USA. - Oecologia 61: 346-351, 1984.

13339 - **FALCON, M.F., FOX, R.L., TRUJILLO, E.E.:** Interactions of soil pH, nutrients and moisture on phytophthora root rot of avocado. - Plant Soil. 81: 165-176, 1984.

13340 - **FANJUL, L., ROSHER, P.H.:** Effects of water stress on internal water relations of apple leaves. - Physiol. Plant. 62: 321-328, 1984.

13341 - **FAPOHUNDA, H.O., AINA, P.O., HOSSAIN, M.M.:** Water use - yield relations for cowpea and maize. - Agr. Water Manage. 9: 219-224, 1984.

13342 - **FARAH, S.M., REGINATO, R.J., NAKAYAMA, F.S.:** Calibration of soil surface neutron moisture meter. - Soil Sci. 138: 235-239, 1984.

13343 - **FARQUHAR, G.D., RICHARDS, R.A.:** Isotopic composition of plant carbon correlates with water-use efficiency of wheat genotypes. - Aust. J. Plant Physiol. 11: 539-552, 1984.

13344 - **FARQUHAR, G.D., WONG, S.C.:** An empirical model of stomatal conductance. - Aust. J. Plant Physiol. 11: 191-209, 1984.

13345 - **FARRIS, M.A.:** Leaf size and shape variation associated with drought stress in *Rumex acetosella* L. (Polygonaceae). - Amer. Midl. Natur. 111: 358-363, 1984.

13346 - **FAWUSI, M.O.A., ORMROD, D.P., EASTHAM, A.M.:** Response to water stress of *Celosia argentea* and *Corchorus olitorius* in controlled environments. - Scientia Hort. (Amsterdam) 22: 163-172, 1984.

*13347 - **FEDDES, R.A.:** Simulation of field water use and crop yield. - In: PENNING de VRIES, F.W.T., VAN LAAR, H.H. (ed.): Simulation of Plant Growth and Crop Production. Pp. 194-209. PUDOC, Wageningen 1982.

13348 - **FELDHAKE, C.M., DANIELSON, R.E., BUTLER, J.D.:** Turfgrass evapotranspiration. II. Responses to deficit irrigation. - Agron. J. 76: 85-89, 1984.

*13349 - **FELIPE, M.R., RODRIQUEZ-PASCUAL, M.L.:** Ultrastructural changes in leaves of maize (*Zea mays* L.) and safflower (*Carthamus tinctoreus* L.) in response to NaCl stress. - In: METZNER, H. (ed.): Photosynthesis and Plant Productivity. Pp. 239-243. Wissenschaftliche Verlagsgesellschaft, Stuttgart 1983.

13350 - **FERERES, E.:** Variability in adaptive mechanisms to water deficits in annual and perennial crop plants. - Bull. Soc. bot. France - Actual. bot. 131: 17-32, 1984.

13351 - **FERGUSON, L., DAVIES, F.S., ISMAIL, M.A., WHEATON, T.A.:** Growth regulator and low-volume irrigation effects on grapefruit quality and fruit-drop. - Scientia Hort. (Amsterdam) 23: 35-40, 1984.

13352 - **FERRARI-ILIOU, R., PHAM THI, A.T., VIEIRA da SILVA, J.:** Effect of water stress on the lipid and fatty acid composition of cotton (*Gossypium hirsutum*) chloroplasts. - Physiol. Plant. 62: 219-224, 1984.

13353 - **FERREE, D.C., CAHOON, G.A., BRAZEE, R.D., FOX, R.D.:** Frost control irrigation under severe conditions. - Ohio Rep. 69: 24-26, 1984.

13354 - **FERRON, F., COUDRET, A.:** Metabolic pathways of photosynthesis in marine algae. - Physiol. vég. 22: 103-113, 1984.

*13355 - FEYEN, J., Van der VEKEN, L., BOSSCHAERTS, L., DECKERS, J.C.: Evaluation
expérimentale de l'irrigation par goutte à goutte des poires dans les
conditions climatique belges. - Rev. Agr. 36: 1641-1660, 1983.

13356 - FICK, G.W.: Simple simulation models for yield prediction applied to
alfalfa in the Northeast. - Agron. J. 76: 235-239, 1984.

13357 - FIEDLER, H.J., POLLEY, H., HÖHNE, H.: Gefässversuch zur Einwirkung verschiedener Stickstoff-Formen auf Kiefernsämlinge bei unterschiedlicher Versorgung
mit Magnesium, Kalium, Bor und Wasser. - Arch. Acker- Pflanzenbau Bodenk.
28: 435-439, 1984.

*13358 - FIELD, C., MERINO, J., MOONEY, H.A.: Compromises between water-use effi-
ciency and nitrogen-use efficiency in five species of California evergreens.
- Oecologia 60: 384-389, 1983.

13359 - FISCUS, E.L., WULLSCHLEGER, S.D., DUKE, H.R.: Integrated stomatal opening
as an indicator of water stress in Zea. - Crop Sci. 24: 245-249, 1984.

13360 - FISCUS, E.L., WULLSCHLEGER, S.D., DUKE, H.R.: Stomatal sensors control the
water supply to Zea mays. - In: Agricultural Electronics - 1983 and Beyond.
Volume I. Pp. 278-285. American Society of Agricultural Engineers, St. Joseph
1984.

13361 - FLETCHER, R.A., NATH, V.: Triadimefon reduces transpiration and increases
yield in water stressed plants. - Physiol. Plant. 62: 422-426, 1984.

13362 - FLINN, J.C., DE DATTA, S.K.: Trends in irrigated-rice yields under inten-
sive cropping at Philippine research stations. - Field Crops Res. 9: 1-15,
1984.

13363 - FLINN, J.C., MAMARIL, C.P., VELASCO, L.E., KAISER, K.: Efficiency of modified
urea fertilizers for tropical irrigated rice. - Fertilizer Res. 5: 157-174,
1984.

13364 - FLORES, H.E., GALSTON, A.W.: Osmotic stress-induced polyamine accumulation
in cereal leaves. I. Physiological parameters of the response. - Plant
Physiol. 75: 102-109, 1984.

13365 - FLORES, H.E., GALSTON, A.W.: Osmotic stress-induced polyamine accumulation
in cereal leaves. II. Relation to amino acid pools. - Plant Physiol. 75:
110-113, 1984.

*13366 - FLORET, C., Le FLOC'H, E., PONTANIER, R.: Phytomasse et production végétale
en Tunisie présaharienne. - Acta oecol. - Oecol. Plant. 4: 133-152, 1983.

*13367 - FLORIA, F.G., GHIORGHȚĂ, G.I.: The influence of the treatment with alky-
lating agents on Papaver somniferum L. in M_1. - Rev. Roum. Biol. - Biol.
vég. 25: 151-156, 1980.

13368 - FLOYD, C.N.: Model experiment on the effect of a plough pan on crop yield
under differing conditions of soil moisture availability. - Soil Tillage
Res. 4: 175-189, 1984.

13369 - FOALE, M.A., WILSON, G.L., COATES, D.B., HAYDOCK, K.P.: Growth and productivity of irrigated Sorghum bicolor (I. Moench) in northern Australia.
II. Low solar altitude as a possible seasonal constraint to productivity
in the tropical dry season. - Aust. J. agr. Res. 35: 229-238, 1984.

13370 - FORD, C.W.: Accumulation of low molecular weight solutes in water-stressed
tropical legumes. - Phytochemistry 23: 1007-1015, 1984.

13371 - **FOREST, F., KALMS, J.-M.:** Influence du régime d'alimentation en eau sur la production du riz pluvial. Simulation du bilan hydrique. - Agron. Trop. 39: 42-49, 1984.

13372 - **FORSCHNER, W., REUTHER, G.:** Photosynthese und Wasserhaushalt von Pelargonium -Stecklingen während der Bewurzelung unter dem Einfluss verschiedener Licht- und CO_2-Bedingungen. - Gartenbauwissenschaft 49: 182-190, 1984.

13373 - **FORSETH, I.N., EHLERINGER, J.R., WERK, K.S., COOK, C.S.:** Field water relations of sonar desert annuals. - Ecology 65: 1436-1444, 1984.

13374 - **FRANCE, J., THORNLEY, J.H.M.:** Mathematical Models in Agriculture. A Quantitative Approach to Problems in Agriculture and Related Sciences. - Butterworth, London - Boston - Durban - Singapore - Sydney - Toronto - Wellington 1984.

*13375 - **FRANCIS, P.E., PIDGEON, J.D.:** A model for estimating soil moisture deficits under cereal crops in Britain. 1. Development. - J. agr. Sci. 98: 651-661, 1982.

*13376 - **FRANCIS, P.E., PIDGEON, J.D.:** A model for estimating soil moisture deficits under cereal crops in Britain. 2. Performance. - J. agr.. Sci. 98: 663-678, 1982.

13377 - **FRANCOIS, L.E.:** Salinity effects on germination, growth, and yield of turnips. - HortScience 19: 82-84, 1984.

13378 - **FRANCOIS, L.E., DONOVAN, T., MAAS, E.V.:** Salinity effects on seed yield, growth, and germination of grain sorghum. - Agron. J. 76: 741-744, 1984.

13379 - **FRANK, A.B., BARKER, R.E., BERDAHL, J.D.:** Pressure-volume characteristics of genotypes of three wheatgrass species. - Crop Sci. 24: 217-220, 1984.

13380 - **FRANK, A.B., BAUER, A.:** Cultivar, nitrogen, and soil water effects on apex development in spring wheat. - Agron. J. 76: 656-660, 1984.

13381 - **FRANK, P.S.,Jr., TRYON, E.H.:** Annual radial increment of oak increased by spray irrigation of campground sewage effluent. - Castanea 49: 86-90, 1984.

*13382 - **FRANK, R., MAREK, M.:** The response of the net photosynthetic rate to irradiance in barley leaves as influenced by nitrogen supply. - Photosynthetica 17: 572-577, 1983.

13383 - **FREER-SMITH, P.H.:** The responses of six broadleaved trees during long-term exposure to SO_2 and NO_2. - New Phytol. 97: 49-61, 1984.

13384 - **FRENCH, R.J., SCHULTZ, J.E.:** Water use efficiency of wheat in a Mediterranean-type environment. I. The relation between yield, water use and climate. - Aust. J. agr. Res. 35: 743-764, 1984.

13385 - **FRENCH, R.J., SCHULTZ, J.E.:** Water use efficiency of wheat in a Mediterranean-type environment. II. Some limitations to efficiency. - Aust. J. agr. Res. 35: 765-775, 1984.

13386 - **FRENKEL, H.:** Reassessment of water quality criteria for irrigation. - In: SHAINBERG, I., SHALHEVET, J. (ed.): Soil Salinity under Irrigation. Processes and Management. Pp. 143-167. Springer-Verlag, Berlin - Heidelberg - New York - Tokyo 1984.

*13387 - **FRESCO, L.F.M.:** An analysis of species response curves and of competition from field data: Some results from heath vegetation. - Vegetatio 48: 175-185, 1982.

13388 - **FRIEDEL, M.H.:** Biomass and nutrient changes in the herbaceous layer of two central Australian mulga shrublands after unusually high rainfall. - Aust. J. Ecol. 9: 27-38, 1984.

13389 - **FUKUTOKU, Y., YAMADA, Y.:** Sources of proline-nitrogen in water-stressed soybean (*Glycine max*). II. Fate of ^{15}N-labelled protein. - Physiol. Plant. 61: 622-628, 1984.

13390 - **FULBRIGHT, T.E., WILSON, A.M., REDENTE, E.F.:** Effects of temporary dehydration on growth of green needlegrass (*Stipa viridula* Trin.) seedlings. - J. Range Manage. 37: 462-464, 1984.

*13391 - **FUNES, F., PEREZ, L., RONDA, A.:** Growth and development of grasses in Cuba. 2. Effect of three cutting intervals on the yield of eight grasses. - Cuban J. agr. Sci. 14: 181-188, 1980.

*13392 - **GACHKOVSKIĬ, V.F.:** O prevrashcheniyakh nativnykh sostoyaniĭ khlorofilla zelenogo lista rasteniĭ pri narusheniyakh ego struktury. [Transformations of native states of chlorophyll of green leaves caused by disruptions of their structure.] - Dokl. Akad. Nauk SSSR 270: 239-242, 1983. [In R.]

13393 - **GAFF, D.F., LOVEYS, B.R.:** Abscisic acid content and effects during dehydration of detached leaves of desiccation tolerant plants. - J. exp. Bot. 35: 1350-1358, 1984.

*13394 - **GALATIS, B., APOSTOLAKOS, P., KATSAROS, C.:** Synchronous organization of two preprophase microtubule bands and final cell plate arragement in subsidiary cell mother cells of some *Triticum* species. - Protoplasma 117: 24-39, 1983.

13395 - **GALES, K., AYLING, S.M., CANNELL, R.Q.:** Effects of waterlogging and drought on winter wheat and winter barley grown on a clay and sandy loam soil. II. Soil and plant water relationships. - Plant Soil 80: 67-78, 1984.

13396 - **GAMBHIR, P.N., PANDA, B.C., PURI, R.K.:** Proton spin-spin relaxation study of moisture & oil in seeds. - Ind. J. exp. Biol. 21: 460-461, 1984.

13397 - **GAMBLE, P.E., BURKE, J.J.:** Effect of water stress on the chloroplast antioxidant system. I. Alterations in glutathione reductase activity. - Plant Physiol. 76: 615-621, 1984.

*13398 - **GARDNER, C.M.K., FIELD, M.:** An avaluation of the success of MORECS, a meteorological model, in estimating soil moisture deficits. - Agr. Meteorol. 29: 269-284, 1983.

*13399 - **GARDNER, W.K., VELTHUIS, R.G., AMOR, R.L.:** Field crop production in Southwest Victoria. I. Area description, current land use and potential for crop production. - J. Aust. Inst. agr. Sci. 49: 60-70, 1983.

*13400 - **GARG, B.K., GARG, O.P.:** Salinity and plant nutrition - effect of sodium carbonate and sodium bicarbonate on the growth and absorption of essential macro-nutrients and sodium in pea (*Pisum sativum* L.). - Proc. Ind. nat. Sci. Acad., Sect. B 46: 694-698, 1980.

*13401 - **GARG, B.K., GARG, O.P.:** Metabolism of germinating moong seeds under saline -alkaline conditions due to sodium carbonate and sodium bicarbonate. - Ind. J. Plant Physiol. 24: 171-174, 1981.

*13402 - **GARG, B.K., GARG, O.P.:** Effect of sodium carbonate and bicarbonate on seed germination. - Geobios 8: 31-33, 1981.

13403 - **GARG, B.K., VYAS, S.P., KATHJU, S., LAHIRI, A.N.:** Influence of repeated
water stress on wheat. - Proc. Ind. Acad. Sci. - Plant Sci. 93: 477-484,
1984.

13404 - **GARG, B.K., VYAS, S.P., KATHJU, S., LAHIRI, A.N.:** Effect of saline waters
on pearl millet under drought. - Ann. Arid Zone 23: 41-45, 1984.

13405 - **GARRATT, J.R.:** The measurement of evaporation by meteorological methods. -
Agr. Water Manage. 8: 99-117, 1984. Also in: SHARMA, M.L. (ed.): Evapotranspiration from Plant Communities. Pp. 99-117. Elsevier Science Publishers, Amsterdam - Oxford - New York - Tokyo 1984.

13406 - **GARRITY, D.P., SULLIVAN, C.Y., WATTS, D.G.:** Changes in grain sorghum stomatal and photosynthetic response to moisture stress across growth stages. -
Crop Sci. 24: 441-446, 1984.

13407 - **GARRITY, D.P., SULLIVAN, C.Y., WATTS, D.G.:** Rapidly determining sorghum
canopy photosynthetic rates with a mobile field chamber. - Agron. J. 76:
163-165, 1984.

13408 - **GÄRTNER, M.:** Anzahl und Lage der Spaltöffnungen als Kenngrösse der ökolo-
gischen Typen. - Wiss. Z. Humboldt-Univ. math.- naturwiss. Reihe 33: 307-309,
1984.

13409 - **GAUSMAN, H.W.:** Evaluation of factors causing reflectance differences between
sun and shade leaves. - Remote Sensing Environ. 15: 177-181, 1984.

13410 - **GAUSMAN, H.W., BURKE, J.J., QUISENBERRY, J.E.:** Use of leaf optical properties in plant stress research. - In: ORY, R.L., RITTIG, F.R. (ed.): Bioregulators: Chemistry and Uses. Pp. 215-233. American Chemical Society,
Washington 1984.

13411 - **GAUSMAN, H.W., QUISENBERRY, J.E., BURKE, J.J., WENDT, C.W.:** Leaf spectral
measurements to screen cotton strains for characters affected by stress. -
Field Crops Res. 9: 373-381, 1984.

13412 - **GEAY, A., VARTANIAN, N., QUEIROZ, O.:** Variations des teneurs en polyamines
et leurs précurseurs au cours de l'adaptation morphogénétique du Colza,
Brassica napus L. var. *oleifera* M. à la sécheresse. - Bull. Soc. Bot.
France, Actual. bot. 131: 99-111, 1984.

13413 - **GEBHARDT, A.:** Beurteilung des Wasserversorgungszustandes von Pflanzenbeständen mittels Thermografie. - Arch. Acker- Pflanzenbau Bodenk. 28: 231-237,
1984.

*13414 - **GEDDES, R.D., SCOTT, H.D., OLIVER, L.R.:** Growth and water use by common
cocklebur (*Xanthium pensylvanicum*) and soybeans (*Glycine max*) under field
conditions. - Weed Sci. 27: 206-212, 1979.

13415 - **GEISLER, D., FERREE, D.C.:** The influence of root pruning on water relations,
net photosynthesis, and growth of young 'Golden Delicious' apple trees. -
J. Amer. Soc. hort. Sci. 109: 827-831, 1984.

13416 - **GEORGE, M.F., BURKE, M.J.:** Supercooling of tissue water to extreme low
temperature in overwintering plants. - Trends biochem. Sci. 9: 211-214,
1984.

*13417 - **GEORGHIOU, K., PSARAS, G., MITRAKOS, K.:** Lettuce endosperm structural changes
during germination under different light, temperature, and hydration conditions. - Bot. Gaz. 144: 207-211, 1983.

13418 - **GERDENITSCH, W.:** Microscopic contributions to the pressure-volume diagram of the cell-water relations as demonstrated with tissue cells. - Protoplasma 119: 35-47, 1984.

13419 - **GERSHENZON, J.:** Changes in the levels of plant secondary metabolites under water and nutrient stress. - In: TIMMERMANN, B.N., STEELINK, C., LOEWUS, F.A. (ed.): Phytochemical Adaptations to Stress. Pp. 273-320. Plenum Press, New York - London 1984.

13420 - **GIBSON, A.C., CALKIN, H.W., NOBEL, P.S.:** Xylem anatomy, water flow, and hydraulic conductance in the fern *Cyrtomium falcatum*. - Amer. J. Bot. 71: 564-574, 1984.

13421 - **GIBSON, T.S., SPEIRS, J., BRADY, C.J.:** Salt-tolerance in plants. II. *In vitro* translation of m-RNAs from salt-tolerant and salt-sensitive plants on wheat germ ribosomes. Responses to ions and compatible organic solutes. - Plant Cell Environ. 7: 579-587, 1984.

13422 - **GILBERTZ, D.A., BARRETT, J.E., NELL, T.A.:** Development of drought-stressed poinsettias. - J. Amer. Soc. hort. Sci. 109: 854-857, 1984.

13423 - **GILLESPIE, T.J., BARR, A.:** Adaptation of a dew estimation scheme to a new crop and site. - Agr. Forest Meteorol. 31: 289-295, 1984.

13424 - **GILMOUR, D.J., HIPKINS, M.F., BONEY, A.D.:** The effect of osmotic and ionic stress on the primary processes of photosynthesis in *Dunaliella tertiolecta*. - J. exp. Bot. 35: 18-27, 1984.

13425 - **GILMOUR, D.J., HIPKINS, M.F., BONEY, A.D.:** The effect of decreasing the external salinity on the primary processes of photosynthesis in *Dunaliella tertiolecta*. - J. exp. Bot. 35: 28-35, 1984.

13426 - **GILMOUR, D.J., HIPKINS, M.F., BONEY, A.D.:** The effect of increasing the external salinity on long and short-term fluorescence parameters in *Dunaliella tertiolecta*. - In: SYBESMA, C. (ed.): Advances in Photosynthesis Research. Volume IV. Pp. 427-430. Martinus Nijhoff / Dr W. Junk Publishers, The Hague - Boston - Lancaster 1984.

*13427 - **GIMMLER, H., LOTTER, G.:** DEAE-dextran induced increase of membrane permeability and inhibition of photosynthesis in *Dunaliella parva*. - Z. Naturforsch. 37c: 609-619, 1982.

*13428 - **GLAIZE, S., ROCHER, J.-P.:** Physiologie des cotylédons de l'embryon de Pommier en relation avec la dormance IV. Différenciation tissulaire et activités photosynthétiques. - Bull. Soc. bot. France, Actual. bot. 130: 127-134, 1983.

13429 - **GLENN, E.P., O'LEARY, J.W.:** Relationship between salt accumulation and water content of dicotyledonous halophytes. - Plant Cell Environ. 7: 253-261, 1984.

13430 - **GOEDHEER, J.C.:** Fluorescence polarization spectra of cells, chromatophores, and chromatophore fractions of *Rhodospirillum rubrum*: dependence upon environmental factors. - Photobiochem. Photobiophys. 7: 91-101, 1984.

13431 - **GOLBERG, A.D., BIERNY, O., RENARD, C.:** Évolution comparée des paramètres hydriques chez *Coffea canephora* Pierre et l'hybride *Coffea arabusta* Capot et Aké Assi. Soumis à deux cycles de sécheresse en conditions contrôlées. - Café Cacao Thé 28: 257-266, 1984.

*13432 - GOLDSTEIN, G., MEINZER, F.: Influence of insulating dead leaves and low temperatures on water balance in an Andean giant rosette plant. - Plant Cell Environ. 6: 649-656, 1983.

13433 - GOLDSTEIN, G., MEINZER, F., MONASTERIO, M.: The role of capacitance in the water balance of Andean giant rosette species. - Plant Cell Environ. 7: 179-186, 1984.

13434 - GOLKIN, K.R., EWEL, K.C.: A computer simulation of the carbon, phosphorus, and hydrologic cycles of a pine flatwoods ecosystem. - Ecol. Model. 24: 113-136, 1984.

13435 - GONCHAROVA, É.A., UDOVENKO, G.V.: Zavisimost' intensivnosti potokov vody v rasteniyakh ot temperaturno-svetovykh usloviǐ i urovnya vodoobespechennos-ti. [Effect of temperature, irradiance and water supply on the rate of water transport in plants.] - Fiziol. Biokhim. kul't. Rast. 16: 176-180, 1984. [In R, ab: E.]

13436 - GONCHAROVA, É.A., UDOVENKO, G.V., NECHIPORENKO, G.A., ZHOLKEVICH, V.N.: Izuchenie vodoobmena plodov kabachka s pomoshch'yu tritievoǐ metki. [Study of water exchange in vegetable marrow fruit with the aid of tritium label.] - Fiziol. Rast. 31: 841-846, 1984. [In R, ab: E.]

*13437 - GOODE, J.E., HIGGS, K.H., HYRYCZ, K.J.: Effects of water stress control in apple trees by misting. - J. hort. Sci. 54: 1-11, 1979.

*13438 - GOREL'KO, N.M.: Produktivnost' fotosinteza yachmenya v zavisimosti ot meteorologicheskikh i agrotekhnicheskikh usloviǐ vozdelyvaniya. [Productivity of barley photosynthesis in dependence on meteorological conditions and agrotechnical management.] - Sbornik nauch. Trudov Belorus. sel'skokhoz. Akad. 102 (Pochvennye Protsessy i Regulirovanie Pitaniya Rasteniǐ): 85-90, 1983. [In R.]

13439 - GORHAM, J., McDONNELL, E., WYN JONES, R.G.: Pinitol and other solutes in salt-stressed *Sesbania aculeata*. - Z. Pflanzenphysiol. 114: 173-178, 1984.

13440 - GORHAM, J., McDONNELL, E., WYN JONES, R.G.: Salt tolerance in the Triticeae: *Leymus sabulosus*. - J. exp. Bot. 35: 1200-1209, 1984.

13441 - GÖRING, H., EHWALD, R., BLEISS, W.: Osmotic relations of growing coleoptile segments. - In: CRAM, W.J., JANÁČEK, K., RYBOVÁ, R., SIGLER, K. (ed.): Membrane Transport in Plants. Pp. 83-89. Academia, Praha 1984.

13442 - GÖRING, H., KOSHUCHOWA, S., MÜNNICH, H., DIETRICH, M.: Stomatal opening and cell enlargement in response to light and phytochrome treatments in primary leaves of red-light-grown seedlings of *Phaseolus vulgaris* L. - Plant Cell Physiol. 25: 683-690, 1984.

13443 - GÓRNY, A.G., PATYNA, H.: The development of the root system in seven spring barley varieties under high and low soil irrigation levels. - Z. Acker-Pflanzenbau 153: 264-273, 1984.

13444 - GOSS, M.J., HOWSE, K.R., VAUGHAN-WILLIAMS, J.M., WARD, M.A., JENKINS, W.: Water use by winter wheat as affected by soil management. - J. agr. Sci. 103: 189-199, 1984.

13445 - GOTOW, K., SHIMAZAKI, K., KONDO, N., SYŌNO, K.: Photosynthesis-dependent volume regulation in guard cell protoplasts from *Vicia faba* L. - Plant Cell Physiol. 25: 671-675, 1984.

*13446 - **GOUDRIAAN, J.:** Potential production processes. - In: PENNING DE VRIES, F.W.T., VAN LAAR, H.H. (ed.): Simulation of Plant Growth and Crop Production. Pp. 98-113, PUDOC, Wageningen 1982.

13447 - **GOUDRIAAN, J., VAN LAAR, H.H., VAN KEULEN, H., LOUWERSE, W.:** Simulation of the effect of increased atmospheric CO_2 on assimilation and transpiration of a closed crop canopy. - Wiss. Z. Humboldt-Univ., math. naturwiss. Reihe 33: 352-356, 1984.

13448 - **GRANT, J.A., RYUGO, K.:** Influence of within-canopy shading on net photosynthetic rate, stomatal conductance, and chlorophyll content of kiwifruit leaves. - HortScience 19: 834-836, 1984.

13449 - **GREACEN, E.L., HIGNETT, C.T.:** Water balance under wheat modelled with limited soil data. - Agr. Water Manage. 8: 291-304, 1984. Also in: SHARMA, M.L. (ed.): Evapotranspiration from Plant Communities. Pp. 291-304. Elsevier Science Publishers, Amsterdam - Oxford - New York - Tokyo 1984.

13450 - **GREEN, A.E., CLOTHIER, B.E., KERR, J.P., SCOTTER, D.R.:** Evapotranspiration from pasture: a comparison of lysimeter and Bowen ratio measurements with Priestly-Taylor estimates. - N. Zeal. agr. Res. 27: 321-327, 1984.

*13451 - **GREEN, T.G.A., JANE, G.T.:** Diurnal patterns of water potential in the evergreen cloud forests of the Kaimai Ranges, North Island, New Zealand. - N. Zeal. J. Bot. 21: 379-389, 1983.

*13452 - **GREEN, T.G.A., JANE, G.T.:** Changes in osmotic potential during bud break and leaf development of *Nothofagus menziesii, Weinmannia racemosa, Quintinia acutifolia* and *Ixerba brexioides*. - N. Zeal. J. Bot. 21: 391-395, 1983.

*13453 - **GREENWAY, H., WATKIN, E.:** Effects of external NaCl, and of changes in turgor pressure and volume, on K^+ concentrations in *Chlorella emersonii*. - Plant Cell Environ. 6: 393-400, 1983.

13454 - **GREGORCZYK, A.:** Zastosowanie terminologii termodynamicznej w gospodarce wodnej roślin. [An adaptation of thermodynamic terminology in the plant water management.] - Wiadomości bot. 28: 191-200, 1984. [In Pol, ab: E.]

13455 - **GREGORIOU, C., RAJ KUMAR, D.:** Effects of irrigation and mulching on shoot and root growth of avocado (*Persea americana* Mill.) and mango (*Magnifera indica* L.). - J. hort. Sci. 59: 109-117, 1984.

13456 - **GREGORY, P.J.:** Water availability and crop growth in arid regions. - Outlook Agr. 13: 208-215, 1984.

13457 - **GREGORY, P.J., SHEPHERD, K.D., COOPER, P.J.:** Effects of fertilizer on root growth and water use of barley in northern Syria. - J. agr. Sci. 103: 429-438, 1984.

*13458 - **GRENNFELT, P., BENGTSON, C., SKÄRBY, L.:** Deposition and uptake of atmospheric nitrogen oxides in a forest ecosystem. - Aquilo Ser. Bot. 19: 208-221, 1983.

13459 - **GREWAL, S.S., SUD, A.D., SINGH, K., SINGH, P.:** Effect of minimal irrigation on profile-water use, water expense and yield of chickpea as modified by seasonal rainfall in Siwalik region. - Ind. J. agr. Sci. 54: 576-581, 1984.

13460 - **GRIEVE, C.M., MAAS, E.V.:** Betaine accumulation in salt-stressed sorghum. - Physiol. Plant. 61: 167-171, 1984.

*13461 - GRIGORYUK, I.A., SHMAT'KO, I.G.: Izmeneniya v vodoobmene semyan i prorostkov pshenitsy pri temperaturnom vozdeĭstvii. [Changes in water relations of wheat seeds and seedlings caused by temperature.] - In: SHMAT'KO, I.G. (ed.): Vodnyĭ Rezhim Rasteniĭ v Svyazi s Deĭstviem Faktorov Sredy. Pp. 6-21, 189-190. Naukova Dumka, Kiev 1983. [In R.]

13462 - GRIMME, K.: Freilanduntersuchungen zum Wasserhaushalt von *Mercurialis perennis* und *Asarum europaeum*. - Flora 175: 249-256, 1984.

*13463 - GRODZINSKI, B., BOESEL, I., HORTON, R.F.: Light stimulation of ethylene release from leaves of *Gomphrena globosa* L. - Plant Physiol. 71: 588-593, 1983.

13464 - GROSE, M.J., PARKER, C.A., SIVASITHAMPARAM, K.: Growth of *Gaeumannomyces graminis* var. *tritici* in soil: effects of temperature and water potential. - Soil Biol. Biochem. 16: 211-216, 1984.

13465 - GROSSMANN, K., JUNG, J.: The influence of new terpenoid analogues of abscisic acid on stomatal movement and leaf senescence. - Z. Acker- Pflanzenbau 153: 14-22, 1984.

13466 - GROSSNICKLE, S.C., REID, C.P.P.: Water relations of Engelmann spruce seedlings on a high-elevation mine site: an example of how reclamation techniques can alter microclimate and edaphic conditions. - Reclamation Reveget.tion Res. 3: 199-221, 1984.

13467 - GU, Z.-M., SHEN, Z.-Y., ZHANG, Z.-L., YAN, J.-Q.: [The effect of cytokinin on the growth of water-stressed wheat coleoptiles.] - Acta phytophysiol. Sin. 10: 353-361, 1984. [In Chin, ab: E.]

13468 - GUERRIER, G.: Selectivité de fixation du sodium au niveau des embryons et des jeunes plantes sensible ou tolerante au NaCl. - Can. J. Bot. 62: 1791-1798, 1984.

13469 - GUINN, G., MAUNEY, J.R.: Fruiting of cotton. I. Effects of moisture status on flowering. - Agron. J. 76: 90-94, 1984.

13470 - GUINN, G., MAUNEY, J.R.: Fruiting of cotton. II. Effects of plant moisture status and active boll load on boll retention. - Agron. J. 76: 94-97, 1984.

13471 - GULYAEV, B.I., SHVEDOVA, O.E.: Ust'ichnyĭ porometer i ego ispol'zovanie dlya otsenki sostoyaniya listovogo apparata. [Stomatal porometer and its use for estimation of the leaf apparatus state.] - Fiziol. Biokhim. kul't. Rast. 16: 504-506, 1984. [In R, ab: E.]

13472 - GURALNICK, L.J., RORABAUGH, P.A., HANSCOM, Z.,III.: Influence of photoperiod and leaf age on Crassulacean acid metabolism in *Portulacaria afra* (L.) Jacq. - Plant Physiol. 75: 454-457, 1984.

13473 - GURALNICK, L.J., RORABAUGH, P.A., HANSCOM, Z.,III.: Seasonal shifts of photosynthesis in *Portulacaria afra* (L.) Jacq. - Plant Physiol. 76: 643-646, 1984.

*13474 - GURICHEVA, N.P., IZMAĬLOVA, N.N., SLEMNEV, N.N., LKHAGVASUREN, S.: *Stellera chamaejasme* (*Thymelaceae*) v stepyakh vostochnogo Khangaya. [*Stellera chamaejasme* (*Thymelaceae*) in the steppes of eastern Khangai.] - Bot. Zh. 68: 453-463, 1983. [In R.]

13475 - GURNEY, R.J., CAMILLO, P.J.: Modelling daily evapotranspiration using remotely sensed data. - J. Hydrol. 69: 305-324, 1984.

*13476 - **GURNEY, R.J., HALL, D.K.**: Satellite-derived surface energy balance estimates in the Alaskan sub-Arctic. - J. Climate appl. Meteorol. 22: 115-125, 1983.

13477 - **GUTSCHICK, V.P.**: Photosynthesis model for C_3 leaves incorporating CO_2 transport, propagation of radiation, and biochemistry 1. Kinetics and their parametrization. - Photosynthetica 18: 549-568, 1984.

13478 - **GUTSCHICK, V.P.**: Photosynthesis model for C_3 leaves incorporating CO_2 transport, propagation of radiation, and biochemistry 2. Ecological and agricultural utility. - Photosynthetica 18: 569-595, 1984.

13479 - **GUY, R.D., WAMPLE, R.L.**: Stable carbon isotope ratios of flooded and non-flooded sunflowers (*Helianthus annuus*). - Can. J. Bot. 62: 1770-1774, 1984.

13480 - **GUY, R.D., WARNE, P.G., REID, D.M.**: Glycinebetaine content of halophytes: improved analysis by liquid chromatography and interpretations of results. - Physiol. Plant. 61: 195-202, 1984.

*13481 - **HABER, M.F., YOUNG, E., FAUST, M.**: Effects of PEG-induced water stress on calcium uptake in peach seedlings. - J. Amer. Soc. hort. Sci. 108: 737-740, 1983.

13482 - **HACK, H.R.B.**: Calculation of the field volumetric water content of cracking clay soils from measurements of gravimetric water content and bulk density. - J. Soil Sci. 35: 299-315, 1984.

13483 - **HAGSTROM, G.R.**: Current management practices for correcting iron deficiency in plants with emphasis on soil management. - J. Plant Nutr. 7: 23-46, 1984.

13484 - **HAJIBAGHERI, M.A., HARVEY, D.M.R., FLOWERS, T.J.**: Photosynthetic oxygen evolution in relation to ion contents in the chloroplasts of *Suaeda maritima*. - Plant Sci. Lett. 34: 353-362, 1984.

13485 - **HÅK, R., NÅTR, L.**: Transpiration of spring barley under nitrogen and phosphorus deficiency during leaf wilting. - Biol. Plant. 26: 11-15, 1984.

13486 - **HÅK, R., NÅTR, L.**: The significance of cuticular transpiration for the calculation of intercellular CO_2 concentration. - Biol. Plant. 26: 74-76, 1984.

*13487 - **HALDER, S., GUPTA, K.**: Effect of RH on sunflower seed viability with special reference to membrane permeability and biochemical changes. - Geobios 8: 6-9, 1981.

*13488 - **HALL, A.E.**: Humidity and plant productivity. - In: RECHCIGL, M.,Jr. (ed.): CRC Handbook of Agricultural Productivity. Volume I. Plant Productivity. Pp. 23-40. CRC Press, Inc., Boca Raton 1982.

13489 - **HALL, A.J., CHIMENTI, C., TRAPANI, N., VILELLA, F., COHEN de HUNAU, R.**: Yield in water-stressed maize genotypes: association with traits measured in seedlings and in flowering plants. - Field Crops Res. 9: 41-57, 1984.

13490 - **HALLDIN, S., SAUGIER, B., PONTAILLER, J.Y.**: Evapotranspiration of a deciduous forest: simulation using routine meteorological data. - J. Hydrol. 75: 323-341, 1984.

13491 - **HALLER, E.**: Effect of the germinating seed environment on crop yields. II. Effect of the water and air regime during germination on cereal yields. - Exp. Agr. 20: 235-243, 1984.

13492 - HÄLLGREN, J.-E.: Photosynthetic gas exchange in leaves affected by air pollutants. - In: KOZIOŁ, M.J., WHATLEY, F.R. (ed.): Gaseous Air Pollutants and Plant Metabolism. Pp. 147-159. Butterworths, London - Boston - Durban - Singapore - Sydney - Toronto - Wellington 1984.

13493 - HAMMES, J.L., PENDLETON, J.W.: Photoperiod-sensitive tropical corn as a potential source of biomass or paper. - Agron. J. 76: 159-160, 1984.

13494 - HAMPP, R., SCHNABL, H.: Adenine and pyridine nucleotide status of isolated *Vicia* guard cell protoplasts during K^+-induced swelling. - Plant Cell Physiol. 25: 1233-1239, 1984.

13495 - HANE, D.C., PUMPHREY, F.V.: Yield-evapotranspiration relationships and seasonal crop coefficients for frequently irrigated potatoes. - Amer. Potato J. 61: 661-668, 1984.

13496 - HANKS, R.J.: Prediction of crop yield and water consumption under saline conditions. - In: SHAINBERG, I., SHALHEVET, J. (ed.): Soil Salinity under Irrigation. Processes and Management. Pp. 272-283. Springer-Verlag, Berlin - Heidelberg - New York - Tokyo 1984.

13497 - HANSEN, S.: Estimation of potential and actual evapotranspiration. - Nordic Hydrol. 15: 205-212, 1984.

*13498 - HANSON, C.L., MORRIS, R.P., WIGHT, J.R.: Using precipitation to predict range herbage production in southwestern Idaho. - J. Range Manage. 36: 766-770, 1983.

*13499 - HARBERD, N.P., EDWARDS, K.J.R.: The effect of a mutation causing alcohol dehydrogenase deficiency on flooding tolerance in barley. - New Phytol. 90: 631-644, 1982.

13500 - HARBINSON, J., WOODWARD, F.I.: Field measurements of the gas exchange of woody plant species in simulated sunflecks. - Ann. Bot. 53: 841-851, 1984.

*13501 - HARDIE, W.J., JOHNSON, J.O., WEAVER, R.J.: The influence of vine water regime on ethephon-enhanced ripening of Zinfandel. - Amer. J. Enol. Viticult. 32: 115-121, 1981.

13502 - HARRIS, C.E., DHANOA, M.S.: The influence of the epidermis on the drying rates of red clover leaf petioles and stems and of Italian ryegrass stems at low water contents. - Grass Forage Sci. 39: 67-74, 1984.

13503 - HARRISON, J.G.: Effect of humidity on infection of field bean leaves by *Botrytis fabae* and on germination of conidia. - Trans. Brit. mycol. Soc. 82: 245-248, 1984.

*13504 - HARZALLAH-SKHIRI, F., GUILLOT-SALOMON, T., SIGNOL, M.: Biochemical and ultrastructural changes in plastids from various alfalfa cultivars growning under salt-stress. - In: WINTERMANS, J.F.G.M., KUIPER, P.J.C. (ed.): Biochemistry and Metabolism of Plant Lipids. Pp. 423-426. Elsevier Biomedical Press. Amsterdam 1982.

13505 - HASEGAWA, P.M., BRESSAN, R.A., HANDA, S., HANDA, A.K.: Cellular mechanisms of tolerance to water stress. - HortScience 19: 371-377, 1984.

13506 - HASHIMOTO, Y., INO, T., KRAMER, P.J., NAYLOR, A.W., STRAIN, B.R.: Dynamic analysis of water stress of sunflower leaves by means of a thermal image processing system. - Plant Physiol. 76: 266-269, 1984.

*13507 - **HATFIELD, J.L.:** Remote sensing estimators of potential and actual crop
yield. - Remote Sensing Environ. 13: 301-311, 1983.

13508 - **HATFIELD, J.L., PINTER, P.J.,Jr., CHASSERAY, E., EZRA, C.E., REGINATO, R.J.,
IDSO, S.B., JACKSON, R.D.:** Effects of panicles on infrared thermometer
measurement of canopy temperature in wheat. - Agr. Forest Meteorol. 32:
97-105, 1984.

13509 - **HATFIELD, J.L., REGINATO, R.J., IDSO, S.B.:** Evaluation of canopy temperature
-evapotranspiration models over various crops. - Agr. Forest Meteorol. 32:
41-53, 1984.

13510 - **HATFIELD, J.L., VAUCLIN, M., VIEIRA, S.R., BERNARD, R.:** Surface temperature
variability patterns within irrigated fields. - Agr. Water Manage. 8: 429-
437, 1984.

13511 - **HATLITLIGIL, M.B., OLSON, R.A., COMPTON, W.A.:** Yield, water use, and
nutrient uptake of corn hybrids under varied irrigation and nitrogen
regimes. - Fertilizer Res. 5: 321- 333, 1984.

13512 - **HAUS, M., WEGMANN, K.:** Glycerol-3-phosphate dehydrogenase (EC 1.1.1.8)
from *Dunaliella tertiolecta*. II. Influence of phosphate esters and different
salts on the enzymatic activity with respect to osmoregulation. - Physiol.
Plant. 60: 289-293, 1984.

13513 - **HAUSER, V.L.:** Neutron meter calibration and error control. - Trans. ASAE
27: 722-728, 1984.

13514 - **HAUSMANN, K., PATTERSON, D.J.:** Contractile vacuole complexes in algae. -
In: WIESSNER, W., ROBINSON, D.G., STARR, R.C. (ed.): Compartments in Algal
Cells and Their Interaction. Pp. 139-146. Springer-Verlag, Berlin - Heidel-
berg - New York - Tokyo 1984.

13515 - **HAVELKA, U.D., ACKERSON, R.C., BOYLE, M.G., WITTENBACH, V.A.:** CO_2-enrichment
effects on soybean physiology. I. Effects of long-term CO_2 exposure. -
Crop Sci. 24: 1146-1150, 1984.

13516 - **HAVELKA, U.D., WITTENBACH, V.A., BOYLE, M.G.:** CO_2-enrichment effects on
wheat yield and physiology. - Crop Sci. 24: 1163-1168, 1984.

13517 - **HAVERKAMP, R., VAUCLIN, M., VACHAUD, G.:** Error analysis in estimating soil
water content from neutron probe measurements: 1. Local standpoint. - Soil
Sci. 137: 78-90, 1984.

*13518 - **HAYASHI, M.:**[Studies on dormancy and germination of rice seed. IX. The
effects of oxygen and moisture upon the release of the rice seed dormancy
and upon the inactivation of inhibitors in the dormant seed.] - Bull. Fac.
Agr. Kagoshima Univ. 30: 1-9, 1980. [In Jap, ab: E.]

13519 - **HEARN, A.B., CONSTABLE, G.A.:** Irrigation for crops in a sub-humid environment. VII. Evaluation of irrigation strategies for cotton. - Irrig. Sci.
5: 75-94, 1984.

13520 - **HEATH, O.V.S.:** Stomatal opening in darkness in the leaves of *Commelina
communis*, attributed to an endogenous circadian rhythm: control of phase. -
Proc. roy. Soc. London, Ser. B 220: 415-422, 1984.

13521 - **HEENAN, D.P., THOMPSON, J.A.:** Growth, grain yield and water use of rice
grown under restricted water supply in New South Wales. - Aust. J. exp. Agr.
anim. Husb. 24: 104-109, 1984.

*13522 - HEGDE, B.A., PATIL, T.M.: Physiological studies on *Parthenium hysterophorus* Linn. under different ecological conditions. - Biovigyanam 6: 15-19, 1980.

13523 - HEIKKILA, J.J., PAPP, J.E.T., SCHULTZ, G.A., BEWLEY, J.D.: Induction of heat shock protein messenger RNA in maize mesocotyls by water stress, abscisic acid, and wounding. - Plant Physiol. 76: 270-274, 1984.

13524 - HEINS, R.D., KARLSSON, M.G., ERWIN, J.E., HAUSBECK, M.K., MILLER, S.H.: Interaction of CO_2 and environmental factors on crop responses. - Acta hort. 162: 21-28, 1984.

*13525 - HELLUM, A.K.: Cone moisture and relative humidity effects on seed release from lodgepole pine cones from Alberta. - Can. J. Forest Res. 12: 102-105, 1982.

*13526 - HENDERSON, R.E., PATRICK, W.H.,Jr.: Soil aeration and plant productivity. - In: RECHCIGL, M.,Jr. (ed.): CRC Handbook of Agricultural Productivity. Volume 1. Plant Productivity. Pp. 51-69. CRC Press, Inc., Boca Raton 1982.

*13527 - HENDERSON, S., PIXTON, S.W.: The influence of the testa on the sorption of water by cocoa beans and some legumes. - J. stored Products Res. 16: 81-84, 1980.

*13528 - HENDRICKS, S.B., VAN DER WOUDE, W.J.: How phytochrome acts - perspectives on the continuing quest. - In: SHROPSHIRE, W.,Jr., MOHR, H. (ed.): Photomorphogenesis. Pp. 3-23. Springer-Verlag, Berlin - Heidelberg - New York - Tokyo 1983.

*13529 - HENRIKSEN, K.: Vanding af asieagurker. [Irrigation of large cucumbers.] - Tidsskr. Planteavl 84: 479-490, 1980. L In Dan, ab: E.]

13530 - HENSON, I.E.: Evidence of a role for abscisic acid in mediating stomatal closure induced by obstructing translocation from leaves of pearl millet (*Pennisetum americanum* (L.) Leeke). - J. exp. Bot. 35: 1419-1432, 1984.

13531 - HENSON, I.E.: Inhibition of abscisic acid accumulation in seedling shoots of pearl millet (*Pennisetum americanum* (L.) Leeke) following induction of chlorosis by norflurazon. - Z. Pflanzenphysiol. 114: 35-43, 1984.

13532 - HENSON, I.E.: The heritability of abscisic acid accumulation in water -stressed leaves of pearl millet (*Pennisetum americanum* (L.) Leeke). - Ann. Bot. 53: 1-11, 1984.

13533 - HENSON, I.E.: Effects of atmospheric humidity on abscisic acid accumulation and water status in leaves of rice (*Oryza sativa* L.). - Ann. Bot. 54: 569-582, 1984.

13534 - HENSON, I.E., MAHALAKSHMI, V., ALAGARSWAMY, G., BIDINGER, F.R.: Leaf abscisic acid content and recovery from water stress in pearl millet (*Pennisetum americanum* (L.) Leeke). - J. exp. Bot. 35: 99-109, 1084.

13535 - HENSON, I.E., MAHALAKSHMI, V., ALAGARSWAMY, G., BIDINGER, F.R.: The effect of flowering on stomatal response to water stress in pearl millet (*Pennisetum americanum* (L.) Leeke). - J. exp. Bot. 35: 219-226, 1984.

*13536 - HERNÁNDEZ-NISTAL, J., ALDASORO, J., RODRIGUEZ, D., MATILLA, A., NICOLÁS, G.: Effect of thiourea on the ionic content and dark fixation of CO_2 in embryonic axes of *Cicer arietinum* seeds. - Physiol. Plant. 57: 273-278, 1983.

13537 - **HERRERA, M.**: Caracterization potencial productiva de las cañeras de provincia Habana. [Characterization of the productive potential of sugarcane
regions in Havana province.] - Cultivos trop. 6: 569-577, 1984. [In Span,
ab: E.]

*13538 - **HESKETH, J.D., LARSON, E.M., GORDON, A.J., PETERS, D.B.**: Internal factors
influencing photosynthesis and respiration. - In: DALE, J.E., MILTHORPE,
F.L. (ed.): The Growth and Functioning of Leaves. Pp. 381-411. Cambridge
University Press, Cambridge - London - New York - New Rochelle - Melbourne
- Sydney 1983.

13539 - **HETRICK, B.A.D., HETRICK, J.A., BLOOM, J.**: Interaction of mycorrhizal
infection, phosphorus level, and moisture stress in growth of field corn. -
Can. J. Bot. 62: 2267-2271, 1984.

*13540 - **HEUCHERT, J.C., MITCHELL, C.A.**: Inhibition of shoot growth in greenhouse
-grown tomato by periodic gyratory shaking. - J. Amer. Soc. hort. Sci.
108: 795-800, 1983.

13541 - **HEW, C.S.**: *Drymoglossum* under water stress. - Amer. Fern J. 74: 37-39,
1984.

13542 - **HIGGS, K.H., JONES, H.G.**: A microcomputer-based system for continuous
measurement and recording fruit diameter in relation to environmental
factors. - J. exp. Bot. 35: 1646-1655, 1984.

13543 - **HIGUCHI, T., YODA, K., TENSHO, K.**: Further evidence for gaseous CO_2 transport in relation to root uptake of CO_2 in rice plant. - Soil Sci. Plant
Nutr. 30: 125-136, 1984.

13544 - **HIJAZI, L.A., TÉKÉTÉ, A., KAHNT, G.**: Effect of rhizosphere temperature and
water management on soybean production. - Z. Acker- Pflanzenbau 153: 438-
445, 1984.

*13545 - **HILDEBRAND, P.D., SUTTON, J.C.**: Weather variables in relation to an epide-
mic of onion downy mildew. - Phytopathology 72: 219-224, 1982.

13546 - **HILDEBRAND, P.D., SUTTON, J.C.**: Interactive effects of the dark period,
humid period, temperature,and light on sporulation of *Peronospora destructor*.
- Phytopathology 74: 1444-1449, 1984.

13547 - **HILDEBRAND, P.D., SUTTON, J.C.**: Effects of weather variables on spore
survival and infection of onion leaves by *Peronospora destructor*. - Can. J.
Plant Pathol. 6: 119-126, 1984.

13548 - **HILDEBRAND, P.D., SUTTON, J.C.**: Relationships of temperature, moisture, and
inoculum density to the infection cycle of *Peronospora destructor*. - Can. J.
Plant Pathol. 6: 127-134, 1984.

13549 - **HILDEBRANDT, P., STOCKBURGER, M.**: Role of water in bacteriorhodopsin's
chromophore: resonance Raman study. - Biochemistry 23: 5539-5548, 1984.

*13550 - **HILLEL, D. (ed.)**: Advances in Irrigation. Volume 1. - Academic Press, New
York - London - Paris - San Diego - San Francisco - São Paulo - Sydney -
Tokyo - Toronto 1982.

13551 - **HILTON, J.R.**: The influence of temperature and moisture status on the photoinhibition of seed germination in *Bromus sterilis* L. by the far-red absorbing
form of phytochrome. - New Phytol. 97: 369-374, 1984.

13552 - **HINCHA, D.K., SCHMIDT, J.E., HEBER, U., SCHMITT, J.M.:** Colligative and non-colligative freezing damage to thylakoid membranes. - Biochim. biophys. Acta 769: 8-14, 1984.

*13553 - **HIPKINS, M.F., FITZSIMONS, P.J., WEYERS, J.D.B.:** The primary processes of photosystem II in purified guard-cell protoplasts and mesophyll-cell protoplasts from *Commelina communis* L. - Planta 159: 554-560, 1983.

13554 - **HIRAI, G.-I., TAKAHASHI, M., TANAKA, O., SHIMAMURA, N., NAKAYAMA, N.:** [Studies on the effects of relative humidity of the atmosphere upon the growth and physiology of rice plant. III. The influence of atmospheric humidity on the rate of photosynthesis.] - Jap. J. Crop Sci. 53: 261-267, 1984. [In Jap, ab: E.]

13555 - **HIRASAWA, T., ARAKI, T., ISHIHARA, K.:** [The relationship between environmental factors and water status in the rice plant. III. On leaf and xylem water potentials of leaves on the different position of a stem.] - Jap. J. Crop Sci. 53: 54-63, 1984. [In Jap, ab: E.]

13556 - **HO, L.C., HURD, R.G., LUDWIG, L.J., SHAW, A.F., THORNLEY, J.H.M., WITHERS, A.C.:** Changes in photosynthesis, carbon budget and mineral content during the growth of the first leaf of cucumber. - Ann. Bot. 54: 87-101, 1984.

*13557 - **HOFFMAN, G.J., OSTER, J.D., ALVES, W.J.:** Evapotranspiration of mature orange trees under drip irrigation in an arid climate. - Trans. ASAE 25: 992-996, 1982.

13558 - **HOFFMANN, R., BISSON, M.A.:** Osmotic regulation of a charophyte in an unusual saline environment. - In: CRAM, W.J., JANACEK, K., RYROVÁ, R., SIGLER, K. (ed.): Membrane Transport in Plants. Pp. 110. Academia, Praha 1984.

*13559 - **HOFMANN, A., BERGER, W.:** Metodische Lösungen für die Quantifizierung vor Beregnungs- und Verlagerungswirkung in Pflanzenproductionsbetrieben. - Arch. Acker- Pflanzenbau Bodenk. 27: 387-393, 1983.

13560 - **HOFMANN, W.C., O'NEILL, M.K., DOBRENZ, A.K.:** Physiological responses to sorghum hybrids and parental lines to soil moisture stress. - Agron. J. 76: 223-228, 1984.

13561 - **HOLDER, C.B., CARY, J.W.:** Soil oxygen and moisture in relation to Russet Burbank potato yield and quality. - Amer. Potato J. 61: 67-75, 1984.

*13562 - **HOLLINGER, D.Y.:** Photosynthesis and water relations of the mistletoe *Phoradendron villosum*, and its host, the California valley oak, *Quercus lobata*. - Oecologia 60: 396-400, 1983.

13563 - **HOLMES, J.W.:** Measuring evapotranspiration by hydrological methods. - Agr. Water Manage. 8: 29-40, 1984. Also in: SHARMA, M.L. (ed.): Evapotranspiration from Plant Communities. Pp. 29-40, Elsevier Science Publishers, Amsterdam - Oxford - New York - Tokyo 1984.

13564 - **HOLOPAINEN, T., KÄRENLAMPI, L.:** Injuries to lichen ultrastructure caused by sulphur dioxide fumigations. - New Phytol. 98: 285-294, 1984.

13565 - **HOOK, D.D.:** Adaptations to flooding with fresh water. - In: KOZLOWSKI, T.T. (ed.): Flooding and Plant Growth. Pp. 265-294. Academic Press, Orlando - San Diego - San Francisco - New York - London - Toronto - Montreal - Sydney - Tokyo - São Paulo 1984.

13566 - **HOOK, J.E., THREADGILL, E.D., LAMBERT, J.R.:** Corn irrigation scheduled by tensiometer and the Lambert model in the humid southeast. - Agron. J. 76: 695-700, 1984.

13567 - HOPMANS, P., DOUGLAS, L.A., CHALK, P.M.: Effects of soil salinity and mineral nitrogen on the acetylene reduction activity of *Trifolium subterraneum* L. - Aust. J. agr. Res. 35: 9-15, 1984.

*13568 - HOPMANS, P., DOUGLAS, L.A., CHALK, P.M., DELBRIDGE, S.G.: Effects of soil moisture, mineral nitrogen and salinity on nitrogen fixation (acetylene reduction) by *Allocasuarina verticillata* (Lam.) L. Johnson seedlings. - Aust. Forest Res. 13: 189-196, 1983.

13569 - HORÁK, J., KOTYK, A., SIGLER, K.: Biochemie Transportních Pochodů. [Biochemistry of Transport Processes.] - Academia, Praha 1984. [In Czech.]

13570 - HORIUCHI, T., NAITO, K.: [Studies on corresponded-relations between plant characters and cultivation methods. III. Comparison of water balance in plants at transplanting among millets.] - Jap. J. Crop Sci. 53: 379-386, 1984. [In Jap, ab: E.]

*13571 - HORLER, D.N.H., DOCKRAY, M., BARBER, J.: The red edge of plant leaf reflectance. - Int. J. remote Sensing 4: 273-288, 1983.

*13572 - HORNE, R., GWALTER, J.: The recovery of rainforest overstorey following logging. I. Subtropical rainforest. - Aust. Forest Res. 13: 29-44, 1982.

13573 - HOROWITZ, M., TAYLORSON, R.B.: Hardseededness and germinability of velvetleaf (*Abutilon theophrasti*) as affected by temperature and moisture. - Weed Sci. 32: 111-115, 1984.

13574 - HORSNELL, L.J.T.: Effect of soil moisture on the response of subterranean clover to lime. - Plant Soil 81: 295-297, 1984.

13575 - HORST, G.L., BEADLE, N.B.: Salinity affects germination and growth of tall fescue cultivars. - J. Amer. Soc. hort. Sci. 109: 419-422, 1984.

*13576 - HORSTMANN, U.: Cultivation of the green alga, *Caulerpa racemosa,* in tropical waters and some aspects of its physiological ecology. - Aquaculture 32: 361-371, 1983.

13577 - HORTON, R.F., SAVILLE, B.J.: Carbon dioxide enrichment, transpiration and l-aminocyclopropane-l-carboxylic acid-dependent ethylene release from oat leaves. - Plant Sci. Lett. 36: 131-135, 1984.

13578 - HOSTALÁCIO, S., VÁLIO, I.F.M.: Desenvolvimento dos frutos de feijão em diferentes regimes de irrigação. [Bean fruit development in different irrigation regimes.] - Pesq. agropec. Bras. 19: 53-57, 1984. [In Port, ab: E.]

13579 - HOSTALÁCIO, S., VÁLIO, I.F.M.: Desenvolvimento de plantes de feijão cv. Goiano Precoce, em diferentes regimes de irrigação. [Bean plant cv. Goiano Precoce development under different irrigation regimes.] - Pesq. agropec. Bras. 19: 211-218, 1984. [In Port, ab: E.]

*13580 - HOUGH, M.N.: A weather-dependent yield model for silage maize. - Agr. Meteorol. 23: 97-114, 1981.

13581 - HOWELL, T.A., DAVIS, K.R., McCORMICK, R.L., YAMADA, H., WALHOOD, V.T., MEEK, D.W.: Water use efficiency of narrow row cotton. - Irrig. Sci. 5: 195-214, 1984.

13582 - HOWELL, T.A., HATFIELD, J.L., RHOADES, J.D., MERON, M.: Response of cotton water stress indicators to soil salinity. - Irrig. Sci. 5: 25-36, 1984.

13583 - HOWELL, T.A., HATFIELD, J.L., YAMADA, H., DAVIS, K.R.: Evaluation of cotton
canopy temperature to detect crop water stress. - Trans. ASAE 27: 84-88,
1984.

13584 - HOY, M.W., OGAWA, J.M., DUNIWAY, J.M.: Effects of irrigation on buckeye rot
of tomato fruit caused by *Phytophthora parasitica*. - Phytopathology 74:
474-478, 1984.

13585 - HSIAO, A.I., McINTYRE, G.I.: Evidence of competition for water as a factor
in the mechanism of root-bud inhibition in milkweed (*Asclepias syriaca*). -
Can. J. Bot. 62: 379-384, 1984.

13586 - HSIAO, T.C., O'TOOLE, J.C., YAMBAO, E.B., TURNER, N.C.: Influence of osmotic
adjustment on leaf rolling and tissue death in rice (*Oryza sativa* L.). -
Plant Physiol. 75: 338-341, 1984.

*13587 - HUBAC, C.: Evolution des sucres solubles de l'amidon, au cours de l'asseche-
ment, puis de la rehumidification, chez des plantes plus ou moins resistantes
a la secheresse (*Carex* et *Gossypium*). Comparaison entre les parties aeriennes
et les parties souterraines. - In: RIEDACKER, A., GAGNAIRE-MICHARD, J. (ed.):
Symposium Physiologie des Racines et Symbioses. Pp. 31-43. Nancy 1978.

*13588 - HUBAC, C., CORNIC, C.: Influence de la photoperiode sur la resistance a la
secheresse du cotonnier: *Gossypium hirsutum* L. var. B.J.A. - Bull. Soc. eco-
physiol. 3: 51-53, 1978.

13589 - HUBAC, C., OUEDRAOGO, M., GUERRIER, D., FERRAN, J.: Rôle de la photopériode
dans l'adaptation à la sécheresse des Cotonniers. - Bull. Soc. bot. France,
Actual bot. 131: 79-88, 1984.

13590 - HUBER, S.C., ROGERS, H.H., MOWRY, F.L.: Effects of water stress on photosynthesis and carbon partitioning in soybean (*Glycine max* (L.) Merr.) plants
grown in the field at different CO_2 levels. - Plant Physiol. 76: 244-249,
1984.

13591 - HUGHES, R.M., COLMAN, R.L., LOVETT, J.V.: Effects of temperature and moisture
stress on germination and seedling growth of four tropical species. - Aust.
J. exp. Agr. anim. Husb. 24: 396-402, 1984.

13592 - HULL, J.C., WOOD, S.G.: Water relations of oak species on and adjacent to
a Maryland serpentine soil. - Amer. Midl. Natur. 112: 224-234, 1984.

13593 - HUNDAL, S.S., De DATTA, S.K.: Water table and tillage effects on root
distribution, soil water extraction, and yield of sorghum grown after wetland rice in a tropical soil. - Field Crops Res. 9: 291-303, 1984.

13594 - HUNDAL, S.S., De DATTA, S.K.: *In situ* water transmission characteristics of
a tropical soil under rice-based cropping systems. - Agr. Water Manage. 8:
384-396, 1984.

13595 - HUNT, E.R.,Jr., WEBER, J.A., GATES, D.M.: Differences between tree species
in hydraulic press calibration of leaf water potential are correlated with
specific leaf area. - Plant Cell Environ. 7: 597-600, 1984.

13596 - HUSSEIN, J.: Laboratory analyses for assessing potential irrigation problems
of basalt-derived vertisols. - Zimbabwe agr. J. 81: 9-16, 1984.

*13597 - HUTCHINSON, C.S., SUTCLIFFE, J.F.: An approach to the teaching of cell water
relations in biology at A-level using the water potential concept. - J.
biol. Educ. 17: 123-130, 1983.

13598 - HUTCHINSON, D.H., OTTEN, L.: Equilibrium moisture content of white beans. Cereal. Chem. 61: 155-158, 1984.

13599 - HUTTUNEN, S., SOIKKELI, S.: Effects of various gaseous pollutants on plant cell ultrastructure. - In: KOZIOŁ, M.J., WHATLEY, F.R. (ed.): Gaseous Air Pollutants and Plant Metabolism. Pp. 117-127. Butterworth, London - Boston - Durban - Singapore - Sydney - Toronto - Wellington 1984.

13600 - HUZULÁK, J., SZABÓ, T., MATEJKA, F.: Influence of meteorological factors on winter wheat, maize and sugar beet canopy temperature. - Biológia (Bratislava) 39: 857-866, 1984.

*13601 - IDSO, S.B.: Carbon Dioxide: Friend of Foe? - IBR Press, Tempe 1982.

13602 - IDSO, S.B.: Atmospheric CO_2 variability: a cause for concern in the reconstruction of past climates. - Speculations Sci. Technol. 7: 37-40, 1984.

13603 - IDSO, S.B.: The case for carbon dioxide. - J. environ. Sci. 27: 19-22, 1984.

13604 - IDSO, S.B., KIMBALL, B.A., CLAWSON, K.L.: Quantifying effects of atmospheric CO_2 enrichment on stomatal conductance and evapotranspiration of water hyacinth via infrared thermometry. - Agr. Forest Meteorol. 33: 15-22, 1984.

13605 - IDSO, S.B., PINTER, P.J.,Jr., REGINATO, R.J., CLAWSON, K.L.: Stomatal conductance and photosynthesis in water hyacinth: Effects of removing water from roots as quantified by a foliage-temperature-based plant water stress index. - Agr. Forest Meteorol. 32: 249-256, 1984.

13606 - IL'INA, L.G.: O selektsii yarovoĭ pshenitsy na zasukhoustoĭchivost'. [Breeding of spring wheat for drought resistance.] - Sel'skokhoz. Biol. 1984 (6): 17-21, 1984. [In R, ab: E.]

13607 - IL'NITSKII, O.A., SHATILOV, I.S., SEMIN, V.S., RADCHENKO, S.S.: Obosnovanie i realizatsiya pochvenno-fiziologicheskogo algoritma upravleniya srokom poliva rasteniĭ. [Principle and realization of soil-physiological management algorythm by plant irrigation timing.] - Sel'skokhoz. Biol. 1984 (3): 31-37, 1984. [In R, ab: E.]

13608 - IL'YASHUK, O,M.: Vzaěmozv'yazok mizh rivnem osmotychnogo tysku v zamykayuchykh klitynakh ta posukhostiĭkistyu *Helianthus annuus* L. [Relationship between the osmotic pressure in guard cells and drought resistance of *Helianthus annuus* L.] - Ukr. bot. Zh. 41 (2): 56-58, 1984. [In Ukr, ab: E.]

13609 - IMAI, K., COLEMAN, D.F., YANAGISAWA, T.: Elevated atmospheric partial pressure of carbon dioxide and dry matter production of cassava (*Manihot esculenta* Crantz). - Jap. J. Crop Sci. 53: 479-485, 1984.

13610 - IMAMUL HUQ, S.M., LARHER, F.: Osmoregulation in higher plants: Effect of maintaining a constant Na:Ca ratio on the growth, ion balance and organic solute status of NaCl stressed cowpea (*Vigna sinensis* L.). - Z. Pflanzenphysiol. 113: 163-176, 1984.

13611 - INDELICATO, S., NICOSIA, O.L.D., TAMBURINO, V.: Wastewater irrigation. Lysimeter investigation on water quality aspects. - Environ. Technol. Lett. 5: 383-388, 1984.

13612 - INDIRA, P., RAMANUJAM, T.: Proline accumulation, relative water content and dry matter in cassava under moisture stress. - Ind. J. agr. Sci. 54: 336-337, 1984.

*13613 - INGALE, A.B., SONAR, K.R.: Effect of presowing soil water treatments and
seed coating with iron compounds on yield and nutrient uptake in upland
rice. - Int. Rice Comm. Newslett. 31: 42-43, 1982.

*13614 - INNES, P., BLACKWELL, R.D., AUSTIN, R.B., FORD, M.A.: The effects of se-
lection for number of ears on the yield and water economy of winter wheat. -
J. agr. Sci. 97: 523-532, 1981.

13615 - INNES, P., BLACKWELL, R.D., QUARRIE, S.A.: Some effects of genetic variation
in drought-induced abscisic acid accumulation on the yield and water use of
spring wheat. - J. agr. Sci. 100: 341-351, 1984.

13616 - INOUE, K., SAKUTARANI, T., UCHIJIMA, Z.: Stomatal resistance of rice leaves
as influenced by radiation intensity and air humidity. - J. agr. Meteorol.
40: 235-242, 1984.

13617 - IOANNOU, N., GROGAN, R.G.: Water requirements for sporangium formation by
Phytophthora parasitica in relation to bioassay in soil. - Plant Dis. 68:
1043-1048, 1984.

13618 - ĬORDANOV, I.T., DILOVA, S.A., STANEV, V.P., PETKOVA, R.A., PANGELOVA, T.K.,
CHICHEV, P.N.: Formirane i Funktsionalna Aktivnost na Fotosintetichniya
Aparat. [Formation and Functional Activity of the Photosynthetic Apparatus.]
- Izdat. B"lgar. Akad. Nauk, Sofiya 1984. [In Bulg, ab: E, R.]

*13619 - ISFAN, D.: Forecasting corn nitrogen fertilizer rate based on soil water
fluctuation. - Commun. Soil Sci. Plant Anal. 12: 1067-1083, 1981.

13620 - ISFAN, D.: Corn yield variation as related to soil water fluctuation and
nitrogen fertilizer. I. Soil water-yield relationships. - Commun. Soil Sci.
Plant Anal. 15: 1147-1161, 1984.

13621 - ISFAN, D.: Corn yield variation as related to soil water fluctuation and
nitrogen fertilizer. II. Soil water-nitrogen-yield relationships. - Commun.
Soil Sci. Plant Anal. 15: 1163.1174, 1984.

13622 - ISHIZAWA, K., ESASHI, Y.: Gaseous factors involved in the enhanced elongation of rice coleoptiles under water. - Plant Cell Environ. 7: 239-245,
1984.

13623 - ISHIZAWA, K., ESASHI, Y.: Osmoregulation in rice coleoptile elongation as
promoted by cooperation between IAA and ethylene. - Plant Cell Physiol. 25:
495-504, 1984.

*13624 - IZMAĬLOVA, N.N., BANNIKOVA, I.A., BERESNEVA, I.A., MAKSIMOVICH, S.V.,
BAYASGALAN, L., CHOĬZHAMTS, B.: Vodoobmen i produktivnost' rastitel'nykh
soobshchestv v svyazi s klimaticheskimi osobennostyami territorii.
[Water exchange and productivity of plant stands in connection with climatic peculiarities of territory.] - In: LAVRENKO, E.M., BANNIKOVA, I.A.
(ed.): Gornaya Lesostep' Vostochnogo Khangaya MNR (Prirodnye Usloviya). Pp.
135-148. Nauka, Moskva 1983. [In R.]

*13625 - JACKSON, D.S., JACKSON, E.A., GIFFORD, H.H.: Soil water in deep Pinaki sands:
some interactions with thinned and fertilised Pinus radiata. - N. Zeal. J.
Forest Sci. 13: 183-196, 1983.

13626 - JACKSON, L.E., BLISS, L.C.: Phenology and water relations of three plant
life forms in a dry tree-line meadow. - Ecology 65: 1302-1314, 1984.

13627 - JACKSON, M.B., DREW, M.C.: Effects of flooding on growth and metabolism of herbaceous plants. - In: KOZLOWSKI, T.T. (ed.): Flooding and Plant Growth. Pp. 47-128. Academic Press, Orlando - San Diego - San Francisco - New York - London - Toronto - Montreal - Sydney - Tokyo - São Paulo 1984.

*13628 - JAGGI, I.K., SINHA, S.B.: Influence of seasonal evaporative demand on evapotranspiration and yield of irrigated wheat. - Ind. J. agr. Sci. 51: 891-895, 1981.

13629 - JAME, Y.W., BIEDERBECK, V.O., NICHOLAICHUK, W., KORVEN, H.C.: Salinity and alfalfa yield under effluent irrigation in southwestern Saskatchewan. - Can. J. Soil Sci. 64: 323-332, 1984.

13630 - JAMIESON, A.P., WILLMER, C.M.: Functional stomata in a variegated leaf chimera of *Pelargonium zonale* L. without guard cell chloroplasts. - J. exp. Bot. 35: 1053-1059, 1984.

13631 - JANA, P.K., SOUNDA, G., SINGH, M.I., SHOW, N., MANDAL, B.B., PATRA, A.P.: Effect of soil-moisture tensions and weed control on yield, evapotranspiration, consumptive-use efficiency and moisture-extraction pattern by soybean. - Ind. J. agr. Sci. 54: 61-63, 1984.

*13632 - JANKOVIČ, M., POPOVIČ, R., STEFANOVIČ, K., DIMITRIJEVIČ, J.: Ekofiziološke karakteristike biljaka i uslovi staništa u zajednici *Chrysopogonetum panonnicum typicum* u Deliblatskoj peščari. [Ecophysiological plant characteristics and habitat conditions in the community *Chrysopogonetum panonnicum typicum* in Deliblatska Peščara.] - In: Drugi Kongres Ekologa Jugoslavije. Pp. 567-584. Savez Društava Ekologa Jugoslavije, Zagreb 1979. [In Croat, ab: E.]

13633 - JARRETT, A.R., MORROW, C.T.: Distribution of frost protection water applied to apple trees during bud and leaf development. - Trans. ASAE 27: 89-92, 98, 1984.

*13634 - JAYAWARDANE, N.S., MEYER, W.S., BARRS, H.D.: Moisture measurement in a swelling clay soil using neutron moisture meters. - Aust. J. Soil Res. 22: 109-117, 1983.

13635 - JELLINGS, A.J., LEECH, R.M.: Anatomical variation in first leaves of nine *Triticum* genotypes, and its relationship to photosynthetic capacity. - New Phytol. 96: 371-382, 1984.

13636 - JENSEN, H.E., MOGENSEN, V.O.: Yield and nutrient content of spring wheat subjected to water stress at various growth stages. - Acta. agr. Scand. 34: 527-533, 1984.

13637 - JENSEN, M., OETTMEIER, W.: Effects of freezing on the structure of chloroplast membranes. - Cryobiology 21: 465-473, 1984.

13638 - JESCHKE, W.D.: Effects of transpiration on potassium and sodium fluxes in root cells and the regulation of ion distribution between roots and shoots of barley seedlings. - J. Plant Physiol. 117: 267-285, 1984.

13639 - JESCHKE, W.D., STEUDLE, E.: Water pathways in barley roots. - In: CRAM, W.J., JANÁCEK, K., RYBOVÁ, R., SIGLER, K. (ed.): Membrane Transport in Plants. Pp. 111-112. Academia, Praha 1984.

*13640 - JHA, K.P., CHANDRA, D., CHALLAIAH : Irrigation requirement of high-yielding rice varieties grown on soils having shallow water-table. - Ind. J. agr. Sci. 51: 732-737, 1981.

13641 - JIANG, S., SHENG, X.-W., QI, Q.-H.: [A comparative study on diurnal changes of photosynthetic rate in *Stipa grandis* community in Nei Monggol region.] - Acta bot. Sin. 26: 644-652, 1984. [In: Chin, ab: E.]

13642 - JOHNSON, R.C., NGUYEN, H.T., CROY, L.I.: Osmotic adjustment and solute accumulation in two wheat genotypes differing in drought resistance. - Crop Sci. 24: 957-962, 1984.

*13643 - JOHNSON, W.C., DAVIS, R.G.: Yield-water relationships of summer-fallowed winter wheat. A precision study in the Texas Panhandle. - Agr. Res. Rep. 5: 1-43, 1980.

13644 - JOHNSTON, M., GROF, C.P.L., BROWNELL, P.F.: Responses to ambient CO_2 concentrations by sodium-deficient C_4 plants. - Aust. J. Plant Physiol. 11: 137-141, 1984.

13645 - JONES, C.S.: The effect of axis splitting on xylem pressure potentials and water movement in the desert shrub *Ambrosia dumosa* (Gray) Payne (Asteraceae). - Bot. Gaz. 145: 125-131, 1984.

13646 - JONES, H., TOMOS, A.D., LEIGH, R.A., WYN JONES, R.G.: The integration of cellular and whole-root hydraulic conductivity in wheat and maize. - In: CRAM, W.J., JANÁČEK, K., RYBOVÁ, R., SIGLER, K. (ed.): Membrane Transport in Plants. Pp. 113-114. Academia, Praha 1984.

*13647 - JONES, H.G.: Estimation of an effective soil water potential at the root surface of transpiring plants. - Plant Cell Environ. 6: 671-674, 1983.

13648 - JONES, H.G., CUMMING, I.G.: Variation of leaf conductance and leaf water potential in apple orchards. - J. hort. Sci. 59: 329-336, 1984.

13649 - JONES, J.B., RAJU, B.C., ENGELHARD, A.W.: Effects of temperature and leaf wetness on development of bacterial spot of geraniums and chrysanthemums incited by *Pseudomonas cichorii*. - Plant Dis. 68: 248-251, 1984.

13650 - JONES, M.B., MUTHURI, F.M.: The diurnal course of plant water potential, stomatal conductance and transpiration in a papyrus (*Cyperus papyrus* L.) canopy. - Oecologia 63: 252-255, 1984.

13651 - JONES, O.R.: Yield, water-use efficiency, and oil concentration and quality of dryland sunflower grown in the Southern High Plains. - Agron. J. 76: 229-235, 1984.

13652 - JONES, P., ALLEN, L.H.,Jr., JONES, J.W., BOOTE, K.J., CAMPBELL, W.J.: Soybean canopy growth, photosynthesis, and transpiration responses to whole-season carbon dioxide enrichment. - Agron. J. 76: 633-637, 1984.

13653 - JONES, P., JONES, J.W., ALLEN, L.H.,Jr., MISHOE, J.E.: Dynamic computer control of closed environmental plant growth chambers. Design and verification. - Trans. ASAE 27: 879-888, 1984.

13654 - JORDAN, P.W., NOBEL, P.S.: Thermal and water relations of roots of desert succulents. - Ann. Bot. 54: 705-717, 1984.

13655 - JORDAN, W.R., SHOUSE, P.J., BLUM, A., MILLER, F.R., MONK, R.L.: Environmental physiology of sorghum. II. Epicuticular wax load and cuticular transpiration. - Crop Sci. 24: 1168-1173, 1984.

*13656 - JOSHI, A.J., IYENGAR, E.R.R.: Effect of salinity on the germination of *Salicornia brachiata* Roxb. - Ind. J. Plant Physiol. 25: 65-70, 1982.

*13657 - JOSHI, S., NIMBALKAR, J.D.: Effect of salt stress on growth and yield in *Cajanus cajan* L. - Plant Soil 74: 291-294, 1983.

13658 - JOSHI, S., NIMBALKAR, J.D.: Salinity effects on photosynthesis and photo-respiration in *Cajanus cajan* (L.). - Photosynthetica 18: 128-133, 1984.

13659 - JOYCE, P.S., PALEG, L.G., ASPINALL, D.: The requirement for low-intensity light in the accumulation of proline as a response to water deficit. - J. exp. Bot. 35: 209-218, 1984.

*13660 - JUANG, Y.D., CHANG, H.S., CHEN, C.Y.: [Studies on the photosynthesis, respiration, and leaf anatomy of several soybean varieties.] - J. Sci. Eng. 17: 263-274, 1980. [In Chin, ab: E.]

13661 - JUDD, M.J., McANENEY, K.J.: Water use by Tamarillos (*Cyphomandra betacea*) within a sheltered orchard environment. - Agr. Forest Meteorol. 32: 31-40, 1984.

13662 - JUNG, Y.-S., TAYLOR, H.M.: Differences in water uptake rates of soybean roots associated with time and depth. - Soil Sci. 137: 341-350, 1984.

*13663 - JUNTTILA, O., STUSHNOFF, C., GUSTA, L.V.: Dehardening in flower buds of saskatoon-berry, *Amelanchier alnifolia*, in relation to temperature, moisture content, and spring bud development. - Can. J. Bot. 61: 164-170, 1983.

13664 - JURY, W.A.: Field scale water and solute transport through unsaturated soils. - In: SHAINBERG, I., SHALHEVET, J. (ed.): Soil Salinity under Irrigation. Processes and Management. Pp. 115-125. Springer-Verlag, Berlin - Heidelberg - New York - Tokyo 1984.

13665 - KACHEL, K., ROTH, D.: Mehrjährige Ergebnisse zum Einfluss der Beregnung auf die Erträge von Winterweizen und Sommergerste auf einer flachgründigen Ton-Schwarzerde. - Arch. Acker- Pflanzenbau Bodenk. 28: 35-43, 1984.

13666 - KAFKAFI, U.: Plant nutrition under saline conditions. - In: SHAINBERG, I., SHALHEVET, J, (ed.): Soil Salinity under Irrigation. Processes and Management. Pp. 319-331. Springer-Verlag, Berlin - Heidelberg - New York - Tokyo 1984.

*13667 - KAĬBIYAĬNEN, L.K., SAZONOVA, T.A,: Dinamika vodnogo obmena sosny. [Dynamics of water regime of pine.] - In: Ekologo-fiziologicheskie Issledovaniya Fotosinteza i Vodnogo Rezhima Rasteniĭ v Polevykh Usloviyakh. Pp. 110-124. Sibirskoe Otdelenie, Akademiya Nauk SSSR, Irkutsk 1983. [In R.]

13668 - KAISER, W.M.: Response of photosynthesis and dark-CO_2-fixation to light, CO_2 and temperature in leaf slices under osmotic stress. - J. exp. Bot. 35: 1145-1155, 1984,

13669 - KAKU, S., IWAYA-INOUE, M., GUSTA, L.V.: Relationship of nuclear magnetic resonance relaxation time to water content and cold hardiness in flower buds of evergreen azalea. - Plant Cell Physiol. 25: 875-882, 1984.

*13670 - KAKU, S., IWAYA-INOUE, M., JEON, K.B.: Effects of temperature on cold acclimation and deacclimation in flower buds of evergreen azaleas. - Plant Cell Physiol. 24: 557-564, 1983.

13671 - KALCHEVA, S.: Optimalni polivni normi na tsarevitsa za silazh p"rva kultura za rajona na Sofiĭskata napoitelna sistema. [Optimal irrigation rates of silage maize first crop in the region of the Sofia irrigation system.] - Rasteniev. Nauki 21(7): 106-114, 1984. [In Bulg, ab: E, R.]

13672 - KALININ, N.I.: Formirovanie élementov struktury urozhaya yarovoĭ pshenitsy
pod vliyaniem temperatury vozdukha i vlazhnosti pochvy. [Formation of spring
wheat productivity elements as affected by air temperature and soil moisture.]
- Sel'skokhoz. Biol. 1984(9): 64-67, 1984. [In R, ab: E.]

13673 - KALLARACKAL, J., MILBURN, J.A.: Specific mass transfer and sink-controlled
phloem translocation in castor bean. - Aust. J. Plant Physiol. 11: 483-490,
1984.

13674 - KALLSEN, C.E., SAMMIS, T.W., GREGORY, E.J.: Nitrogen and yield as related
to water use of spring barley. - Agron. J. 76: 59-64, 1984.

*13675 - KALMS, J.M., IMBERNON, J.: Modalités d'alimentation en eau du riz pluvial:
bilan des recherches méthodologiques effectuées à Bouaké en Côte d'Ivoire. -
Agron. trop. 38: 198-205, 1983.

*13676 - KÄMPF, A.N., LANGE, P.: Der Einfluss der Blattdüngung auf morphologische
und anatomische Merkmale bei Bromeliaceen. - Angew. Bot. 57: 9-18, 1983.

13677 - KANDA, A., IMAI, K., MORIYA, T., HANDA, S., TERAJIMA, T.: A gas exchange
measurement system and its application to rice and cassava leaves. - Jap.
J. Crop Sci. 53: 472-478, 1984.

*13678 - KANDPAL, R.P., APPAJI RAO, N.: Alterations in the amount of soluble proteins
and activities of acid phosphatase and nucleotide pyrophosphatase in Ragi
(Eleusine coracana) and mung bean (Vigna radiata) seedlings subjected to
water stress. - Ind. J. exp. Biol. 20: 856-858, 1982.

*13679 - KANDPAL, R.P., APPAJI RAO, N.: Water stress induced alterations in the
properties of ornithine aminotransferase from Ragi (Eleusine coracana)
leaves. - Biochem. int. 5: 297-302, 1982.

13680 - KANDPAL, R.P., APPAJI RAO, N.: Water stress induced alterations in ornithine
aminotransferase of ragi (Eleusine coracana): Protection by proline against
heat inactivation and denaturation by urea and guanidinium chloride. -
J. Biosci. 6: 61-67, 1984.

13681 - KANDPAL, R.P., APPAJI RAO, N.: Water stress-induced desensitization of
aspartate transcarbamylase from Ragi (Eleusine coracana) seedlings. -
Biochem. int. 9: 307-314, 1984.

*13682 - KANDPAL, R.P., VAIDYANATHAN, C.S., KUMAR, M.U., SASTRY, K.S.K., APPAJI RAO,
N.: Alterations in the activities of the enzymes of proline metabolism in
Ragi (Eleusine coracana) leaves during water stress. - J. Biosci. 3: 361-
370, 1981.

13683 - KANG, S., JUDEL, G.K.: Einfluss von NaCl-Salinität auf die CO₂-Assimilation
und auf den Einbau von ¹⁴C in verschiedene Inhaltsstoffe bei jungem Sommer-
weizen. - Z. Pflanzenernähr. Bodenk. 147: 565-571, 1984.

13684 - KANWAR, R.S., JOHNSON, H.P., FENTON, T.E.: Determination of crop production
loss due to inadequate drainage in a large watershed. - Water Resour. Bull.
20: 589-597, 1984.

13685 - KARATELA, Y.Y., GILL, L.S.: Stomatogenesis in the leaves of Murraya koenigii
(L.) Spreng. (Rutaceae). - Feddes Repertorium 95: 347-349, 1984.

13686 - KARATELA, Y.Y., GILL, L.S.: Epidermal structure and stomatal ontogeny in
some Gymnosperms from Nigeria. - Feddes Repertorium 95: 351-354, 1984.

*13687 - KARKANIS, P.G.: Determining field capacity and wilting point using soil saturation by capillary rise. - Can. agr. Eng. 25: 19-21, 1983.

*13688 - KARMANOV, V.G., SOLOV'EV, E.V.: Novye metody i izuchenie mnogofunktsional'-nykh vzaimosvyazeĭ fotosinteza. [New methods and study of multi-functional interrelationships of photosynthesis.] - Trudy prikl. Bot. Genet. Selektsii 72(2): 46-52, 1982. [In R, ab: E.]

13689 - KARUKSTIS, K.K., SAUER, K.: Energy transfer and distribution in the red alga Porphyra perforata studied using picosecond fluorescence spectroscopy. - Biochim. biophys. Acta 766: 141-147, 1984.

*13690 - KAŠE, M., ČATSKÝ, J.: Calculator-assisted measurements of photosynthetic, respiration and photorespiration rates in a closed gas exchange system. - Biol. Plant. 25: 139-146, 1983.

13691 - KATERJI, N., DAUDET, F., VALANCOGNE, C.: Contribution des réserves profondes du sol au bilan hydrique des cultures. Détermination et importance. - Agronomie 4: 779-787, 1984.

13692 - KATERJI, N., HALLAIRE, M.: Les grandeurs de référence utilisables dans l'étude de l'alimentation en eau des cultures. - Agronomie 4: 999-1008, 1984.

13693 - KATERJI, N., HALLAIRE, M., DURAND, R.: Transfert hydrique dans le végétal. III. - Simulation de l'influence des paramètres du couvert sur l'évolution diurne du potentiel hydrique foliaire. - Acta oecol. - Oecol. Plant. 5: 107-117, 1984.

*13694 - KAUFMANN, M.R.: A canopy model (RM-CWU) for determining transpiration of subalpine forests. I. Model development. - Can. J. Forest Res. 14: 218-226, 1983.

*13695 - KAUFMANN, M.R.: A canopy model (RM-CWU) for determining transpiration of subalpine forests. II. Consumptive water use in two watersheds. - Can. J. Forest Res. 14: 227-232, 1983.

13696 - KAUFMANN, M.R.: Effects of weather and physiographic conditions on temperature and humidity in subalpine watersheds of the Fraser Experimental Forest. - USDA Forest Serv. Res. Paper RM-251: 1-9, 1984.

13697 - KECK, T.J., WAGENET, R.J., CAMPBELL, W.F., KNIGHTON, R.E.: Effects of water and salt stress on growth and acetylene reduction in alfalfa. - Soil Sci. Soc. Amer. J. 48: 1310-1316, 1984.

*13698 - KEITEL, A., ARNDT, U.: Ozoninduzierte Turgeszenzverluste bei Tabak (Nicotiana tabacum var. Bel W3) - ein Hinweis auf schnelle Permeabilitätsveränderungen der Zellmembranen. - Angew. Bot. 57: 193-204, 1983.

13699 - KELLER, T., HÄSLER, R.: The influence of a fall fumigation with ozone on the stomatal behavior of spruce and fir. - Oecologia 64: 284-286, 1984.

13700 - KELLIHER, F.M., BLACK, T.A., BARR, A.G.: Estimation of twig xylem water potential in young Douglas-fir trees. - Can. J. Forest Res. 14: 481-487, 1984.

13701 - KENERLEY, C.M., PAPKE, K., BRUCK, R.I.: Effect of flooding on development of Phytophthora root rot in Fraser fir seedlings. - Phytopathology 74: 401-404, 1984.

13702 - KERCHER, J.R., AXELROD, M.C.: A process model of fire ecology and succession in a mixed-conifer forest. - Ecology 65: 1725-1742, 1984.

*13703 - KEREN, R., MEIRI, A., KALO, Y.: Plant spacing effect on yield of cotton
          irrigated with saline waters. - Plant Soil 74: 461-465, 1983.

13704 - KERSHAW, K.A.: Seasonal photosynthetic capacity changes in lichens:
          a provisional mechanistic interpretation. - Lichenologist 16: 145-171,
          1984.

*13705 - KETCHIE, D.O., LOPUSHINSKY, W.: Composition of root pressure exudate from
          conifers. - Pacific Nortwest Forest Range Exp. Sta. Res. Rep. Note PNW-395:
          1-6, 1981.

*13706 - KEYS, A.J.: Effects of environmental conditions on the general metabolism
          of sugars, amino acids and organic acids. - In: METZNER, H. (ed.): Photo-
          syntnesis and Plant Productivity. Pp. 319-325. Wissenschaftliche Verlags-
          gesellschaft, Stuttgart 1983.

13707 - KHADDAR, V.K., RAY, N., GUPTA, R.K., SHRIVASTAVA, M.K.: Effect of salt
          concentrations and moisture regimes on physiological characteristics of
          tomato plants. - Z. Acker- Pflanzenbau 153: 1-13, 1984.

*13708 - KHAN, A.R., DATTA, B.: Effect of aggregate size on water uptake by peanut
          seeds. - Soil Tillage Res. 3: 171-184, 1983.

13709 - KHAN, M.A., UNGAR, I.A.: The effect of salinity and temperature on the
          germination of polymorphic seeds and growth of *Atriplex triangularis*
          Willd. - Amer. J. Bot. 71: 481-489, 1984.

13710 - KHAN, S.M., WILSON, A.M.: Nonstructural carbohydrates and dehydration
          tolerance of blue grama seedlings. - Agron. J. 76: 637-642, 1984.

13711 - KHANNA-CHOPRA, R., KOUNDAL, K.R., SINHA, S.K.: A simple technique of
          studying water deficit effects on nitrogen fixation in nodules without
          influencing the whole plant. - Plant Physiol. 76: 254-256, 1984.

13712 - KHATIBU, A.I., LAL, R., JANA, R.K.: Effects of tillage methods and mulching
          on erosion and physical properties of a sandy clay loam in an equatorial
          warm humid region. - Field Crops Res. 8: 239-254, 1984.

13713 - KIM, J.H., LEE-STADELMANN, O.Y.: Water relations and cell wall elasticity
          quantities in *Phaseolus vulgaris* leaves. - J. exp. Bot. 35: 841-858, 1984.

13714 - KING, R.W.: Water uptake in relation to pre-harvest sprouting damage in
          wheat: grain characteristics. - Aust. J. agr. Res. 35: 337-345, 1984.

*13715 - KING, R.W., GALE, M.D., QUARRIE, S.A.: Effects of NORIN 10 and Tom Thumb
          dwarfing genes on morphology, physiology and abscisic acid production in
          wheat. - Ann. Bot. 51: 201-208, 1983.

13716 - KING, R.W., RICHARDS, R.A.: Water uptake in relation to pre-harvest
          sprouting damage in wheat: ear characteristics. - Aust. J. agr. Res. 35:
          327-336, 1984.

13717 - KINGSBURY, R.W., EPSTEIN, E., PEARCY, R.W.: Physiological responses to
          salinity in selected lines of wheat. - Plant Physiol. 74: 417-423, 1984.

13718 - KIRKHAM, M.B.: Water relations of drought-resistant and drought-sensitive
          wheat cultivars sprinkled with saline water. - Irrig. Sci. 5: 137-146,
          1984.

13719 - KIRKHAM, M.B., SUKSAYRETRUP, K., WASSOM, C.E., KANEMASU, E.T.: Canopy
          temperature of drought-resistant and drought-sensitive genotypes of maize. -
          Maydica 29: 287-303, 1984.

13720 - **KIRSCHBAUM, M.U.F., FARQUHAR, G.D.:** Temperature dependence of whole-leaf photosynthesis in *Eucalyptus pauciflora* Sieb. ex Spreng. - Aust. J. Plant Physiol. 11: 519-538, 1984.

*13721 - **KITANO, M., EGUCHI, H., MATSUI, T.:** Analysis of heat balance of leaf with reference to stomatal responses to environmental factors. - Biotronics 12: 19-27, 1983.

13722 - **KLAGHOFER, E., STENITZER, E.:** Zur Frage der Bestimmung des Wasserbedarfes von Pflanzenbeständen aus Lysimetermessungen. - Z. Kulturtech. Flurbereinig. 25: 6-11, 1984.

13723 - **KLEINSTÄUBER, G.:** Automatische Erfassung und Registrierung der individuellen Wasserzugabe bei Gefässkulturversuchen mit Vegetationstischen. - Arch. Acker-Pflanzenbau Bodenk. 28: 351-356, 1984.

*13724 - **KLEPPER, B., RICKMAN, R.W., TAYLOR, H.M.:** Farm management and the function of field crop root systems. - Agr. Water Manage. 7: 115-141, 1983.

13725 - **KNAPP, A.K.:** Post-burn differences in solar radiation, leaf temperature and water stress influencing production in a lowland tallgrass prairie. - Amer. J. Bot. 71: 220-227, 1984.

13726 - **KNOTH, A., WIENCKE, C.:** Dynamic changes of protoplasmic volume and of fine structure during osmotic adaptation in the intertidal red alga *Porphyra umbilicalis.* - Plant Cell Environ. 7: 113-119, 1984.

13727 - **KOBATA, T.:** [Measurement of relative water content in rice plant.] - Jap. J. Crop Sci. 53: 526-527, 1984. [In Jap, ab: E.]

13728 - **KOBATA, T., TAKAMI, S.:** [Estimation of the leaf water potential in rice by the pressure chamber technique.] - Jap. J. Crop Sci. 53: 290-298, 1984. [In Jap, ab: E.]

13729 - **KOBRIGER, J.M., KLIEWER, W.M., LAGIER, S.T.:** Effects of wind on water relations of several grapevine cultivars. - Amer. J. Enol. Viticult. 35: 164-169, 1984.

13730 - **KOBRIGER, J.M., TIBBITTS, T.W., BRENNER, M.L.:** Injury, stomatal conductance, and abscisic acid levels of pea plants following ozone plus sulfur dioxide exposures at different times of the day. - Plant Physiol. 76: 823-826, 1984.

*13731 - **KOLAROV, D., BROSHCHILOVA, M.:** Prouchvaniya v"rkhu produktivnostta na *P.× eur.* cv. I-214 i *P.× eur.* cv. Bachelieri i intenzivnostta na fotosintezata im v zavisimost ot vlazhnostta na pochvata i mineralnoto khranene. [Productivity of the hybrid, *Populus euramericana,* cultivars I-214 and Bachelieri, and their photosynthetic rate as a function of soil moisture and mineral nutrition.] - Gorskostop. Nauka 17(5): 24-31, 1980. [In Bulg, ab: E, R.]

*13732 - **KOLOMEÏCHENKO, A.A.:** K voprosu ob êffektivnom ispol'zovanii FAR na oroshaemykh zemlyakh. [Effective utilization of PhAR on irrigated soils.] - In: Biologicheskie i Agrotekhnicheskie Osnovy Oroshaemogo Zemledeliya. Pp. 242-246. Nauka, Moskva 1983. [In R.]

*13733 - **KONDRATEV, K.Ya., FEDCHENKO, P.P.:** Spektral'naya Otrazhatel'naya Sposobnost' i Raspoznavanie Rastitel'nosti. [Spectral Reflectance Ability and Recognition of Vegetation.] - Gidrometeoizdat, Leningrad 1982. [In R, ab: E.]

*13734 - **KOPYTOVA, L.D.**: Osobennosti vodnogo rezhima lugovykh rasteniĭ zapadnogo uchastka BAM. [Peculiarities of the water regime of pasture plants of the western part of BAM.] - In: Ėkologo-Fiziologicheskie Issledovaniya Foto-sinteza i Vodnogo Rezhima Rasteniĭ v Polevykh Usloviyakh. Pp. 133-141, 165. Sibirskoe Otdelenie, Akademiya Nauk SSSR, Irkutsk 1983. [In R.]

*13735 - **KOSTYUK, V.I.**: Svyaz' fotochimicheskoĭ aktivnosti khloroplastov s nekoto-rymi fiziologo-biokhimicheskimi pokazatelyami u rasteniĭ kartofelya. [Relation between photochemical activity of chloroplasts and some physio-logical and biochemical indices in potato plants.] - Trudy prikl. Bot. Genet. Selektsii 72(2): 68-76, 1982. [In R, ab: E.]

*13736 - **KOTOV, M.M., KOTOVA, L.I.**: Otsenka zasukhoustoĭchivosti derev'ev po reaktsii khvoi na zasukhu. [The estimation of the conifers drought-resistance by their needles reaction on the drought.] - Izv. Akad. Nauk SSSR, Ser. biol. 1982: 867-873, 1982. [In R, ab: E.]

*13737 - **KOTOV, M.M., ZABIYAKINA, L.A., POLUSHINA, G.I.**: Otsenka zasukhoustoĭchi-vosti i zharoustoĭchivosti sosny obyknovennoĭ. [The estimation of drought-resistance and heat-resistance of *Pinus silvestris*.] - Izv. Akad. Nauk SSSR, Ser. biol. 1981: 266-274, 1981. [In R, ab: E.]

13738 - **KOWALIK, P.J.**: Mathematical modelling of energy forest growth: an outline. In: Swed. Univ. agr. Sci. Rep. 15 (PERTTU, K. (ed.): Ecolody and Manage-ment of Forest Biomass Production Systems): 429-459, 1984.

13739 - **KOWALIK, P.J., ECKERSTEN, H.**: Water transfer from soil through plants to the atmosphere in willow energy forest. - Ecol. Model. 26: 251-284, 1984.

13740 - **KOZHUSHKO, N.N.**: Vodouderzhivayushchaya sposobnost' razlichnykh vidov pshenitsy i ikh dikorastushchikh sorodicheĭ i ee svyaz' s genomnym sostavom i zasukhoustoĭchivost'yu. [Water-holding capacity of different species of *Triticum* and *Aegilops* and its relation to genome composition and drought resistance.] - Sel'skokhoz. Biol. 1984(11): 14-17, 1984. [In R, ab: E.]

13741 - **KOZIOL, M.J., WHATLEY, F.R.**: Gaseous Air Pollutants and Plant Metabolism. Butterworth, London - Boston - Durban - Singapore - Sydney - Toronto - Wellington 1984.

*13742 - **KOZLOWSKI, T.T. (ed.)**: Water Deficits and Plant Growth. Volume VII. Addi-tional Woody Crop Plants. - Academic Press, New York - London - Paris - San Diego - San Francisco - São Paulo - Sydney - Tokyo - Toronto 1983.

13743 - **KOZLOWSKI, T.T.**: Plant responses to flooding of soil. - Bioscience 34: 162-167, 1984.

13744 - **KOZLOWSKI, T.T. (ed.)**: Flooding and Plant Growth. (Physiological Ecology. A Series of Monographs, Texts, and Treatises.). - Academic Press, Orlando - San Diego - San Francisco - New York - London - Toronto - Montreal - Sydney - Tokyo - São Paulo 1984.

13745 - **KOZLOWSKI, T.T.**: Extent, causes, and impacts of flooding. - In: KOZLOWSKI, T.T. (ed.): Flooding and Plant Growth. Pp. 1-7. Academic Press, Orlando - San Diego - San Francisco - New York - London - Toronto - Montreal - Sydney - Tokyo - São Paulo 1984.

13746 - **KOZLOWSKI, T.T.**: Responses of woody plants to flooding. - In: KOZLOWSKI, T.T. (ed.): Flooding and Plant Growth. Pp. 129-163. Academic Press, Orlando - San Diego - San Francisco - New York - London - Toronto - Montreal - Sydney - Tokyo - São Paulo 1984.

13747 - KOZLOWSKI, T.T., PALLARDY, S.G.: Effect of flooding on water, carbohydrate and mineral relations. - In: KOZLOWSKI, T.T. (ed.): Flooding and Plant Growth. Pp. 165-193. Academic Press, Orlando - San Diego - San Francisco - New York - London - Toronto - Montreal - Sydney - Tokyo - São Paulo 1984.

13748 - KRAFT, H.: Verdunstungsleistung von Limnophyten bei der Abwasserreinigung in hydrobotanischen Kläranlagen. - Z. Wasser- Abwasser- Forsch. 17: 12-15, 1984.

13749 - KRAJÍČKOVÁ, A., MEJSTŘÍK, V.: The effect of fly ash particles on the plugging stomata. - Environ. Pollut., Ser. A 36: 83-93, 1984.

*13750 - KRAMER, P.J.: Water and plant productivity or yield. - In: RECHCIGL, M.,Jr. (ed.): CRC Handbook of Agricultural Productivity. Volume 1. Plant Productivity. Pp. 41-47. CRC Press, Inc., Boca Raton 1982.

13751 - KRAMPITZ, M.J., FOCK, H.P.: $^{14}CO_2$ assimilation and carbon flux in the Calvin cycle and the glycollate pathway in water-stressed sunflower and bean leaves. - Photosynthetica 18: 329-337, 1984.

13752 - KRAMPITZ, M.J., KLUG, K., FOCK, H.P.: Rates of photosynthetic CO_2 uptake, photorespiratory CO_2 evolution and dark respiration of water-stressed sunflower and bean leaves. - Photosynthetica 18: 322-328, 1984.

*13753 - KRASAVTSEV, O.A., RAZNOPOLOV, O.N., KHVALIN, N.N.: Ottok vody iz pereokhlazhdennykh zimuyushchikh butonov. [Water efflux from supercooled wintering flower primordia.] - Fiziol. Rast. 30: 1025-1032, 1983. [In R, ab: E.]

*13754 - KRAUSE, C.R.: Diagnosis of ambient air pollution injury to red maple leaves. - Scanning Electron Microscopy 3: 203-206, 1981.

*13755 - KRAUSE, G.H.: Umweltstress: Wirkung extremer Umweltbedingungen auf die Photosynthese. - Ber. Deut. bot. Ges. 96: 255-257, 1983.

13756 - KREIL, W., MUNDEL, G.: Untersuchungen über die Evapotranspiration von Grasland auf Grundwasserstandorten. 5. Mitteilung: Quantifizierung des Einflusses untersuchungen Faktoren in bezug auf Meliorationsmassnahmen und Graslanderträge. - Arch. Acker- Pflanzenbau Bodenk. 28: 75-81, 1984.

13757 - KRETSCHMER, H.: Relativer Blattwassergehalt und Stressindex als Kriterium des Wasserversorgungszustandes bei Zuckerrüben unter Feldbedingungen. - Wiss. Z. Humboldt-Univ., math. naturwiss. Reihe 33: 381-382, 1984.

13758 - KRETSCHMER, H., SCHÄFER, W.: Adaptationsverhalten von Zuckerrüben nach Wasserdefizitperioden auf einem Diluvilstandort. - Arch. Acker- Pflanzenbau Bodenk. 28: 295-303, 1984.

13759 - KRIEBITZSCH, W.-U., GRIMME, K., REGEL, J.: Über den H_2O-Gaswechsel von *Mercurialis perennis* im submontanen Kalkbuchenwald. I. Die Bedeutung steuernder Parameter für den Tagesgang der Transpiration in verschiedenen Phasen der Vegetationsentwicklung. - Flora 175: 257-272, 1984.

13760 - KRIEDEMANN, P.E., SANDS, R.: Salt resistance and adaptation to root-zone hypoxia in sunflower. - Aust. J. Plant Physiol. 11: 287-301, 1984.

13761 - KROECKEL, L., STOLP, H.: Influence of soil water potential on respiration and nitrogen fixation of *Azotobacter vinelandii*. - Plant Soil 79: 37-49, 1984.

*13762 - KRSTIČ, B.: Uticaj abiotskih i biotskih činilaca na fotosintezu pšenice.
[Effects of abiotic and biotic factors upon photosynthesis in wheat.] -
Posebna Izd. Srpska Akad. Nauka Umetn. 536, Od. prirod.-mat. Nauka 53:
21-38, 1981. [In Serb, ab: E.]

*13763 - KRSTIČ, B.: Uticaj nekih činilaca na proces asimilacije CO_2 kod šećerne
repe. [Effect of some factors on CO_2 assimilation process in sugar beet.] -
Posebna Izd. Srpska Akad. Nauka Umetn. 538, Od. prirod.-mat. Nauka 54 (BELIČ,
J. (ed.): Fiziologija Šećerne Repe.): 17-38, 1981. [In Serb, ab: E.]

13764 - KUBIK, M., PŁONKA, A.: Abscisic acid induced decay of strawberry transpiration. - Physiol. Plant. 60: 539-542, 1984.

13765 - KUBOTA, F., TANAKA, N.: [Growth retardation and wilting of corn seedlings
(Zea mays L.) by aboveground and underground environmental factors.] -
Jap. J. Crop Sci. 53: 145-152, 1984. [In Jap, ab: E.]

13766 - KŮDELA, V., KOVÁČIKOVÁ, E.: Vliv vláhových podmínek v roce výsevu na vyzi-
mování jetele lučního. [The influence of moisture conditions in the year
of sowing on the red clover winter killing.] - Rost. Výroba (Praha) 30:
387-392, 1984. [In Czech, ab: E, R.]

*13767 - KUGANATHAN, A., PALANIAPPAN, S.: Effect of antitranspirants on growth and
production of grain sorghum. - Turrialba 29: 69-74, 1979.

*13768 - KUGANATHAN, A., PALANIAPPAN, S.: Effect of antitranspirants on soil and
plant water status in grain sorghum. - Acta agron. Acad. Sci. Hung. 29:
401-409, 1980.

*13769 - KULANDAIVELU, G., NOORUDEEN, A.M., SAMPATH, P., PERIYANAN, S.: A modified
leaf chamber assembly to determine the rate of CO_2 exchange for upper and
lower sides of leaves. - Experientia 39: 1179-1180, 1983.

13770 - KULIEVA, L.K., OBRUCHEVA, N.V., PROKOF'EV, A.A.: O fiziologicheskoĭ zre-
losti sozrevayushchikh semyan tonkovoloknistogo khlopchatnika. [On physiological maturity of developing seeds of Gossypium barbadense L.] - Fiziol.
Rast. 31: 928-933, 1984. [In R, ab: E.]

13771 - KUMAR, A., SINGH, P., SINGH, D.P., SINGH, H., SHARMA, H.C.: Differences in
osmoregulation in Brassica species. - Ann. Bot. 54: 537-541, 1984.

*13772 - KUO, C.G., CHEN, B.W.: Physiological responses of tomato cultivars to
flooding. - J. Amer. Soc. hort. Sci. 105: 751-755, 1980.

*13773 - KURKIN, K.A., MEDVEDEVA, A.S.: Ékologicheskie mekhanizmy transformatsii
fitotsenozov kratkopoemnykh lugov pod vozdeĭstviem orosheniya, udobreniya
i rezhimov otchuzhdeniya. [Ecological mechanisms of transformation in the
phytocoenoses of the short-flooded meadows under the influence of irrigation, fertilization and alienation.] - Ékologiya 1983(6): 3-8, 1983.
[In R.]

13774 - KUTÍK, J., ŠESTÁK, Z., VOLFOVÁ, A.: Ontogenetic changes in the internal
limitations to bean-leaf photosynthesis. 8. Primary leaf blade characteristics and chloroplast number, size and ultrastructure. - Photosynthetica
18: 1-8, 1984.

*13775 - KUZNETSOVA, I.V., VINOGRADOVA, G.B.: Vlazhnost' zavyadaniya rasteniĭ
v uplotnennykh gorizontakh pochv. [Wilting moisture of plants in the soil
compact horizons.] - Pochvovedenie 1983 (5): 58-64, 1983. [In R, ab: E.]

13776 - **KUZNETSOVA, I.V., VINOGRADOVA, G.B.:** Vliyanie plotnosti illyuvial'nogo gorizonta dernovo-podzolistykh pochv na dostupnost' vody rasteniyam (po rezul'tatam model'nogo opyta). [Water availability influenced by the density of illuvial horizons in sod-podzolic soil (from results of the model experiments).] - Pochvovedenie 1984(2): 55-60, 1084. [In R, ab: E.]

13777 - **LADIPO, D.O., GRACE, J., SANFORD, A.P., LEAKEY, R.R.B.:** Clonal variation in photosynthetic and respiration rates and diffusion resistances in the tropical hardwood *Triplochiton scleroxylon* K. Schum. - Photosynthetica 18: 20-27, 1984.

13778 - **LAFFRAY, D., VAVASSEUR, A., GARREC, J.-P., LOUGUET, P.:** Moist air effects on stomatal movements and related ionic content in dark conditions. Study on *Pelargonium* x *hortorum* and *Vicia faba*. - Physiol. vég. 22: 29-36, 1984.

*13779 - **LAHAV, E., KALMAR, D.:** Determination of the irrigation regimen for an avocado plantation in spring and autumn. - Aust. J. agr. Res. 34: 717-724, 1983.

13780 - **LAKHANOV, A.P.:** Fiziologo-biochemicheskie osobennosti ustoĭchivosti rastenii zernovykh bobovykh kul'tur pri deĭstvii zamorozkov. [Physiological and biochemical changes in plants of grain legumes as affected by frosts.] - Sel'skokhoz. Biol. 1984(10): 7-11, 1984. [In R, ab: E.]

13781 - **LAKSO, A.N., BIERHUIZEN, J.F., MARTAKIS, G.F.P.:** Light responses of photosynthesis and transpiration of two tomato cultivars under ambient and altered CO_2 and O_2. - Scientia Hort. (Amsterdam) 23: 119-128, 1984.

13782 - **LAKSO, A.N., GEYER, A.S., CARPENTER, S.G.:** Seasonal osmotic relations in apple leaves of different ages. - J. Amer. Soc. hort. Sci. 109: 544-547, 1984.

*13783 - **LANCASTER, J.E., MANN, J.D.:** Water relationships of *Lupinus luteus*. - In: Proceedings of Soil and Plant Water Symposium. Pp. 49-54. Palmerston North 1976.

*13784 - **LANDSBERG, J.J.:** Limits to apple yields imposed by weather. - In: HURD, R.G., BISCOE, P.V., DENNIS, C. (ed.): Opportunities for Increasing Crop Yields. Pp. 161-180. Pitman Publishing Ltd., London 1979.

*13785 - **LANDSBERG, J.J.:** The use of models in interpreting plant response to weather. - GRACE, J., FORD, E.D., JARVIS, P.G. (ed.): Plants and their Atmospheric Environment. Pp. 369-389. Blackwell Scientific Publications, Oxford - London - Edinburgh - Boston - Melbourne 1980.

13786 - **LANDSBERG. J.J.. McMURTRIE. R.:** Water use by isolated trees. - Agr. Water Manage 8: 223-242, 1984. Also in: SHARMA, M.L. (ed.): Evapotranspiration from Plant Communities. Pp. 223-242. Elsevier Science Publishers, Amsterdam - Oxford - New York - Tokyo 1984.

13787 - **LANE, L.J., ROMNEY, E.M., HAKONSON, T.E.:** Water balance calculations and net production of perennial vegetation in the northern Mojave Desert. - J. Range Manage. 37: 12-18, 1984.

13788 - **LANGENHEIM, J.H., OSMOND, C.B., BROOKS, A., FERRAR, P.J.:** Photosynthetic responses to light in seedlings of selected Amazonian and Australian rain-forest tree species. - Oecologia 63: 215-224, 1984.

13789 - **LANKES, C., LENZ, F.:** Veränderungen der stomatären Diffusionswiderstände von Sauerkirschblättern unter verschiedenen Umweltbedingungen. - Gartenbauwissenschaft 49: 199-204, 1984.

13790 - **LAPOINTE, B.E., RICE, D.L., LAWRENCE, J.M.:** Responses of photosynthesis, respiration, growth and cellular constituents to hypo-osmotic shock in the red alga *Gracilaria tikvahiae*. - Comp. Biochem. Physiol. 77 A: 127-132, 1984.

*13791 - **LARSON, D.W.:** Differential wetting in some lichens and mosses: the role of morphology. - Bryologist 84: 1-15, 1981.

*13792 - **LARSON, D.W.:** The pattern of production within individual *Umbilicaria* lichen thalli. - New Phytol. 94: 409-419, 1983.

13793 - **LASCANO, R.J., Van BAVEL, C.H.M.:** Root water uptake and soil water distribution: test of an availability concept. - Soil Sci. Soc. Amer. J. 48: 233-237, 1984.

*13794 - **LAURIDSON, T.C., WILSON, R.G., HADERLIE, L.C.:** Effect of moisture stress on Canada thistle (*Cirsium arvense*) control. - Weed Sci. 31: 674-680, 1983.

13795 - **LAUTENSCHLAGER, K.:** Osmotic potential and gas exchange in healthy and diseased silver firs (*Abies alba* Mill.). - Europ. J. Forest Pathol. 14: 359-372, 1984.

*13796 - **LAWN, R.J.:** Response of four grain legumes to water stress in south-eastern Queensland. IV. Interaction with sowing arrangement. - Aust. J. agr. Res. 34: 661-669, 1983.

13797 - **LAYNE, R.E.C., TAN, C.S.:** Long-term influence of irrigation and tree density on growth, survival, and production of peach. - J. Amer. Soc. hort. Sci. 109: 795-799, 1984.

*13798 - **LECHOWICZ, M.J.:** Ecological trends in lichen photosynthesis. - Oecologia 53: 330-336, 1982.

13799 - **LeDREW, H.D., DAYNARD, T.B., MULDOON, J.F.:** Relationships among hybrid maturity, environment, dry matter yield and moisture content of whole -plant corn. - Can. J. Plant Sci. 64: 565-573, 1984.

*13800 - **LEE, D.R.:** Water affects on plant growth and development. - In: METZNER, H. (ed.): Photosynthesis and Plant Productivity. Pp. 305-314. Wissenschaftliche Verlagsgesellschaft, Stuttgart 1983.

13801 - **LEE, D.R., ATKEY, P.T.:** Water loss from the developing caryopsis of wheat (*Triticum aestivum*). - Can. J. Bot. 62: 1319-1326, 1984.

13802 - **LEE, R.F., MARAIS, L.J., TIMMER, L.W., GRAHAM, J.H.:** Syringe injection of water into the trunk: A rapid diagnostic test for citrus blight. - Plant Dis. 68: 511-513, 1984.

*13803 - **LEECH, R.M., BAKER, N.R.:** The development of photosynthetic capacity in leaves. - In: DALE, J.E., MILTHORPE, F.L. (ed.): The Growth and Functioning of Leaves. Pp. 271-307. Cambridge University Press, Cambridge - London - New York - New Rochelle - Melbourne - Sydney 1983.

13804 - **LEES, G.L.:** Cuticle and cell wall thickness: Relation to mechanical strength of whole leaves and isolated cells from some forage legumes. - Crop Sci. 24: 1077-1081, 1984.

13805 - **LEE-STADELMANN, O.Y., BUSHNELL, W.R., STADELMANN, E.J.:** Changes of plasmolysis form in epidermal cells of *Hordeum vulgare* infected by *Erysiphe graminis*: evidence for increased membrane-wall adhesion. - Can. J. Bot. 62: 1714-1723, 1984.

*13806 - **LEGEL, S.**: Untersuchungen zum Anbau und zur Ertragsleistung von Alexandriner-
klee (*Trifolium alexandrinum* L.) in Abhängigkeit von Vegetationsstadium,
Aufwuchs und Bewässerung in Ägypten. Teil 2: Ertragsleistungen. - Beitr.
trop. Landwirt. Veterinärmed. 21: 355-363, 1983.

13807 - **LEHMAN, W.F., RUTGER, J.N., ROBINSON, F.E., KADDAH, M.**: Value of rice
characteristics in selection for resistance to salinity in an arid environment. - Agron. J. 76: 366-370, 1984.

13808 - **LEHNBERG, W., SCHRAMM, W.**: Mass culture of brackish-water-adapted seaweeds
in sewage-enriched seawater I. Productivity and nutrient accumulation. -
Hydrobiologia 116/117: 276-281, 1984.

13809 - **LE MASSON, B., PAULIN, A.**: Influence des séjours à sec l'évolution de l'état
hydrique et de la fuite des solutés chez l'oeillet coupé. - Physiol. vég. 22:
425-435, 1984.

13810 - **Le RUDULIER, D., BERNARD, T., GOAS, G., HAMELIN, J.C.**: Osmoregulation in
Klebsiella pneumoniae: enhancement of anaerobic growth and nitrogen fixation
under stress by proline betaine, γ-butyrobetaine, and other related compounds.
- Can. J. Microbiol. 30: 299-305, 1984.

13811 - **Le RUDULIER, D., STROM, A.R., DANDEKAR, A.M., SMITH, L.T., VALENTINE, R.C.**:
Molecular biology of osmoregulation. - Science 224: 1064-1068, 1984.

13812 - **LETEY, J., VAUX, H.J.,Jr., FEINERMAN, E.**: Optimum crop water application as
affected by uniformity of water infiltration. - Agron. J. 76: 435-441, 1984.

*13813 - **LEVI, Y., BERNER, T., COHEN, Y.**: CO_2 exchange and growth rate of the loess
soil crusts algae in the Negev desert of Israel. - In: SHUVAL, H.I. (ed.):
Development in Arid Zone Ecology and Environmental Quality. Pp. 43-48.
Balaban International Science Services, Philadelphia 1981.

*13814 - **LEVITT, J.**: Injury and repair during rehydration droughted cabbage leaves. -
Carnegie Inst. Washington Year Book 82: 87, 1983.

*13815 - **LEVY, G.**: Comportement de jeunes plants d'Epicéa commun en sol à engorge-
ment temporaire de surface: influence de divers facteurs du milieu. - Ann.
Sci. forest. 38: 3-30, 1981.

13816 - **LIAO, Z.-Z., PAN, R.-C.**: [Fusicoccin-ethylene interaction in the growth of
etiolated pea seedlings.] - Acta bot. Sin. 26: 272-279, 1984. [In Chin,
ab: E.]

13817 - **LICHTENTHALER, H.K., BUSCHMANN, C.**: Das Waldsterben aus Botanischer Sicht.
G. Braun, Karlsruhe 1984.

13818 - **LIETH, J.H., REYNOLDS, J.F.**: A model of canopy irradiance in relation to
changing leaf area in a phytotron-grown snap bean (*Phaseolus vulgaris* L.)
crop. - Int. J. Biometeorol. 28: 61-71, 1984.

13819 - **LIN, S.-H., HUANG, Y.-X., CHEN, Z.-L., HAN, R.-Z., YAO, Y.-Q.**: [The effect
of several pollutants on the transpiration of rice plants.] - Acta bot.
Sin. 26: 85-93, 1984. [In Chin, ab: E.]

13820 - **LIN, T.-S., CRANE, J.C., RYUGO, K., POLITO, V.S., DeJONG, T.M.**: Comparative
study of leaf morphology, photosynthesis, and leaf conductance in selected
Pistacia species. - J. Amer. Soc. hort. Sci. 109: 325-330, 1984.

13821 - **LINK, S.O., NASH, T.H.,III.**: The influence of water on CO_2 exchange in the
lichen *Parmelia praesignis* Nyl. - Oecologia 64: 204-210, 1984.

13822 - LINK, S.O., NASH, T.H.,III.: A mathematical description of the effect of resaturation on net photosynthesis in the lichen, *Parmelia praesignis* Nyl. - New Phytol. 96: 257-262, 1984.

13823 - LIVINGSTON, N.J., BLACK, T.A., BEAMES, D., DUNSWORTH, B.G.: An instrument for measuring the average stomatal conductance of conifer seedlings. - Can. J. Forest Res. 14: 512-517, 1984.

13824 - LONG, F.L.: A field system for automatically measuring soil water potential. - Soil Sci. 137: 227-230, 1984.

13825 - LONGSTRETH, D.J., BOLAÑOS, J.A., SMITH, J.E.: Salinity effects on photosynthesis and growth in *Alternanthera philoxeroides* (Mart.) Griseb. - Plant Physiol. 75: 1044-1047, 1984.

13826 - LOPUSHINSKY, W., KAUFMANN, M.R.: Effects of cold soil on water relations and spring growth of Douglas-fir seedlings. - Forest Sci. 30: 628-634, 1984.

13827 - LÖSCH, R.: Die Ökologie der mainfränkischen Kalktrockenrasen. - Abhandl. naturwiss. Vereins Würzburg 21/22: 1-14, 1981/1984.

13828 - LÖSCH, R.: Jahreszeitlicher Verlauf der diurnalen Säureschwankungen bei Semperviven von unterschiedlichen Standorten auf Teneriffa. - In: Mitteilungsband Botaniker-Tagung in Wien. Pp. 3-4. Deutsche Botanische Gesellschaft, Vereinigung für Angewandte Botanik, Wien 1984.

13829 - LÖSCH, R.: Evolution funktioneller Standortpassungen als Grundlage differenzierter Nischenbesiedelung bei der Makaronesischen Semperviven-Gattung *Aichryson*. - In: Mitteilungsband Botaniker-Tagung in Wien. Pp. 15-16. Deutsche Botanische Gesellschaft, Vereinigung für Angewandte Botanik, Wien 1984.

13830 - LÖSCH, R.: Plant water relations. - Progress Bot. 46: 38-55, 1984.

13831 - LOVATT, C.J., BATES, L.M.: Early effects of excess boron on photosynthesis and growth of *Cucurbita pepo*. - J. exp. Bot. 35: 297-305, 1984.

*13832 - LOVETT, J.V., ORCHARD, P.W.: Morphological and anatomical changes induced in sunflower by chlormequat and their possible significance. - In: Proceedings of the VIIth International Sunflower Conference. Volume II. Pp. 323-331. Krasnodar 1976.

13833 - LOVEYS, B.R.: Diurnal changes in water relations and abscisic acid in field-grown *Vitis vinifera* cultivars. III. The influence of xylem-derived abscisic acid on leaf gas exchange. - New Phytol. 98: 563-573, 1984.

13834 - LOVEYS, B.R.: Abscisic acid transport and metabolism in grapevine (*Vitis vinifera* L.). - New Phytol. 98: 575-582, 1984.

13835 - LOVEYS, B.R., DÜRING, H.: Diurnal changes in water relations and abscisic acid in field-grown *Vitis vinifera* cultivars. II. Abscisic acid changes under semi-arid conditions. - New Phytol. 97: 37-47, 1984.

*13836 - LUDLOW, M.M.: External factors influencing photosynthesis and respiration. - In: DALE, J.E., MILTHORPE, F.L. (ed.): The Growth and Functioning of Leaves. Pp. 347-380. Cambridge University Press, Cambridge - London - New York - New Rochelle - Melbourne - Sydney 1983.

*13837 - LUDLOW, M.M., BJÖRKMAN, O.: Paraheliotropic leaf movement as a protective mechanism against drought-induced damage to primary photosynthetic reactions. - Carnegie Inst Washington Year Book 82: 89-91, 1983.

13838 - **LUDLOW, M.M., BJÖRKMAN, O.:** Paraheliotropic leaf movement in Siratro as a protective mechanism against drought-induced damage to primary photosynthetic reactions: damage by excessive light and heat. - Planta 161: 505-518, 1984.

*13839 - **LÜTTGE, U.:** Transport functions of leaves. - In: DALE, J.E., MILTHORPE, F.L. (ed.): The Growth and Functioning of Leaves. Pp. 413-448. Cambridge University Press, Cambridge - London - New York - New Rochelle - Melbourne - Sydney 1983.

13840 - **LÜTTGE, U., NOBEL, P.S.:** Day-night variations in malate concentration, osmotic pressure, and hydrostatic pressure in *Cereus validus*. - Plant Physiol. 75: 804-807, 1984.

13841 - **LUXMOORE, R.J., SHARMA, M.L.:** Evapotranspiration and soil heterogeneity. - Agr. Water Manage. 8: 279-289, 1984. Also in: SHARMA, M.L. (ed.): Evapotranspiration from Plant Communities. Pp. 279-289. Elsevier Science Publishers, Amsterdam - Oxford - New York - Tokyo 1984.

13842 - **LYLES, L., TATARKO, J., DICKERSON, J.D.:** Windbreak effects on soil water and wheat yield. - Trans. ASAE 27: 69-72, 1984.

13843 - **LYNCH, J., LÄUCHLI, A.:** Potassium transport in salt-stressed barley roots. - Planta 161: 295-301, 1984.

*13944 - **LYUTOVA, M.I.:** Sravnenie ustoĭchivosti reaktsii Khilla k vysokoĭ temperature i gidroliticheskim fermentam u dvukh vidov vysshikh rasteniĭ. [Comparison of the resistance of Hill reaction to heat and to hydrolytic enzymes in two higher plant species.] - Fiziol. Rast. 30: 1194-1200, 1983. [In R, ab: E.]

13845 - **MacDONALD, J.D.:** Salinity effects on the susceptibility of chrysanthemum roots to *Phytophthora cryptogea*. - Phytopathology 74: 621-624, 1984.

*13846 - **MACHADO, R.C.R., SOUZA, H.M.F., MORENO, M.A., ALVIM, P.de T.:** Variáveis relacionadas com a tolerância de gramíneas forrageiras ao déficit hídrico. [Variables associated with the tolerance to water deficit in forage grasses.] - Pesq. agropec. Bras. 18: 603-608, 1983. [In Port, ab: E.]

13847 - **MACKAY, M.A., NORTON, R.S., BOROWITZKA, L.J.:** Organic osmoregulatory solutes in cyanobacteria. - J. gen. Microbiol. 130: 2177-2191, 1984.

*13848 - **MACOVSCHI, E.:** The biostructured water. - Rev. Roum. Biochim. 17: 39-49, 1980.

13849 - **MACROBBIE, E.A.C.:** Effects of light/dark on anion fluxes in isolated guard cells of *Commelina communis* L. - J. exp. Bot. 35: 707-726, 1984.

13850 - **MACROBBIE, E.A.C.:** Light effects on ion fluxes in guard cells. - CRAM, W.J., JANÁČEK, K., RYBOVÁ, R., SIGLER, K. (ed.): Membrane Transport in Plants. Pp. 129-134. Academia, Praha 1984.

13851 - **MADHAVAN, S., SMITH, B.N.:** Phosphoenolpyruvate carboxylase in guard cells of several species as determined by an indirect, immunofluorescent technique. - Protoplasma 122: 157-161, 1984.

13852 - **MADORE, M., GRODZINSKI, B.:** Stimulation of net photosynthesis in cucumber leaf discs: effect of K-glyoxylate and KCl. - J. exp. Bot. 35: 941-947, 1984.

*13853 - **MAGALHÃES, A.A.,de, CHOUDHURY, M.M., MILLAR, A.A., ALBUQUERQUE, M.M., de:** Efeito do déficit de água no solo sobre o ataque de *Macrophomina phaseolina* em feijão. [Effect of soil water deficit on the attact of *Macrophomina phaseolina* on beans.] - Pesq. agropec. Bras. 17: 407-411, 1982. [In Port, ab: E.]

13854 - **MAGAN, N., LACEY, J.:** Water relations of some *Fusarium* species from infected wheat ears and grain. - Trans. Brit. mycol. Soc. 83: 281-285, 1984.

13855 - **MAGID, A.H.A., BIERHUIZEN, J.F.:** Growth, photosynthesis and yield of some snapbean (*Phaseolus vulgaris* L.) cultivars I. Gas exchange measurements. - Gartenbauwissenschaft 49: 1-6, 1984.

*13856 - **MAGNUSSEN, S.:** Wirkungen von Licht, Wasser und Nährstoffen auf junge Küsten-tannen (*Abies grandis* Lindley). - Allg. Forst. Jagdzeit. 154(1): 10-20, 1983.

*13857 - **MAGNUSSEN, S., PESCHL, A.:** Die Einwirkung verschiedener Beschattungsgrade auf die Photosynthese und die Transpiration junger Weiss- und Küstentannen. - Allg. Forst. Jagdzeit. 152(5): 82-93, 1981.

13858 - **MAGOMEDOV, Z.G.:** Kolichestvennye i kachestvennye izmeneniya v list'yakh vinogradnogo rasteniya pri razlichnykh usloviyakh osveshcheniya. [Qualita-tive and quantitative changes in grape leaves under various illumination conditions.] - Sel'skokhoz. Biol. 1984(8): 81-84, 1984. [In R, ab: E.]

*13859 - **MAHAMA, A., SILVY, A.:** Influence de la teneur en eau sur la radiosensibilite des semences d'*Hibiscus cannabinus* L. - I. Role des differentes etats de l'eau. - Environ. exp. Bot. 22: 233-242, 1982.

13860 - **MAHEY, R.K., CHEEMA, S.S.:** Evaluating methods of nitrogen application for efficiency of water use and yield of rainfed barley. - J. agr. Sci. 102: 81-83, 1984.

13861 - **MAHEY, R.K., FFYFN, J., WYSEURE, G.:** A numerical analysis of irrigation treatments of barley with respect to drainage losses and crop response. - Trans. ASAE 27: 1805-1810, 1984.

*13862 - **MAHON, J.D., HOBBS, S.L.A., SALMINEN, S.O.:** Characteristics of pea leaves and their relationships to photosynthetic CO_2 exchange in the field. - Can. J. Bot. 61: 3283-3292, 1983.

13863 - **MAHRER, Y., NAOT, O., RAWITZ, E., KATAN, J.:** Temperature and moisture regi-mes in soils mulched with transparent polyethylene. - Soil Sci. Soc. Amer. J. 48: 362-367, 1984.

13864 - **MAIER-MAERCKER, U.:** Turgor-mediated leaf movements in analogy with stomatal function and under the general aspect of water flux through the plant. I. Microautoradiographic localization of ^{86}Rb and ^{43}K in the laminar pulvinus of *Phaseolus vulgaris* L. - J. Plant Physiol. 115: 405-418, 1984.

13865 - **MAIER-MAERCKER, U.:** Turgor-mediated leaf movements in analogy with stomatal function and under the general aspect of water flux through the plant. II. Rhythmic transport of ^{86}Rb and ^{43}K in *Trifolium repens* L. and *Oxalis aceto-sella* L. - J. Plant Physiol. 115: 419-425, 1984.

13866 - **MAITI, R.K., PRASADA RAO, K.E., RAJU, P.S., HOUSE, L.R.:** The glossy trait in sorghum: its characteristics and significance in crop improvement. - Field Crops Res. 9: 279-289, 1984.

13867 - **MAJUMDAR, D.K., MANDAL, M.:** Effect of irrigation based on pan evaporation and nitrogen levels on the yield and water use in wheat. - Ind. J. agr. Sci. 54: 613-614, 1984.

13868 - **MAKINO, A., MAE, T., OHIRA, K.:** Changes in photosynthetic capacity in rice leaves from emergence through senescence. Analysis from ribulose-1,5-bis-phosphate carboxylase and leaf conductance. - Plant Cell Physiol. 25: 511-521, 1984.

*13869 - **MALEK, L.**: The effect of drying on *Spirodela polyrhiza* turion germination. - Can. J. Bot. 59: 104-105, 1981.

*13870 - **MALI, C.V., VARADE, S.B.**: Response of rice to varying soil moisture conditions in vertisols. I. Studies on soil pH and nutrient availability. - Z. Acker- Pflanzenbau 150: 129-139, 1981.

*13871 - **MALI, C.V., VARADE, S.B.**: Response of rice to varying soil moisture conditions in vertisols. II. Studies on nutrient content in rice plants. - Z. Acker- Pflanzenbau 150: 140-146, 1981.

*13872 - **MALIK, Y.S., MEHROTRA, N., PANDITA, M.L., JAISWAL, R.C.**: Effects of soil salinity levels on the yield and quality of radish, carrot and turnip. - Ind. J. agr. Sci. 53: 861-862, 1983.

*13873 - **MALLICK, S., KALRA, N., NAGARAJARAO, Y.**: Field water balance and water use efficiency of wheat under shallow water table condition. - Ind. J. Agron. 26: 435-441, 1981.

13874 - **MANDAL, B.K., CHATTERJEE, B.N.**: Growth and yield performance of selected rice varieties during cooler months under two water regimes. - Ind. J. Agron. 29: 94-100, 1984.

13875 - **MANDAL, B.K., GHOSH, T.K.**: Efficacy of mulches in the reduction of irrigation requirement of groundnut. - Ind. J. agr. Sci. 54: 446-449, 1984.

*13876 - **MANDOLI, D.F., BOYER, J.S., BRIGGS, W.R.**: Fiber-optic capacity of plant tissues varies as a function of tissue water status. - Carnegie Inst. Washington Year Book 82: 40-42, 1983.

*13877 - **MANSFIELD, T.A., DAVIES, W.J.**: Abscisic acid and water stress. - Biochem. Soc. Transactions 11: 557-560, 1983.

13878 - **MANSFIELD, T.A., FREER-SMITH, P.H.**: The role of stomata in resistance mechanisms. - In: KOZIOŁ, M.J., WHATLEY, F.R. (ed.): Gaseous Air Pollutants and Plant Metabolism. Pp. 131-146. Butterworths, London - Boston - Durban - Singapore - Sydney - Toronto - Wellington 1984.

13879 - **MANSOUR, K.S., HALLET, J.N., LECOCQ, F.M.**: Effets d'une période de déshydratation sur le métabolisme des ARN et des protéines chez une mousse revivis- cente, le *Polytrichum formosum* Hedw. - Bull. Soc. bot. France, Actual. bot. 131: 113-124, 1984.

*13880 - **MARANO, B., MATTEI, F.**: Behaviour of sorghum hybrid (*Sorghum bicolor* (L.) Moench x *Sorghum sudanese* (Piper) Stapf) in presence of high plant density. II - Chemical composition and nutritive value. - Agrochimica 21: 370-378, 1977.

*13881 - **MARANO, B., MATTEI, F.**: Behaviour of a sorghum hybrid (*Sorghum bicolor* L., Moench x *Sorghum sudanese* Piper, Stapf.) in presence of high plant density. I - Competition and productivity. - Ann. Fac. Sci. agr. Univ. Napoli Portici, Ser. IV 12: 217-230, 1978.

13882 - **MARGARIS, N.S., ARIANOUSTOU-FARRAGITAKI, M., OECHEL, W.C. (ed.)**: Being Alive on Land (Tasks for Vegetation Science. Volume 13). - Dr W. Junk Publishers, The Hague 1984.

13883 - **MARIE, B.A., ORMROD, D.P.**: Tomato plant growth with continuous exposure to sulphur dioxide and nitrogen dioxide. - Environ. Pollut. Ser. A 33: 257-265, 1984.

*13884 - **MARIGO, G., LÜTTGE, U., SMITH, J.A.C.:** Cytoplasmic pH and the control of
crassulacean acid metabolism. - Z. Pflanzenphysiol. 109: 405-413, 1983.

13885 - **MARKHART, A.H.,III.:** Amelioration of chilling-induced water stress by abscisic acid-induced changes in root hydraulic conductance. - Plant Physiol. 74:
81-83, 1984.

13886 - **MARKHART, A.H.,III., SMIT-SPINKS, B.:** Comparison of J-14 hydraulic press
with the Scholander pressure bomb for measuring leaf water potential. -
HortScience 19: 52-54, 1984.

13887 - **MARÓTI, I., TUBA, Z., CSIK, M.:** Changes of chloroplast ultrastructure and
carbohydrate level in *Festuca, Achillea* and *Sedum* during drought and after
rocovery. - J. Plant Physiol. 116: 1-10, 1984.

*13888 - **MARQUES, J.B.B., LIN, S.S.:** Efeito de espacamento entre fileiras população
de plantas e irrigação sobre o rendimento de sementes de soja. [The effect
of spacing, plant population and irrigation on soybean seed yield.] - Pesq.
agropec. Bras. 17: 733-739, 1982. [In Port, ab: E.]

13889 - **MARTIN, A.J.:** Testing volume equation accuracy with water displacement
techniques. - Forest Sci. 30: 41-50, 1984.

13890 - **MARTIN, C.E., WARNER, D.A.:** The effects of desiccation on concentrations
and a/b ratios of chlorophyll in *Leucobryum glaucum* and *Thuidium delicatulum*.
New Phytol. 96: 545-550, 1984.

13891 - **MARTIN, G.E.,II., OUTLAW, W.H.,Jr., ANDERSON, L.C., JACKSON, S.G.:** Photosynthetic electron transport in guard cells of diverse species. - Plant
Physiol. 75: 336-337, 1984.

*13892 - **MARTINEZ, M.R., DULAY-LEONG, N., AQUINO, R.F.:** Resistance of the blue-green
alga *Gloeotrichia natans* to desiccation. - Kalikasan (Phillippine J. Biol.)
12: 182-186, 1983.

13893 - **MARTY, J.-R., CABELGUENNE, M., PUECH, J.:** Perspectives de valorisation d'un
milieu par des assolements de grandes cultures: essais d'optimisation
technico-économique. I. - Elaboration d'un modèle de choix d'assolement. -
Agronomie 4: 871-884, 1984.

13894 - **MARTY, J.-R., CABELGUENNE, M., PUECH, J.:** Perspectives de valorisation d'un
milieu par des assolements de grandes cultures: essais d'optimisation
technico-économique. II. -. Exemples d'assolements: résultats techniques et
agronomiques. - Agronomie 4: 915-925, 1984.

13895 - **MASIUNAS, J.B., CARPENTER, P.L.:** The effect of temperature and polyethylene
glycol-induced osmotic stress on radicle growth of *Lespedeza stipulacea,
Lolium multiflorum,* and *Bouteloua curtipendula*. - J. Amer. Soc. hort. Sci.
109: 67-69, 1984.

13896 - **MASON, W.K., SMALL, D.R., PRITCHARD, K.E.:** Effects of irrigation and soil
management for fodder crops on root zone conditions in a red-brown earth. -
Aust. J. Soil Res. 22: 207-218, 1984.

13897 - **MATEJKA, F., HUZULÁK, J.:** Dependence of the leaf water potential of forest
trees on the atmospheric evaporative demand. - Biológia (Bratislava) 39:
25-30, 1984.

*13898 - **MATSUI, T., EGUCHI, H., KOUTAKI, M.:** Effects of environmental factors on
leaf temperature in a temperature controlled room. III. Difference in leaf
temperature between adaxial and abaxial surfaces in cotyledon of *Cucurbita
maxima* Duch. - Environ. Control Biol. 19: 69-72, 1981.

*13899 - MATSUI, T., EGUCHI, H., SOEJIMA, Y.: Evaluation of artificial light for
plants on the basis of transpiration model. - Environ. Control Biol. 19:
25-34, 1981.

13900 - MATTHEWS, M.A., BOYER, J.S.: Acclimation of photosynthesis to low leaf water
potentials. - Plant Physiol. 74: 161-166, 1984.

13901 - MATTHEWS, M.A., Van VOLKENBURGH, E., BOYER, J.S.: Acclimation of leaf growth
to low water potentials in sunflower. - Plant Cell Environ. 7: 199-206,
1984.

13902 - MATZIRIS, D.I.: Genetic variation in morphological and anatomical needle
characteristics in the black pine of Peloponnesos. - Silvae Genet. 33: 164-
169, 1984.

13903 - MAUK, C.S., BREEN, P.J., MACK, H.J.: Flowering-pattern and yield components
at inflorescence nodes of snap bean as affected by irrigation and plant
density. - Scientia Hort. (Amsterdam) 23: 9-19, 1984.

13904 - MAWSON, B.T., FRANKLIN, A., FILION, W.G., CUMMINS, W.R.: Comparative studies
of fluorescence from mesophyll and guard cell chloroplasts in *Saxifraga
cernua*. Analysis of fluorescence kinetics as a function of excitation intensity. - Plant Physiol. 74: 481-486, 1984.

*13905 - MAYEUX, H.S.,Jr., JORDAN, W.R., MEYER, R.E., MEOLA, S.M.: Epicuticular wax
on goldenweed (*Isocoma* spp.) leaves: Variation with species and season. -
Weed Sci. 29: 389-393, 1981.

13906 - MAZOR, L., PERL, M., NEGBI, M.: Changes in some ATP-dependent activities in
seeds during treatment with polyethylene glycol and during the redrying
process. - J. exp. Bot. 35: 1119-1127, 1984.

*13907 - McANENEY, K.J., JUDD, M.J.: Observations on kiwifruit (*Actinidia chinensis*
Planch.) root exploration, root pressure, hydraulic conductivity, and water
uptake. - N. Zeal. J. agr. Res. 26: 507-510, 1983.

13908 - McBURNEY, T., COSTIGAN, P.A.: Rapid oscillations in plant water potential
measured with a stem psychrometer. - Ann. Bot. 54: 851-853, 1984.

13909 - McBURNEY, T., COSTIGAN, P.A.: The relationship between plant water potential
and transpiration rate in young cabbage plants growing in wet soil. - J. exp.
Bot. 35: 1032-1038, 1984.

13910 - McBURNEY, T., COSTIGAN, P.A.: The relationship between stem diameter and
water potentials in stems of young cabbage plants. - J. exp. Bot. 35: 1787-
1793, 1984.

*13911 - McCLENDON, J.H.: Water relations curves for plant cells: toward a realistic
Hofler diagram for textbooks. - What's New Plant Physiol. 13(5): 17-20, 1982.

13912 - McCONNELL, D.B., RUGABER, P., SHEEHAN, T.J., HENNY, R.J.: Light levels alter
anatomy of *Aphelandra squarrosa* 'Dania'. - J. Amer. Soc. hort. Sci. 109:
298-301, 1984.

13913 - McCOY, E.L., BOERSMA, L., UNGS, M.L., AKRATANAKUL, S.: Toward understanding
soil water uptake by plant roots. - Soil Sci. 137: 69-77, 1984.

13914 - McCREE, K.J., KALLSEN, C.E., RICHARDSON, S.G.: Carbon balance of sorghum
plants during osmotic adjustment to water stress. - Plant Physiol. 76:
898-902, 1984.

13915 - McDONALD, G.K., SUTTON, B.G., ELLISON, F.W.: The effect of sowing date,
irrigation and cultivar on the growth and yield of wheat in the Namoi River
Valley, New South Wales. - Irrig. Sci. 5: 123-135, 1984.

13916 - McGOWAN, M., BLANCH, P., GREGORY, P.J., HAYCOCK, D.: Water relations of
winter wheat. 5. The root system and osmotic adjustment in relation to crop
evaporation. - J. agr. Sci. 102: 415-425, 1984.

13917 - McGRANAHAN, G.H., SMALLEY, E.B.: Influence of moisture, temperature, leaf
maturity, and host genotype on infection of elms by *Stegophora ulmea*. -
Phytopathology 74: 1296-1300, 1984.

13918 - McHUGHEN, A., SWARTZ, M.: A tissue-culture derived salt-tolerant line of
flax (*Linum usitatissimum*). - J. Plant Physiol. 117: 109-117, 1984.

13919 - McILROY, I.C.: Terminology and concepts in natural evaporation. - Agr. Water
Manage. 8: 77-98, 1984. Also in: SHARMA, M.L. (ed.): Evapotranspiration from
Plant Communities. Pp. 77-98. Elsevier Science Publishers, Amsterdam -
Oxford - New York - Tokyo 1984.

13920 - McINTYRE, G.I., BOYER, J.S.: The effect of humidity, root excision, and
potassium supply on hypocotyl elongation in dark-grown seedlings of
Helianthus annuus. - Can. J. Bot. 62: 420-428, 1984.

13921 - McINTYRE, G.I., QUICK, W.A.: Control of sprout growth in the potato. Effect
of humidity and water supply. - Can. J. Bot. 62: 2140-2145, 1984.

13922 - McIVOR, J.G.: Leaf growth and senescence in *Urochloa mosambicensis* and
U. oligotricha in a seasonally dry tropical environment. - Aust. J. agr.
Res. 35: 177-187, 1984.

13923 - McKEE, W.H.,Jr., HOOK, D.D., DeBELL, D.S., ASKEW, J.L.: Growth and nutrient
status of loblolly pine seedlings in relation to flooding and phosphorus. -
Soil Sci. Soc. Amer. J. 48: 1438-1442, 1984.

13924 - McKEON, G.M.: Field changes in germination requirements: effect of natural
rainfall on potential germination speed and light requirement of *Stylosanthes
humilis, Stylosanthes hamata* and *Digitaria ciliaris*. - Aust. J. agr. Res. 35:
807-819, 1984.

13925 - McLACHLAN, K.D.: Effects of drought, aging and phosphorus status on leaf
acid phosphatase activity in wheat. - Aust. J. agr. Res. 35: 777-787, 1984.

13926 - McLAUGHLIN, N.B., DAVIDSON, H.R.: A differential fluid manometer for
hydraulic lysimeters. - Agron. J. 76: 160-162, 1984.

*13927 - McNAUGHTON, K.G.: Net interception losses during sprinkler irrigation. -
Agr. Meteorol. 24: 11-27, 1981.

*13928 - McNAUGHTON, K.G., CLOTHIER, B.E., KERR, J.P.: Evaporation from land surfaces.
- In: MURRAY, D.L., ACKROYD, P. (ed.): Physical Hydrology. New Zealand
Experience. Pp. 97-119. N.Z. Hydrological Society, Wellington North 1979.

13929 - McNAUGHTON, K.G., JARVIS, P.G.: Using the Penman-Monteith equation predicti-
vely. - Agr. Water Manage. 8: 263-278, 1984. Also in: SHARMA, M.L. (ed.):
Evapotranspiration from Plant Communities. Pp. 263-278. Elsevier Science
Publishers, Amsterdam - Oxford - New York - Tokyo 1984.

13930 - McPARTLAND, J.M., SCHOENEWEISS, D.F.: Hyphal morphology of *Botryosphaeria
dothidea* in vessels of unstressed and drought-stressed stems of *Betula alba*.
Phytopathology 74: 358-362, 1984.

*13931 - McWHORTER, C.G.: Effect of temperature and relative humidity on transloca-
         tion of ^{14}C-metriflufen in johnsongrass (Sorghum halepense) and soybean
         (Glycine max). - Weed Sci. Soc. Amer. J. 29: 87-93, 1981.

13932 - MECKEL, L., EGLI, D.B., PHILLIPS, R.E., RADCLIFFE, D., LEGGETT, J.E.:
        Effect of moisture stress on seed growth in soybean. - Agron. J. 76: 647-
        650, 1984.

*13933 - MEEK, B.D., EHLIG, C.F., STOLZY, L.H., GRAHAM, L.E.: Furrow and trickle
         irrigation: effects on soil oxygen and ethylene and tomato yield. - Soil
         Sci. Soc. Amer. J. 47: 631-635, 1983.

13934 - MEHLHORN, R.J., BLUMWALD, E., PACKER, L.: ESR methods for studies of osmo-
        regulation in the cyanobacterium Synechococcus 6311. - In: CRAM, W.J.,
        JANÁČEK, K., RYBOVÁ, R., SIGLER, K. (ed.): Membrane Transport in Plants.
        Pp. 115-116. Academia, Praha 1984.

13935 - MEI, H.-S., THIMANN, K.V.: The relation between nitrogen deficiency and
        leaf senescence. - Physiol. Plant. 62: 157-161, 1984.

13936 - MEINERS, T.M., SMITH, D.W., SHARIK, T.L., BECK, D.E.: Soil and plant water
        stress in an Appalachian oak forest in relation to topography and stand
        age. - Plant Soil 80: 171-179, 1984.

13937 - MEINZER, F., GOLDSTEIN, G., JAIMES, M.: The effect of atmospheric humidity
        on stomatal control of gas exchange in two tropical coniferous species. -
        Can. J. Bot. 62: 591-595, 1984.

13938 - MEIRI, A.: Plant response to salinity: experimental methodology and appli-
        cation to the field. - In: SHAINBERG, I., SHALHEVET, J. (ed.): Soil Sali-
        nity under Irrigation. Processes and Management. Pp. 284-297. Springer
        -Verlag, Berlin - Heidelberg - New York - Tokyo 1984.

*13939 - MENOUX-BOYER, Y., BURIOL, G., De PARCEBAUX, S.: Use of two non-destructive
         methods for the measurement of plant water content: ecophysiological study
         of adaptive mechanisms in response to environmental stresses. - In:
         METZNER, H. (ed.): Photosynthesis and Plant Productivity. Pp. 101-104.
         Wissenschaftliche Verlagsgesellschaft, Stuttgart 1983.

13940 - MERRIEN, A., BLANCHET, R.: Aspects agronomiques de la résistance à la
        sécheresse chez le Tournesol (Helianthus annuus L.). - Bull. Soc. bot.
        France, Actual. bot. 131: 45-50, 1984.

13941 - METZGER, J.D., KENG, J.: Effects of dimethipin, a defoliant and dessicant,
        on stomatal behavior and protein synthesis. - J. Plant Growth Regulation
        3: 141-156, 1984.

13942 - MICHALOV, J.: Transport processes throught root tissue. - In: CRAM, W.J.,
        JANÁČEK, K., RYBOVÁ, R., SIGLER, K. (ed.): Membrane Transport in Plants.
        Pp. 51-52. Academia, Praha 1984.

13943 - MICHELS, P., FEYEN, J.: Automatic control of irrigation in greenhouses by
        simulation of the water balance in the root zone. - J. agr. Eng. Res. 29:
        223-230, 1984.

13944 - MIGLIACCIO, F., MAAS, E.V., OGATA, G.: Phosphate absorption, fluxes, and
        symplasmic transport in osmotically-shocked Zea mays roots. - J. exp. Bot.
        35: 8-17, 1984.

*13945 - MILLER, C.A., DAVIS, D.D.: Response of pinto bean plants exposed to O_3, SO_2,
         or mixtures at varying temperatures. - HortScience 16: 548-550, 1981.

*13946 - **MILLER, S.S.:** Effect of preharvest antitranspirant sprays on the size and quality of 'Delicious' apples at harvest. - J. Amer. Soc. hort. Sci. 104: 204-207, 1979.

13947 - **MILLER, T.E., WING, J.S., HUETE, A.R.:** The agricultural potential of selected C_4 plants in arid environments. - J. arid Environ. 7: 275-286, 1984.

*13948 - **MINASHINA, N.G., KOCHETKOVA, G.N.:** Soli v pochvennom rastvore i vlazhnost' zavyadaniya rasteniĭ pri oroshenii peschanykh pochv khloridnymi i sul'fatnymi vodami. [Salts in the soil solution and wilting point in sandy soils irrigated by chloride and sulphate waters.] - Pochvovedenie 1983(5): 90-100, 1983. [In R, ab: E.]

13949 - **MIRECKI, R.M., TERAMURA, A.H.:** Effects of ultraviolet-B irradiance on soybean V. The dependence of plant sensitivity on the photosynthetic photon flux density during and after leaf expansion. - Plant Physiol. 74: 475-480, 1984.

13950 - **MITCHELL, P.D., JERIE, P.H., CHALMERS, D.J.:** The effects of regulated water deficits on pear tree growth, flowering, fruit growth, and yield. - J. Amer. Soc. hort. Sci. 109: 604-606, 1984.

*13951 - **MITSCH, W.J.:** Ecological models for management of freshwater wetlands. - In: JØRGENSEN, S.E., MITSCH, W.J. (ed.): Application of Ecological Modelling in Environmental Management. Part B. Pp. 283-310. Elsevier Science Publishers, Amsterdam 1983.

*13952 - **MITSCH, W.J., DAY, J.W.,Jr., TAYLOR, J.R., MADDEN, C.:** Models of North American freshwater wetlands. - Int. J. Ecol. environ. Sci. 8: 109-140, 1982.

13953 - **MITSCH, W.J., RUST, W.G.:** Tree growth responses to flooding in a bottomland forest in northeastern Illinois. - Forest Sci. 30: 499-510, 1984.

13954 - **MIYAMOTO, S., DAVIS, J., PIELA, K.:** Water use, growth and rubber yields of four guayule selections as related to irrigation regimes. - Irrig. Sci. 5: 95-103, 1084.

13955 - **MIYAMOTO, S., PIELA, K., DAVIS, J., FENN, L.B.:** Salt effects on emergence and seedling mortality of guayule. - Agron. J. 76: 295-300, 1984.

13956 - **MOHR, H.:** "Baumsterben" als pflanzenphysiologisches Problem. - Biol. unserer Zeit 14(4): 103-110, 1984.

*13957 - **MOKHAMED, T,:** Proekten poliven rezhim na lyutserna. [A planned irrigation regime of alfalfa]-Rasteniev. Nauki 20(7): 127-132, 1983. [In Bulg, ab: R, E.]

13958 - **MOLLOV, M.Z., VELICHKOV, D.K., TSONEV, Ts.D., STOYANOVA, Z.P., DEKOV, I.Ch.:** Aminoacid composition of transpired water from maize, bean and sunflower plants. - Dokl. Bolg. Akad. Nauk 37(5): 649-651, 1984.

13959 - **MONSON, R.K.:** A field study of photosynthetic temperature acclimation in *Carex eleocharis* Bailey. - Plant Cell Environ. 7: 301-308, 1984.

*13960 - **MONTEITH, J.L., ELSTON, J.:** Performance and productivity of foliage in the field. - In: DALE, J.E., MILTHORPE, F.L. (ed.): The Growth and Functioning of Leaves. Pp. 499-518. Cambridge University Press, Cambridge - London - New York - New Rochelle - Melbourne - Sydney 1983.

*13961 - **MOONEY, H.A., FIELD, C., YANES, C.V., CHU, C.:** Environmental controls on stomatal conductance in a shrub of the humid tropics. - Proc. nat. Acad. Sci. USA 80: 1295-1297, 1983.

*13962 - **MOONEY, H.A., GULMON, S.L.:** The determinants of plant productivity - natural versus man-modified communities. - In: MOONEY, H.A., GORDON, M. (ed.): Disturbance and Ecosystems Components of Response. Pp. 146-158. Springer -Verlag, Berlin 1983.

13963 - **MORAN, R.J., O'SHAUGHNESSY, P.J.:** Determination of the evapotranspiration of *E. regnans* forested catchments using hydrological measurements. - Agr. Water Manage. 8: 57-76, 1984. Also in: SHARMA, M.L. (ed.): Evapotranspiration from Plant Communities. Pp. 57-76. Elsevier Science Publishers, Amsterdam - Oxford - New York - Tokyo 1984.

13964 - **MORESHET, S., GREEN, G.C.:** Seasonal trends in hydraulic conductance of field-grown 'Valencia' orange trees. - Scientia Hort. 23: 169-180, 1984.

13965 - **MORGAN, J.A.:** Interaction of water supply and N in wheat. - Plant Physiol. 76: 112-117, 1984.

*13966 - **MORGAN, J.M.:** Osmoregulation as a selection criterion for drought tolerance in wheat. - Aust. J. agr. Res. 34: 607-614, 1983.

13967 - **MORGAN, J.M.:** Modelling environmental effects on crop productivity. - In: PEARSON, C.J. (ed.): Control of Crop Productivity. Pp. 289-304. Academic Press, Sydney - Orlando - San Diego - Petaluma - New York - London - Toronto - Montreal - Tokyo 1984.

13968 - **MORGAN, J.M.:** Osmoregulation and water stress in higher plants. - Annu. Rev. Plant Physiol. 35: 299-319, 1984.

13969 - **MORGAN, J.M., KING, R.W.:** Association between loss of leaf turgor, abscisic acid levels and seed set in two wheat cultivars. - Aust. J. Plant Physiol. 11: 143-150, 1984.

13970 - **MORIN, J., RAWITZ, E., HOOGMOED, W.B., BENYAMINI, Y.:** Tillage practices for soil and water conservation in the semi-arid zone. III. Runoff modeling as a tool for conservation tillage design. - Soil Tillage Res. 4: 215-224, 1984.

13971 - **MORISON, J.I.L., GIFFORD, R.M.:** Plant growth and water use with limited water supply in high CO_2 concentrations. I. Leaf area, water use and transpiration. - Aust. J. Plant Physiol. 11: 361-374, 1984.

13972 - **MORISON, J.I.L., GIFFORD, R.M.:** Plant growth and water use with limited water supply in high CO_2 concentrations. II. Plant dry weight, partitioning and water use efficiency. - Aust. J. Plant Physiol. 11: 375-384, 1984.

13973 - **MORIZET, J., CRUIZIAT, P., CHATENOUD, J., PICOT, P., LECLERCQ, P.:** Essai d'amélioration de la résistance à la sécheresse du tournesol (*Helianthus annuus*) par croisement interspécifique avec une espèce sauvage (*Helianthus agrophyllus*). Réflexions sur les méthodes utilisées et les premiers résultats obtenus. - Agronomie 4: 577-585, 1984.

13974 - **MORIZET, J., CRUIZIAT, P., TOGOLA, D.:** Quelques données et réflexions sur les mécanismes de résistance à la sécheresse à partir des exemples du Maïs et du Tournesol. - Bull. Soc. bot. France, Actual. bot. 131: 33-43, 1984.

13975 - MORIZET, J., ROBELIN, M., BAUCHER, G.: Résultats de 18 années d'observations
lysimétriques sous climat limagnais. II. - Etude des relations entre l'eau
et la production végétale. - Agronomie 4: 407-416, 1984.

*13976 - MOROHASHI, Y., MATSUSHIMA, H.: Appearance and disappearance of cyanide-resis-
tant respiration in *Vigna mungo* cotyledons during and following germination
of the axis. - Plant Physiol. 73: 82-86, 1983.

13977 - MORTON, F.I.: What are the limits on forest evaporation? - J. Hydrol. 74:
373-398, 1984.

13978 - MOTT, K.A., O'LEARY, J.W.: Stomatal behavior and CO_2 exchange characteristics
in amphistomatous leaves. - Plant Physiol. 74: 47-51, 1984.

13979 - MOUGOU, A., LEMEUR, R., SCHALCK, J.: Water stress induction using polyethylene glycol in nutrient film technique. Effects on the photosynthetic parameters of three tomato species and one hybrid. - Acta oecol. - Oecol. Plant.
5: 375-385, 1984.

13980 - MOUSSEAU, M.: CO_2 and H_2O exchanges in response to alteration of photoperiod
in *Anagallis arvensis*, a long-day flowering plant. - Can. J. Bot. 62:
1880-1883, 1984.

*13981 - MOUTONNET, P., BRANDY-CHERRIER, M., CHAMBON, J.: Possibilites d'utilisation
des tensiometres pour l'automation de l'irrigation. - Plant Soil 59: 335-
345, 1981.

13982 - MUELLER, P.A., DENGLER, N.G.: Leaf development in the anisophyllous shoots
of *Pellionia daveauana* (*Urticaceae*). - Can. J. Bot. 62: 1158-1170, 1984.

13983 - MÜLLER, H.J. (ed.): Ökologie (Studienreihe Biowissenschaften). - VEB Gustav
Fischer Verlag, Jena 1984.

13984 - MUNNS, R., PASSIOURA, J.B.: Hydraulic resistance of plants. III. Effects of
NaCl in barley and lupin. - Aust. J. Plant Physiol. 11: 351-359, 1984.

13985 - MUNNS, R., PASSIOURA, J.B.: Effect of prolonged exposure to NaCl on the
osmotic pressure of leaf xylem sap from intact, transpiring barley plants. -
Aust. J. Plant Physiol. 11: 497-507, 1984.

*13986 - MUÑOZ, C.E., DAVIES, F.S., SHERMAN, W.B.: Hydraulic conductivity and ethyle-
ne production in detached flowering peach shoots. - HortScience 17: 226-228,
1982.

13987 - MUNRO, D.S.: Summer soil moisture content and the water table in a forested
wetland peat. - Can. J. Forest Res. 14: 331-335, 1984.

*13988 - MUNTAJABUDDIN, K., BORULKAR, D.N.: Consumptive use of water by rice and its
relationship with climate. - Ind. J. agr. Sci. 53: 913-917, 1983.

13989 - MURALI, N.S., SAXE, H.: Effects of ultraviolet-C radiation on net photosynthesis, transpiration and dark respiration of *Spathiphyllum wallisii*. -
Physiol. Plant. 60: 192-196, 1984.

13990 - MURPHY, G.J.P.: Metabolism of R,S-[2-^{14}C]abscisic acid by non-stressed and
water-stressed detached leaves of wheat (*Triticum aestivum* L.). - Planta
160: 250-255, 1984.

13991 - MURRAY, A.J.S.: Light affects the deposition of NO_2 to the *flacca* mutant of
tomato without affecting the rate of transpiration. - New Phytol. 98:
447-450, 1984.

13992 - **MURRAY, D.R. (ed.)**: Seed Physiology. Volume 1. Development. - Academic Press, Sydney - Orlando - San Diego - Petaluma - New York - London - Toronto - Montreal - Tokyo 1984.

13993 - **MURRAY, D.R. (ed.)**: Seed Physiology. Volume 2. Germination and Reserve Mobilization. - Academic Press, Sydney - Orlando - San Diego - Petaluma - New York - London - Toronto - Montreal - Tokyo 1984.

13994 - **MYERS, B.A., NEALES, T.F.**: Seasonal changes in the water relations of *Eucalyptus behriana* F. Muell. and *E. microcarpa* (Maiden) Maiden in the field. - Aust. J. Bot. 32: 495-510, 1984.

13995 - **MYERS, R.J.K., FOALE, M.A., DONE, A.A.**: Responses of grain sorghum to varying irrigation frequency in the Ord Irrigation Area. II. Evapotranspiration, water use efficiency and root distribution of different cultivars. - Aust. J. agr. Res. '35: 31-42, 1984.

13996 - **MYERS, R.J.K., FOALE, M.A., DONE, A.A.**: Responses of grain sorghum to varying irrigation frequency in the Ord Irrigation Area. III. Water relations. - Aust. J. agr. Res. 35: 43-52, 1984.

13997 - **NABLE, R.O., BROWNELL, P.F.**: Effect of sodium nutrition and light upon the concentrations of alanine in leaves of C_4 plants. - Aust. J. Plant Physiol. 11: 319-324, 1984.

13998 - **NAGARAJARAO, Y.**: Behaviour of leaf water potentials of wheat and mungbean under field conditions as a function of available soil moisture and relative humidity. - Z. Pflanzenernähr. Bodenk. 147: 37-48, 1984.

*13999 - **NAGY, J., VINCZE, L.**: Az öntözés, a növényállomány és a trágyázás hatása a cukorrépa gyökér- és fehércukor termésére. [Effect of irrigation, plant stand and fertilization on root and white sugar yield in sugar beet.] - Növénytermelés 32: 565-573, 1983. [In Hung, ab: E.]

*14000 - **NAÏDENOVA, K.A., FILIPENKO, N.K., BOBROVSKIĬ, N.A.**: Produktivnost' kul'turnykh pastbishch v svyazi s primeneniem mineral'nykh udobreniĭ i orosheniya. [Productivity of cultured pastures affected by fertilization and irrigation.'] - In: Pochvennye Protsessy i Regulirovanie Pitaniya Rasteniĭ. Pp. 36-39, 119. Ministerstvo Sel'skogo Khozyaĭstva SSSR, Gorki 1983. [In R.]

14001 - **NAIK, G.R., PATIL, T.M., HEGDE, B.A.**: Effect of pre-planting treatment of growth regulators on photosynthetic productivity in sugarcane grown under saline conditions. - Biovigyanam 10: 143-147, 1984.

14002 - **NAKAYAMA, F.S., BUCKS, D.A.**: Crop water stress index, soil water, and rubber yield relations for the guayule plant. - Agron. J. 76: 791-794, 1984.

*14003 - **NAKHUTSRISHVILI, G.Sh., CHERNUSKA, A.**: Ėkologicheskiĭ analiz vliyaniya vypasa na vysokogornye luga Tsentral'nogo Kavkaza. [Ecological analysis of the influence of grazing on alpine meadows in Central Kaukasus.] - Dokl. Akad. Nauk SSSR 267: 503-505, 1982. [In R.]

*14004 - **NASRUDINOVA, R.I., SHCHERBATYUK, A.S.**: Vodnyĭ rezhim sosny obyknovennoĭ v lesostepi Predbaĭkal'ya. [Water regime of Scots pine in the forest steppe of the Bajkal region.] - In: Ėkologo-fiziologicheskie Issledovaniya Fotosinteza i Vodnogo Rezhima Rasteniĭ v Polevykh Usloviyakh. Pp. 102-110, 164. Sibirskoe Otdelenie, Akademiya Nauk SSSR, Irkutsk 1983. [In R.]

*14005 - NASRUDINOVA, R.I., SHCHERBATYUK, A.S., BURYAKOV, B.M.: Vliyanie vodnogo rezhima na fotosintez khvoi sosny obyknovennoĭ. [Effect of water regime on photosynthesis of needles of Scots pine.] - In: Ėkologo-fiziologicheskie Issledovaniya Fotosinteza i Vodnogo Rezhima Rasteniĭ v Polevykh Usloviyakh. Pp. 125-132, 164-165. Sibirskoe Otdelenie, Akademiya Nauk SSSR, Irkutsk 1983. [In R.]

*14006 - NASYROV, Yu.S.: Genetic modification of the CO_2 carboxylation reactions as a factor improving efficiency of photosynthesis. - Ind. J. Plant Physiol. 24: 26-36, 1981.

14007 - NÁTR, L., PAZOUREK, J.: Anatomical characteristics from physiological point of view. - In: Biogenez, Struktury i Funktsii Fotosinteticheskogo Apparata. Pp. 30. ÚEB ČSAV, Praha 1984.

14008 - NAVARA, J.: Das Wachstum der oberirdischen Organe und die Wasseraufnahme bei beginnender Fruchtbarkeit von Apfelbäumen (Malus domestica Borkh.). - Biológia (Bratislava) 39: 493-505, 1984.

14009 - NELSON, N.D., EHLERS, P.: Comparative carbon dioxide exchange for two Populus clones grown in growth room, greenhouse, and field environments. - Can. J. Forest Res. 14: 924-932, 1984.

*14010 - NELSON, N.D., ISEBRANDS, J.G.: Late-season photosynthesis and photosynthate distribution in an intensively-cultured Populus nigra x laurifolia clone. - Photosynthetica 17: 537-549, 1983.

14011 - NESTEROVICH, N.D., LUCHKOV, A.I.: Sravnitel'naya morfologo-anatomicheskaya kharakteristika khvoi seyantsev tui zapadnoĭ, vyrashchennykh na pochvakh s razlichnym rezhimom gruntovykh vod. [Comparative morphological and anatomical characterization of Thuja occidentalis L. seedlings grown on soils with different groundwater levels.]- Dokl. Akad. Nauk BSSR 28: 1119-1122, 1984. [In R., ab: E.]

14012 - NICOLAS, M.E., GLEADOW, R.M., DALLING, M.J.: Effects of drought and high temperature on grain growth in wheat. - Aust. J. Plant Physiol. 11: 553-566, 1984.

*14013 - NIKOLOV, G.: Optimizirane i prognozirane na napoyavaneto na pamuka. [Optimizing and predicting cotton irrigation.] - Rasteniev. Nauki 20(7): 78-87, 1983. [In Bulg, ab: R, E.]

14014 - NILSEN, E.T., SHARIFI, M.R., RUNDEL, P.W.: Comparative water relations of phreatophytes in the Sonoran Desert of California. - Ecology 65: 767-778, 1984.

*14015 - NILSSON, T.: The influence of soil type, nitrogen and irrigation on yield, quality and chemical composition of cauliflower. - Swedish J. agr. Res. 10: 65-75, 1980.

14016 - NOBEL, P.S.: Productivity of Agave deserti: measurement by dry weight and monthly prediction using physiological responses to environmental parameters. - Oecologia 64: 1-7, 1984.

14017 - NOBEL, P.S., CALKIN, H.W., GIBSON, A.C.: Influences of PAR, temperature and water vapor concentration on gas exchange by ferns. - Physiol. Plant. 62: 527-534, 1984.

14018 - NOBEL, P.S., HARTSOCK, T.L.: Physiological responses of Opuntia ficus-indica to growth temperature. - Physiol. Plant. 60: 98-105, 1984.

14019 - **NOBLE, C.L., HALLORAN, G.M., WEST, D.W.**: Identification and selection for salt tolerance in lucerne (*Medicago sativa* L.). - Aust. J. agr. Res. 35: 239-252, 1984.

14020 - **NOITSAKIS, B., BERGER, A.**: Relations hydriques chez *Dactylis glomerata* et *Dichanthium ischaemum* cultivés sous deux régimes hydriques contrastés. - Acta oecol. - Oecol. Plant. 5: 75-88, 1984.

14021 - **NOKS, P.P., VENEDIKTOV, P.S., KONONENKO, A.A., RUBIN, A.B., GARAB, D., FALUDI-DANIÉL, A.**: Termolyuminestsentsiya i êlektricheskaya polyrizatsiya v khloroplastakh. [Thermoluminescence and electric polarization of chloroplasts.] - Mol. Biol. (Moskva) 18: 766-775, 1984. [In R, ab: E.]

*14022 - **NORBY, R.J., KOZLOWSKI, T.T.**: Flooding and SO_2 stress interaction in *Betula papyrifera* and *B. nigra* seedlings. - Forest Sci. 29: 739-750, 1983.

14023 - **NORLYN, J.D., EPSTEIN, E.**: Variability in salt tolerance of four triticale lines at germination and emergence. - Crop Sci. 24: 1090-1092, 1984.

*14024 - **NORWINE, J., GREEGOR, D.H.**: Vegetation classification based on advanced very high resolution radiometer (AVHRR) satellite imagery. - Remote Sensing Environ. 13: 69-87, 1983.

*14025 - **NOSONENKO, N.A.**: Temperaturnaya ustoĭchivost'v'yashchikhsya roz v Krymu. [Temperature resistance of climbing roses in the Crimea.] - Byull. gosud. Nikitskogo bot. Sada 1979(3): 69-72, 1979. [In R, ab: E.]

14026 - **NOWAK, R.S., CALDWELL, M.M.**: A test of compensatory photosynthesis in the field: implications for herbivory tolerance. - Oecologia 61: 311-318, 1984.

*14027 - **NUKAYA, A.**: Salt tolerance studies in muskmelons and other vegetables. - Tech. Bull. Dept. hort. Fac. agr. Shizuoka Univ. 8: 1-97, 1983.

14028 - **NUKAYA, A., MASUI, M., ISHIDA, A.**: Salt tolerance of muskmelons as affected by diluted sea water applied at different growth stages in nutrient solution culture. - J. Jap. Soc. hort. Sci. 53: 168-175, 1984.

14029 - **NUKAYA, A., MASUI, M., ISHIDA, A.**: Salt tolerance of muskmelons as affected by various salinities. - J. Jap. Soc. hort. Sci. 52: 420-428, 1984.

14030 - **NULSEN, R.A.**: Evapotranspiration of four major agricultural plant communities in the south-west of Western Australia measured with large ventilated chambers. - Agr. Water Manage. 8: 191-202, 1984. Also in: SHARMA, M.L. (ed.): Evapotranspiration from Plant Communities. Pp. 191-202. Elsevier Science Publishers, Amsterdam - Oxford - New York - Tokyo 1984.

14031 - **NUNES, M.A., DIAS, M.A., CORREIA, M.M., OLIVEIRA, M.M.**: Further studies on growth and osmoregulation of sugar beet leaves under low salinity conditions. - J. exp. Bot. 35: 322-331, 1984.

14032 - **NUNGESSER, D., KLUGE, M., TOLLE, H., OPPELT, W.**: A dynamic computer model of the metabolic and regulatory processes in Crassulacean acid metabolism. - Planta 162: 204-214, 1084.

*14033 - **OATES, B.R., MURRAY, S.N.**: Photosynthesis, dark respiration and desiccation resistance of the intertidal seaweeds *Hesperophycus harveyanus* and *Pelvetia fastigiata* f. *gracilis*. - J. Phycol. 19: 371-380, 1983.

14034 - **OBERBAUER, S.F., STRAIN, B.R.**: Photosynthesis and successional status of Costa Rican rain forest trees. - Photosynthesis Res. 5: 227-232, 1984.

*14035 - OECHEL, W.C., HASTINGS, S.J.: The effects of fire on photosynthesis in
         chaparral resprouts. - In: KRUGER, F.J., MITCHELL, D.T., JARVIS, J.U.M. (ed.)
         (ed.): Mediterranean-Type Ecosystems. The Role of Nutrients. Pp. 274-285.
         Springer-Verlag, Berlin - Heidelberg - New York 1983.

 14036 - OERTLI, J.J.: Water relations in cell walls and cells in the intact plant. -
         Z. Pflanzenernähr. Bodenk. 147-197, 1984.

*14037 - OHTAKI, E.: Turbulent transport of carbon dioxide over a paddy field. -
         Boundary-Layer Meteorol. 19: 315-336, 1980.

^14038 - OHTAKI, E., MATSUI, T.: Infrared device for simultaneous measurement of
         fluctuations of atmospheric carbon dioxide and water vapor. - Boundary-Layer
         Meteorol. 24: 109-119, 1982.

 14039 - OKAMOTO, H., KATOU, K., MIZUNO, A., KOJIMA, H., MATSUMURA, Y., ONO, Y.:
         Osmotic and hormonal control of electrogenic ion pumps and growth in pea
         hypocotyl segments by pressurized intra-organ perfusion. - In: CRAM, W.J.,
         JANÁČEK, K., RYBOVÁ, R., SIGLER, K. (ed.): Membrane Transport in Plants. Pp.
         447-452. Academia, Praha 1984.

 14040 - OKAZAKI, Y., SHIMMEN, T., TAZAWA, M.: Turgor regulation in a brackish
         charophyte, *Lamprothamnium succinctum*. I. Artificial modification of intra-
         cellular osmotic pressure. - Plant Cell Physiol. 25: 565-571, 1984.

 14041 - OKAZAKI, Y., SHIMMEN, T., TAZAWA, M.: Turgor regulation in a brackish
         charophyte, *Lamprothamnium succinctum*. II. Changes in K^+, Na^+ and Cl^- concen-
         trations, membrane potential and membrane resistance during turgor regula-
         tion. - Plant Cell Physiol. 25: 573-581, 1984.

 14042 - OKUSANYA, O.T., UNGAR. I.A.: The growth and mineral composition of three
         species of *Spergularia* as affected by salinity and nutrients at high salini-
         ty. - Amer. J. Bot. 71: 439-447, 1984.

 14043 - OLIVARES, E., URICH, R., MONTES, G., CORONEL, I., HERRERA, A.: Occurrence of
         Crassulacean acid metabolism in *Cissus trifoliata* L. (*Vitaceae*). - Oecologia
         61: 358-362, 1984.

 14044 - OLIVER, M.J., BEWLEY, J.D.: Plant desiccation and protein synthesis. IV. RNA
         synthesis, stability, and recruitment of RNA into protein synthesis during
         desiccation and rehydration of the desiccation-tolerant moss, *Tortula rura-
         lis*. - Plant Physiol. 74: 21-25, 1984.

 14045 - OLIVER, M.J., BEWLEY, J.D.: Plant desiccation and protein synthesis. V.
         Stability of poly(A)$^-$ and poly(A)$^+$ RNA during desiccation and their synthesis
         upon rehydration in the desiccation-tolerant moss *Tortula ruralis* and the
         intolerant moss *Cratoneuron filicinum*. - Plant Physiol. 74: 917-922, 1984.

 14046 - OLIVER, M.J., BEWLEY, J.D.: Plant desiccation and protein synthesis. VI.
         Changes in protein synthesis elicited by desiccation of the moss *Tortula
         ruralis* are effected at the translational level. - Plant Physiol. 74: 923-
         927, 1984.

*14047 - OLIVIER, H.: Irrigation and plant productivity. - In: RECHCIGL, M., Jr. (ed.)
         (ed.): CRC Handbook of Agricultural Productivity. Volume 1. Plant Produc-
         tivity. Pp. 263-272. CRC Press, Inc., Boca Raton 1982.

 14048 - OLSSON, K.A., CARY, P.R., TURNER, D.W.: Fruit crops. - In: PEARSON, C.J.
         (ed.): Control of Crop Productivity. Pp. 219-234. Academic Press, Sydney -
         Orlando - San Diego - Petamula - New York - London - Toronto - Montreal -
         Tokyo 1984.

14049 - **OLSZYK, D.M., TINGEY, D.T.:** Fusicoccin and air pollutant injury to plants. Evidence for enhancement of SO_2 but not O_3 injury. - Plant Physiol. 76: 400-402, 1984.

14050 - **OLUDIMU, O.:** Irrigation and agricultural development: some conceptual and empirical considerations. - J. environ. Manage. 19: 175-184, 1984.

14051 - **OMAR, M.H., MEHANNA, A.M.:** Measurements and estimates of potential evapotranspiration over Egypt. - Agr. Forest Meteorol. 31: 117-129, 1984.

14052 - **OMASA, K., HASHIMOTO, Y., AIGA, I.:** Image instrumentation of plants exposed to air pollutants. 1. Quantification of physiological information included in thermal infrared image. - Res. Rep. Nat. Inst. Environ. Studies 1984 (66): 69-78, 1984.

14053 - **OMASA, K., ONOE, M.:** Measurement of stomatal aperture by digital image processing. - Plant Cell Physiol. 25: 1379-1388, 1984.

14054 - **O'NEILL, P.E., JACKSON, T.J., BLANCHARD, B.J., WANG, J.R., GOULD, W.I.:** Effects of corn stalk orientation and water content on passive microwave sensing of soil moisture. - Remote Sensing Environ. 16: 55-67, 1984.

*14055 - **O'NEILL, S.D.:** Role of osmotic potential gradients during water stress and leaf senescence in *Fragaria virginiana*. - Plant Physiol. 72: 931-937, 1983.

*14056 - **O'NEILL, S.D.:** Osmotic adjustment and the development of freezing resistance in *Fragaria virginiana*. - Plant Physiol. 72: 938-944, 1983.

14057 - **ONG, C.K.:** The influence of temperature and water deficits on the partitioning of dry matter in groundnut (*Arachis hypogaea* L.). - J. exp. Bot. 35: 746-755, 1984.

*14058 - **ORAM, R.N.:** Ecotypic differentiation for dormancy levels in oversummering buds of *Phalaris aquatica*. - Bot. Gaz. 144: 544-551, 1983.

14059 - **ORAM, R.N., FREEBAIRN, R.D.:** Genetic improvement of drought survival ability in *Phalaris aquatica* L. - Aust. J. exp. Agr. anim. Husb. 24: 403-409, 1984.

14060 - **ORCHARD, P.:** Waterlogging studies on sunflower and sorghum. - Aust. Field Crops Newslett. 1984: 271-272, 1984.

*14061 - **ORCHARD, P.W.:** The effects of transient waterlogging on mesophytic plants: a review. - In: SMITH, R.J., RIXON, A.J. (ed.): Rural Drainage in Northern Australia. Pp. 19-37. Darling Downs Institute of Advanced Education, Toowoomba 1982.

*14062 - **ORCHARD, P.W.:** Changes in the soil environment during waterlogging of sunflower and sorghum under controlled conditions: I. Growth and yield in relation to stage of development and nutrient availability. - In: SMITH, R. J., RIXON, A.J. (ed.): Rural Drainage in Northern Australia. Pp. 67-74. Darling Downs Institute of Advanced Education, Toowoomba 1982.

14063 - **ORCHARD, P.W., JESSOP, R.S.:** The response of sorghum and sunflower to short--term waterlogging. I. Effects of stage of development and duration of waterlogging on growth and yield. - Plant Soil 81: 119-132, 1984.

*14064 - **ORCHARD, P.W., LOVETT, J.V.:** Chlormequat induced drought avoidance in sunflowers. - In: Proceedings of the VIIth International Sunflower Conference. Volume II. Pp. 332-343. Krasnodar 1976.

*14065 - **ORCHARD, P.W., SO, H.B.:** The effect of transient waterlogging on the yield
of sunflower and sorghum. - In: Agronomy Australia 1982. (Proceedings of the
Second Australian Agronomy Conference.) Pp. 309. Australian Society of
Agronomy, Wagga Wagga 1982.

*14066 - **ORMROD, D.P.:** Effects of environmental conditions on the mineral demand on
the plant organs. - In: METZNER, H. (ed.): Photosynthesis and Plant Productivity. Pp. 316-318. Wissenschaftliche Verlagsgesellschaft, Stuttgart 1983.

 14067 - **ORY, R.L., RITTIG, F.R. (ed.):** Bioregulators, Chemistry and Uses. (ACS
Symposium Series, Volume 257.) American Chemical Society, Washington 1984.

*14068 - **OSMOND, C.B.:** Carbon cycling and stability of the photosynthetic apparatus
in CAM. - In: TING, I.P., GIBBS, M. (ed.): Crassulacean Acid Metabolism. Pp.
112-127. Amer. Soc. Plant Physiol., Rockville 1982.

 14069 - **OSONUBI, O.:** Effect of water stress on leaf and root growth, and water uptake of *Gmelina arborea* Roxb. seedlings. - Biol. Plant 26: 246-252, 1984.

*14070 - **O'TOOLE, J.C., BALDIA, E.P.:** Water deficits and mineral uptake in rice. -
Crop Sci. 22: 1144-1150, 1982.

 14071 - **O'TOOLE, J.C., HSIAO, T.C., NAMUCO, O.S.:** Panicle water relations during
water stress. - Plant Sci. Lett. 33: 137-143, 1984.

*14072 - **O'TOOLE, J.C., NAMUCO, O.S.:** Role of panicle excertion in water stress induced sterility. - Crop Sci. 23: 1093-1097, 1983.

 14073 - **O'TOOLE, J.C., PADILLA, J.L.:** Water deficits and nitrogen uptake as affected
by water table depth in rice (*Oryza sativa* L.). - Plant Soil 80: 127-132, 1904.

 14074 - **O'TOOLE, J.C., TURNER, N.C., NAMUCO, O.P., DINGKUHN, M., GOMEZ, K.A.:** Comparison of some crop water stress measurement methods. - Crop Sci. 24: 1121-
1128, 1984.

 14075 - **OUEDRAOGO, M., TREMOLIÈRES, A., HUBAC, C.:** Change in fatty acids composition
during water stress in cotton plants. Relation with drought resistance induced by far red light. - Z. Pflanzenphysiol. 114: 239-245, 1984.

 14076 - **OUTLAW, W.H., Jr., TARCZYNSKI, M.C.:** Guard cell starch biosynthesis regulated by effectors of ADP-glucose pyrophosphorylase. - Plant Physiol. 74:
424-429, 1984.

 14077 - **OUTLAW, W.H., Jr., TARCZYNSKI, M.C., MILLER, W.I.:** Histological compartmentation of phosphate in *Vicia faba* L. leaflet. Possible significance to
stomatal functioning. - Plant Physiol. 74: 430-433, 1984.

 14078 - **OVERDIECK, D.:** CO_2-Gaswechsel und Transpiration von Waldschattenpflanzen bei
unterschiedlichen Strahlungsqualitäten. - Ber. Deut. bot. Ges. 97: 351-357,
1984.

*14079 - **ÖZTÜRK, M., MERT, H.H.:** Water relations and germination of seeds of *Inula
graveolens* (L.) Desf. - Biotronics 12: 11-17, 1983.

 14080 - **PAEZ, A., HELLMERS, H., STRAIN, B.R.:** Carbon dioxide enrichment and water
stress interaction on growth of two tomato cultivars. - J. agr. Sci. 102:
687-693, 1984.

*14081 - **PAHLICH, E., GRIEB, B.:** Turgor pressure and proline accumulation in water
stressed plants. - Angew. Bot. 57: 295-299, 1983.

14082 - **PAHLICH, E., STADERMANN, T.**: The thermodynamic activity of proline in ternary solutions of different water potentials. - J. Plant Physiol. 115: 91-96, 1984

*14083 - **PÄIVÖKE, A.**: The long-term effects of zinc on the growth and development, chlorophyll content and nitrogen fixation of the garden pea. - Ann. bot. Fenn. 20: 205-213, 1983.

*14084 - **PÄIVÖKE, A.**: The long-term effects of lead and arsenate on the growth and development, chlorophyll content and nitrogen fixation of the garden pea. - Ann. bot. Fenn. 20: 297-306, 1983.

14085 - **PALOSCIA, S., PAMPALONI, P.**: Microwave remote sensing of plant water stress. - Remote Sensing Environ. 16: 249-255, 1984.

14086 - **PALTA, J.A.**: Influence of water deficits on gas-exchange and the leaf area development of cassava cultivars. - J. exp. Bot. 35: 1441-1449, 1984.

14087 - **PANDEY, R., GANAPATHY, P.S.**: Effects of sodium chloride stress on callus cultures of *Cicer arietinum* L. cv. BG-203: growth and ion accumulation. - J. exp. Bot. 35: 1194-1199, 1984.

14088 - **PANDEY, R.K., HERRERA, W.A.T., PENDLETON, J.W.**: Drought response of grain legumes under irrigation gradient: I. Yield and yield components. - Agron. J. 76: 549-553, 1984.

14089 - **PANDEY, R.K., HERRERA, W.A.T., PENDLETON, J.W.**: Drought response of grain legumes under irrigation gradient: II. Plant water status and canopy temperature. - Agron. J. 76: 553-557, 1984.

14090 - **PANDEY, R.K., HERRERA, W.A.T., VILLEGAS, A.N., PENDLETON, J.W.**: Drought response of grain legumes under irrigation gradient: III. Plant growth. - Agron. J. 76: 557-560, 1984.

14091 - **PAPADOPOULOS, I.**: Effect of sulphate waters on soil salinity, growth and yield of tomatoes. - Plant Soil 81: 353-361, 1984.

14092 - **PAQUIN, R.**: Influence of the environment on cold hardening and winter survival of forage and cereal species with consideration of proline as a metabolic marker of hardening. - In: MARGARIS, N.S., ARIANOUSTOU-FARAGGITAKI, M., OECHEL, W.C.: Being Alive on Land. Pp. 137-154. Dr W. Junk publishers, The Hague - Boston - Lancaster 1984.

14093 - **PARAMESWARAN, K.V.M., GRAHAM, R.D., ASPINALL, D.**: Studies on the nitrogen and water relations of wheat. II. Effects of varying nitrogen and water supply on growth and grain yield. - Irrig. Sci. 5: 105-121, 1984.

*14094 - **PARIHAR, J.S., BAIJAL, B.D.**: Effect of salinity on nitrogen metabolism in berseem (*Trifolium alexandrium* L.). - Acta bot. Ind. 10: 282-287, 1982.

*14095 - **PARK, I.-K., SASHARA, T., TSUNODA, S.**: Response of four rice varieties to three regimes of soil moisture tension under two levels of fertilization. - Tohoku J. agr. Res. 29: 63-69, 1978.

*14096 - **PARK, I.-K., TSUNODA, S.**: [Photosynthetic responses of rice varieties to low temperature treatment with special reference to the accumulation of soluble sugars and starch in the leaves.] - Jap. J. Breed. 33: 404-410, 1983. [In Jap, ab: E.]

14097 - **PARKHURST, D.F.**: Mesophyll resistance to photosynthetic carbon dioxide uptake in leaves: dependence upon stomatal aperture. - Can. J. Bot. 62: 163-165, 1984.

14098 - **PARSONS, L.R., HOWE, T.K.:** Effects of water stress on the water relations of
        of *Phaseolus vulgaris* and the drought resistant *Phaseolus acutifolius.* -
        Physiol. Plant. 60: 197-202, 1984.

14099 - **PARTON, W.J., SINGH, J.S.:** Adapting a biomass simulation model to a tropical
        grassland. - Ecol. Model. 23: 151-163, 1984.

14100 - **PARYS, E., ROMANOWSKA, E., POSKUTA, J.:** Does the embryonic axis affect res-
        piration in pea cotyledons? - J. Plant Physiol. 115: 153-159, 1984.

14101 - **PASSIOURA, J.B.:** Hydraulic resistance of plants. I. Constant or variable? -
        Aust. J. Plant Physiol. 11: 333-339, 1984.

14102 - **PASSIOURA, J.B., MUNNS, R.:** Hydraulic resistance of plants. II. Effects of
        rooting medium, and time of day, in barley and lupin. - Aust. J. Plant.
        Physiol. 11: 341-350, 1984.

14103 - **PASTOR, J., POST, W.M.:** Calculating Thornthwaite and Mather's actual evapo-
        transpiration using an approximating function. - Can. J. Forest Res. 14:
        466-467, 1984.

*14104 - **PATAKY, S., BÁLINT, J., MARÓTI, I.:** Anatomical comparison of the flag and
        second leaves of two *Triticum aestivum* cv. species. - Acta biol. Szeged. 29:
        45-65, 1983.

14105 - **PATE, F.M., SNYDER, G.H.:** Effect of water table and nitrogen fertilization
        on tropical grasses grown on organic soil. - Trop. Grasslands 18: 74-78, 1984.

14106 - **PATE, J.S.:** The carbon and nitrogen nutrition of fruit and seed - case
        studies of selected grain legumes. - In: MURRAY, D.R. (ed.): Seed Physiology.
        Volume I. Development. Pp. 41-82. Academic Press, Sydney - Orlando - San
        Diego - Petaluma - New York - London - Toronto - Montreal - Tokyo 1984.

14107 - **PATEL, C.L., GHILYAL, B.P., TOMAR, V.S.:** Nutrient flow rates in rice roots
        under varying drainage conditions. - Plant Soil 77: 243-252, 1984.

*14108 - **PATEL, J.C., SINGH, R.M.:** Yield and nutrient uptake of sunflower (*Helianthus
        annuus* L.) as affected by irrigation, mulch and cycocel. - Ind. J. Agron. 28:
        205-210, 1983.

*14109 - **PATIL, T.M., HEGDE, B.A.:** Influence of water stress on relative rate of
        photosynthesis and translocation of photosynthate in the leaves of *Parthenium
        hysterophorus.* - Ind. bot. Rep. 2: 9-12, 1983.

*14110 - **PATIL, T.M., MIRAJKAR, P.B., HEGDE, B.A., JOSHI, G.V.:** Influence of soil
        salinity on morphology, rate of carbon assimilation, photosynthetic products
        and enzyme activities in a sorghum hybrid CSH-5. - Ind. J. Plant Physiol. 26:
        155-162, 1983.

*14111 - **PATIL, V.M., SOMAWANSHI, R.B.:** Correction of iron chlorosis in sugarcane on
        saline calcareous soil. - Commun. Soil Sci. Plant Anal. 14: 471-480, 1983.

14112 - **PATRICK, J.W.:** Photosynthate unloading from seed coats of *Phaseolus vulgaris*
        L. Control by tissue water relations. - J. Plant Physiol. 115: 297-310, 1984.

14113 - **PAUL, E.M.M.:** The response to temperature of leaf area in tomato genotypes.
        I. Cell size and number in relation to the area of a leaf. - Euphytica 33:
        347-354, 1984.

14114 - **PAUL, N.D., AYRES, P.G.:** Effects of rust and post-infection drought on photo-
        synthesis, growth and water relations in groundsel. - Plant Pathol. 33: 561-
        569, 1984.

14115 - **PAVELKOVÁ, A., ČURIOVÁ, S.:** Vliv extrémního vysoušení semen hrachu na jejich biologickou hodnotu. [The effect of extreme desiccation of pea seeds on their biological value.] - Sborník ÚVTIZ, Genet. Šlecht. (Praha) 20: 251-256, 1984. [In Czech, ab: R, E.]

14116 - **PAVLIK, B.M.:** Seasonal changes of osmotic pressure, symplasmic water content and tissue elasticity in the blades of dune grasses growing *in situ* along the coast of Oregon. - Plant Cell Environ. 7: 531-539, 1984.

14117 - **PAW U, K.T.:** A theoretical basis for the leaf equivalence point temperature. - Agr. Meteorol. 30: 247-256, 1984.

14118 - **PAW U, K.T., DAUGHTRY, C.S.T.:** A new method for the estimation of diffusive resistance of leaves. - Agr. Forest Meteorol. 33: 141-155, 1984.

14119 - **PEACOCK, C.H., DUDECK, A.E.:** Physiological response of St. Augustinegrass to irrigation scheduling. - Agron. J. 76: 275-279, 1984.

*14120 - **PEARCE, A.J., ROWE, L.K.:** Rainfall interception in a multi-storied, evergreen mixed forest: estimated using Gash's analytical model. - J. Hydrol. 49: 341-353, 1981.

*14121 - **PEARCY, R.W., BJÖRKMAN, O.:** Physiological effects. - In: LEMON, E.R. (ed.): CO_2 and Plants. The Response of Plants to Rising Levels of Atmospheric Carbon Dioxide. Pp. 65-105. American Association for the Advancement of Science, Washington 1983.

14122 - **PEARCY, R.W., EHLERINGER, J.:** Comparative ecophysiology of C_3 and C_4 plants. - Plant Cell Environ. 7: 1-13, 1984.

14123 - **PEARCY, R.W., USTIN, S.L.:** Effects of salinity on growth and photosynthesis of three California tidal marsh species. - Oecologia 62: 68-73, 1984.

14124 - **PEARSON, C.J. (ed.):** Control of Crop Productivity. - Academic Press, Sydney - Orlando - San Diego - Petaluma - New York - London - Toronto - Montreal - Tokyo 1984.

14125 - **PEARSON, C.J., HALL, A.J.:** Maize and pearl millet. - In: PEARSON, C.J. (ed.): Control of Crop Productivity. Pp. 141-158. Academic Press, Sydney - Orlando - San Diego - Petaluma - New York - London - Toronto - Montreal - Tokyo 1984.

14126 - **PEARSON, C.J., LARSON, E.M., HESKETH, J.D., PETERS, D.B.:** Development and source-sink effects on single leaf and canopy carbon dioxide exchange in maize. - Field Crop Res. 9: 391-402, 1984.

14127 - **PEARSON, C.J., MUIRHEAD, W.A.:** Nitrogen uptake. - In: PEARSON, C.J. (ed.): Control of Crop Productivity. Pp. 73-86. Academic Press, Sydney, Orlando - San Diego - Petaluma - New York - London - Toronto - Montreal - Tokyo 1984.

14128 - **PEARSON, R.G., ROSS, B.E.:** Growth rate and bending properties of selected loblolly pines. - Wood Fiber Sci. 16: 37-47, 1984.

14129 - **PELLETIER, B., MOUSSEAU, M.:** Effet du fractionnement d'un éclairement d'appoint sur la croissance, les bilans carbonés et la transpiration de jeunes plants de tomate de serre (*Lycopersicon esculentum* Mill., var. "Lucy"). - Agronomie 4: 987-992, 1984.

*14130 - **PEMADASA, M.A.:** Differential abaxial and adaxial stomatal responses to indole-3-acetic acid in *Commelina communis* L. - New Phytol. 90: 209-219, 1982.

14131 - **PENNAZIO, S., ROGGERO, P.:** Effects of salicylate on stomatal resistance in detached tobacco leaves. - Biol. Plant. 26: 455-460, 1984.

*14132 - **PENNING de VRIES, F.W.T.**: Systems analysis and models of crop growth. - In:
PENNING de Vries, F.W.T., VAN LAAR, H.H. (ed.): Simulation of Plant Growth
and Crop Production. Pp. 9-19. PUDOC, Wageningen 1982.

14133 - **PENNING de VRIES, F.W.T., AKKERSDIJK, J.W.J., VAN OORSCHOT, J.L.P.**: An error
in measuring respiration and photosynthesis due to transpiration. - Photosynthetica 18: 146-149, 1984.

*14134 - **PENNING de VRIES, F.W.T., VAN LAAR, H.H.**: Simulation of growth processes and
the model BACROS. - In: PENNING de VRIES, F.W.T., VAN LAAR, H.H. (ed.): Simulation of Plant Growth and Crop Production. Pp. 114-135. PUDOC, Wageningen
1982.

14135 - **PERETZ, J., EVANS, R.G., PROEBSTING, E.L.**: Leaf water potentials for management of high frequency irrigation on apples. - Trans. ASAE 27: 437-442, 1984.

14136 - **PESCI, P., BEFFAGNA, N.**: Inhibiting effect of fusicoccin on abscisic acid-induced proline accumulation in barley leaves. - Plant Sci. Lett. 36: 7-12,
1984.

14137 - **PETERSON, D.L., BAZZAZ, F.A.**: Photosynthetic and growth responses of silver
maple (*Acer saccharinum* L.) seedlings to flooding. - Amer. Midl. Natur. 112:
261-272, 1984.

14138 - **PETRASOVITS, I.**: Water utilization in agriculture. - Acta agron. Acad. Sci.
Hung. 33: 269-275, 1984.

14139 - **PETRASOVITS, I.**: Az öntözővíz minösége és a növénytermesztés. [Quality of
irrigation water and crop growing.] - Növénytermelés 33: 357-366, 1984. [In
Hung, ab: E.]

*14140 - **PETROVA, L., DEKOV, I., KOZAROVA, M.**: Vliyanie na razlichnite s"otnosheniya
na makro- i mikroelementi v khranitelniya substrat v"rkhu transpiratsiyata i
formite na vodata pri lyutsernata. [Effect of the various macro- and trace
element ratios in nutrient solution on transpiration and some other characteristics of alfalfa water regime.] - Fiziol. Rast. (Sofia) 9 (4): 71-77,
1983. [In Bulg, ab: R, E.]

14141 - **PFEUFER, B., KRUG, H.**: Effects of high CO_2-concentrations on vegetables. -
Acta Hort. 162: 37-44, 1984.

14142 - **PHAM THI, A.T.**: Action de la sécheresse sur les lipides polaires des feuilles
de Cotonnier (*Gossypium hirsutum* L.). - Bull. Soc. bot. France, Actual. bot.
131: 89-97, 1984.

14143 - **PHOGAT, B.S., SINGH, D.P., SINGH, P.**: Responses of cowpea (*Vigna unguiculata*
(L.) Walp.) and mung bean (*Vigna radiata* (L.) Wilczek) to irrigation. I. Effects on soil-plant water relations, evapotranspiration, yield and water use
efficiency. - Irrig. Sci. 5: 47-60, 1984.

14144 - **PHOGAT, B.S., SINGH, D.P., SINGH, P.**: Responses of cowpea (*Vigna unguiculata*
(L.) Walp.) and mung bean (*Vigna radiata* (L.) Wilczek) to irrigation. II. Effects on CO_2 exchange, radiation characteristics and growth. - Irrig. Sci.
5: 61-72, 1984.

14145 - **PIGOTT, C.D.**: Effect of photoperiod and water supply on apical abscision of
long-shoots of *Tilia cordata* Mill. - New Phytol. 97: 575-581, 1984.

*14146 - **PIRANI, V.**: Il clima e le rese produttive del girasole nell'Italia centrale.
[Climate and yields of sunflower (*Helianthus annuus* L.) in central Italy.] -
Riv. Agron. 15: 107-114, 1981. [In Ital, ab: E.]

*14147 - **PIRANI, V.:** Il clima e le rese produttive della barbabietola da zucchero in provincia di Ancona. [The climate and productive yields of sugat beet in the province of Ancona.] - Riv. Agron. 14: 275-286, 1980. [In Ital, ab: E.]

14148 - **PITCAIRN, C.E.R., GRACE, J.:** The effect of wind on provenances of *Molinia caerulea* L. - Ann. Bot. 54: 135-143, 1984.

14149 - **PITT, R.E.:** Forage drying in relation to pan evaporation. - Trans. ASAE 27: 1933-1937, 1944, 1984.

14150 - **PITT, R.E.:** Stochastic theory of forage drying as related to pan evaporation. - Agr. Forest Meteorol. 32: 197-215, 1984.

14151 - **PLESKANKA, J., SLAVÍK, B.:** Model of water vapour transport in intercellular spaces of leaf mesophyll. - In: CRAM, W.J., JANÁČEK, K., RYBOVÁ, R., SIGLER, K. (ed.): Mambrane Transport in Plants. Pp. 159-160. Academia, Praha 1984.

14152 - **PLHÁK, F.:** Dark transpiration rate and water deficit as growth limiting factors in alfalfa plants. - Biol. Plant. 26: 441-447, 1984.

14153 - **POCARD, J.-A., BERNARD, T., GOAS, G., Le RUDULIER, D.:** Restauration partielle, par la glycine bétaïneet la proline bétaïne, de l'activité fixatrice d'azote de jeunes plantes de *Medicago sativa* L. soumises à un stress hydrique. - C.R. Acad. Sci. (Paris), Sér. III 298: 477-480, 1984.

*14154 - **POHLHEIM, F.:** Untersuchungen an *Antirrhinum majus wettsteinii*, einer an vegetativ haploiden Pflanzen entstandenen Sprossvariante. - Zentralbl. Biol. 97: 53-67, 1978.

14155 - **POKORNY, F.A., WETZSTEIN, H.Y.:** Internal porosity, water availability, and root penetration of pine bark particles. - HortScience 19: 447-449, 1984.

14156 - **POKORNÝ, J., LIPTAJ, T.:** Effects of metal ions on the hydration of membrane phospholipids. - In: CRAM, W.J., JANÁČEK, K., RYBOVÁ, R., SIGLER, K. (ed.): Membrane Transport in Plants. Pp. 505-506. Academia, Praha 1984.

14157 - **POLIMBETOVA, F.A., OMAROVA, É.I., BOGDANOVA, E.D., KHUSAINOVA, G.K., KUDYSHEVA, K.K.:** Fiziologicheskie osobennosti morozoustoïchivykh mutantov pshenitsy. [Physiological peculiarities of frost-resistant wheat mutants.] - Fiziol. Biokhim. kul't. Rast. 16: 369-375, 1984. [In R, ab: E.]

14158 - **POLONENKO, D.R., MAYFIELD, C.I., DUMBROFF, E.B.:** Microbial responses to salt-induced osmotic stress. IV. A model of a root region. - Plant Soil 80: 363-371, 1984.

*14159 - **POLSON, A., Van der MERWE, K.J.:** The effect of weak electrical potential gradients on the transport of water in broad bean plants under stress. - Experientia 39: 576-577, 1983.

14160 - **POLUEKTOV, R.A., NEUSYPINA, T.A., KÜNKEL, K., WEIRAUCH, M., SCHÄFER, W.:** Modell der Ertragsbildung von Winterweizen bei limitiertem Wasserregime. - Arch. Acker- Pflanzenbau Bodenk. 28: 21-33, 1984.

14161 - **POMERANZ, Y., BOLTE, L.C.:** Time-dependent moisture gradients in conditioned wheat, determined by electrical methods. - Cereal Chem. 61: 559-561, 1984.

*14162 - **POMEROY, M.K., PIHAKASKI, S.J., ANDREWS, C.J.:** Membrane properties of isolated winter wheat cells in relation to icing stress. - Plant Physiol. 72: 535-539, 1983.

14163 - **POND, E.C., MENGE, J.A., JARRELL, W.M.:** Improved growth of tomato in salinized soil by vesicular-arbuscular mycorrhizal fungi collected from saline soils. - Mycologia 76: 74-84, 1984.

14164 - **PONNAMPERUMA, F.N.**: Effects of flooding on soils. - In: KOZLOWSKI, T.T. (ed.): Flooding and Plant Growth. Pp. 9-45. Academic Press, Orlando - San Diego - San Francisco - New York - London - Toronto - Montreal - Sydney - Tokyo - São Paulo 1984.

14165 - **POPOVA, Ĭ., DIMOVA, R.**: Vliyanie na sroka na seitba, g"stotata na poseva i polivaneto v"rkhu bratimostta na tsarevitsata. [Influence of seeding date, plant density and irrigation on maize tillering.] - Rasteniev. Nauki 21 (5): 92-100, 1984. [In Bulg, ab: R, E.]

*14166 - **POPOVIĆ, R., DIMITRIJEVIĆ, J., JANKOVIĆ, M.M.**: Ekofiziološka istraživanja vegetacije Deliblatske peščare. I. Dinamika i intenzitet transpiracije i količine vode u listovima biljaka livadsko-stepske i šumske zajednice. [Ecophysiological studies of the vegetation of Deliblato sands. I. Dynamics and intensity of plant transpiration and water content in the meadow-steppe and forest community.] - Ekologija 18: 15-42, 1983. [In Croat, ab: E.]

*14167 - **POPOVIĆ, R., DIMITRIJEVIĆ, J., JANKOVIĆ, M.M.**: Ekofiziološka istraživanja vegetacije Deliblatske peščare. II. Hidraturni odnosi nekih biljaka livadsko--stepske i šumske vegetacije. [Ecophysiological studies of the vegetation of Deliblato sands. II. Water relations in some plants of the meadow-steppe and forest vegetation.] - Ekologija 18: 93-106, 1983, [In Croat, ab: E.]

*14168 - **POPOVIĆ, R., JANKOVIĆ, M.M., DIMITRIJEVIĆ, J.**: Ovnovne karakteristike vodnog režima nekih vrsta biljaka zajednice *Querco-Carpinetum serbicum* Rud. na Fruškoj gori. [Basic characteristics of water regime in some plant species of the community *Querco-Carpinetum serbicum* Rud. on the mountain Fruška gora.] - Arh. biol. Nauka 31: 13-30, 1979. [In Serb, ab: E.]

*14168 - **POPOVIĆ, Ž.**: Prinosnost gajenih biljaka i njena veza sa pojedinim fiziološkim procesima. [Productivity of cultivated plants and its connection with some physiological processes.] - Agrohemija 1982: 389-438, 1982. [In Croat, ab: E.]

14170 - **POSPÍŠILOVÁ, J., SOLÁROVÁ, J.**: Ontogenetic changes in response of adaxial and abaxial epidermal conductances to water stress. - Biol. Plant. 26: 49-55, 1984.

14171 - **POSPÍŠILOVÁ, J., SOLÁROVÁ, J.**: Ontogenetic changes in stomatal conductance. - In: CRAM, W.J., JANÁČEK, K., RYBOVÁ, R., SIGLER, K. (ed.): Membrane Transport in Plants. Pp. 141-146. Academia, Praha 1984.

14172 - **POSPÍŠILOVÁ, J., SOLÁROVÁ, J.**: Regulation der Transpiration und Photosynthese-rate über die stomatäre Leitfähigkeit bei unterschiedlichen äusseren und pflanzeninteren Bedingungen. - In: Erhöhung der Wassernutzung im Boden und durch die Pflanze zur Steigerung und Stabilisierung der Erträge in der Pflanzenproduktion. Pp. 49-50. Forschungszentrum für Bodenfruchtbarkeit Müncheberg, Bereich Jena, Jena 1984.

14173 - **POSPÍŠILOVÁ, J., SOLÁROVÁ, J. (ed.)**: Water-in-Plants Bibliography. Volume 9 1983. References no. 11193-12911 / ABD-ZUR. Dr W. Junk Publishers, Dordrecht - Boston - Lancaster 1984.

14174 - **POTTOSIN, I.I., CHAMOROVSKIĬ, S.K., KONONENKO, A.A., USPENSKAYA, N.Ya.**: Vliyanie degidratatsii na perenos élektrona ot svyazannykh s membranoĭ tsitokhromov *c* na bakteriokhlorofill fotosinteticheskogo reaktsionnogo tsentra v khromatoforakh purpurnykh serobakterii. [Effects of dehydration on electron transfer between membrane-bound cytochrome *c* proteins and photosynthetic reaction centres of chromatophores of purple sulphur bacteria.] - Mol. Biol. (Moskva) 18: 821-830, 1984. [In R, ab: E.]

14175 - **POTTS, M., BOWMAN, M.A., MORRISON, N.S.**: Control of matric water potential (ψ_m) in immobilised cultures of cyanobacteria. - FEMS Microbiol. Lett. 24: 193-196, 1984.

*14176 - POTTS, M., OCAMPO-FRIEDMANN, R., BOWMAN, M.A., TÖZÜN, B.: *Chroococcus* S24 and *Chroococcus* N41 (cyanobacteria): morphological, biochemical and genetic characterization and effects of water stress on ultrastructure. - Arch. Microbiol. 135: 81-90, 1983.

14177 - POTVIN, M.A., WERNER, P.A.: Seasonal patterns in water relations of two species of goldenrods (*Solidago*) grown on an experimental soil moisture gradient. - Bull. Torrey bot. Club 111: 171-178, 1984.

14178 - POWLES, S.B.: Photoinhibition of photosynthesis induced by visible light. - Annu. Rev. Plant Physiol. 35: 15-44, 1984.

14179 - PRANGE, R.K., ORMROD, D.P., PROCTOR, J.T.A.: Effect of front age on frond elongation, gas exchange, and water relations in the ostrich fern (*Matteuccia struthiopteris*). - Can. J. Bot. 62: 2094-2100, 1984.

*14180 - PRASAD, U.K., SINGH, Y.: Effect of soil-moisture regimes and nitrogen levels on the growth, leaf-water potential, sucrose content and yield of sugarbeet. - Ind. J. agr. Sci. 53: 948-958, 1983.

*14181 - PREGNALL, A.M.: Release of dissolved organic carbon from the estuarine intertidal macroalga *Enteromorpha prolifera*. - Mar. Biol. 73: 37-42, 1983.

14182 - PREVOT, L., BERNARD, R., TACONET, O., VIDAL-MADJAR, D., THONY, J.L.: Evaporation from a bare soil evaluated using a soil water transfer model and remotely sensed surface soil moisture data. - Water Resour. Res. 20: 311-316, 1984.

14183 - PRIEHRADNÝ, S.: Deviácia kriviek sýtenia pletív vodou u jačmeňa infikovaného múčnatkou. [Deviation of the curves of tissue saturation by water in powdery mildew-infected barley.] - Biológia (Bratislava) 39: 471-482, 1984. [In Slov, ab: E, R.]

14184 - PRIEHRADNÝ, S.: Zmeny obsahu vody v rastlinách jačmeňa infikovaných *Erysiphe graminis* DC. [Changes of water content in barley plants infected by *Erysiphe graminis* DC.] - Biológia (Bratislava) 39: 483-491, 1984. [In Slov, ab: E, R.]

14185 - PRIEHRADNÝ, S.: Die Reaktion anfälliger und resistenter Gerstensorten auf pilzliche Krankheitserreger. IV. Wasser- und Trockensubstanzgehalt. - Phytopathol. Z. 111: 271-282, 1984.

14186 - PROEBSTING, E.L., DRAKE, S.R., EVANS, R.G.: Irrigation management, fruit quality, and storage life of apple. - J. Amer. Soc. hort. Sci. 109: 229-232, 1984.

*14187 - PROEBSTING, E.L., PERETZ, J.: Plant response to methods of irrigation. - In: Water, Energy and Economic Alternatives. Pp. 155-161. Irrigation Association, Silver Spring 1982.

14188 - PUNJA, Z.K., JENKINS, S.F.: Influence of temperature, moisture, modified gaseous atmosphere, and depth in soil on eruptive sclerotial germination of *Sclerotium rolfsii*. - Phytopathology 74: 749-754, 1984.

14189 - PUNTHAKEY, J.F., McFARLAND, M.J., WORTHINGTON, J.W.: Stomatal responses to leaf water potentials of drip irrigated peach (*Prunus persica*). - Trans ASAE 27: 1442-1450, 1984.

14190 - PURSE, J.G.: Phloem exudate of *Perilla crispa* and its effects on flowering of *P. crispa* shoot explants. - J. exp. Bot. 35: 227-238, 1984.

14191 - PURVIS, A.C.: Importance of water loss in the chilling injury of grapefruit stored at low temperature. - Scientia Hort. 23: 261-267, 1984.

14191 - PURVIS, A.C.: Importance of water loss in the chilling injury of grapefruit
        stored at low temperature. - Scientia Hort. 23: 261-267, 1984.

14192 - PUTA, H., SCHMIDT, M.: Anwendung mathematischer Optimierungsverfahren zur
        Steuerung des Pflanzenwachstums. - Wiss. Z. Humboldt-Univ., math. naturwiss.
        Reihe 33: 375-377, 1984.

14193 - QUAMME, H.A., STUSHNOFF, C.: Resistance to environmental stress. - In: MOORE,
        J.N., JANICK, J. (ed.): Methods in Fruit Breeding. Pp. 242-266. Purdue Uni-
        versity Press, West Lafayette 1984.

14194 - QUARRIE, S.A., LISTER, P.G.: Evidence of plastid control of abscisic acid
        accumulation in barley (Hordeum vulgare L.). - Z. Pflanzenphysiol. 114: 295-
        308, 1984.

14195 - RADCHENKO, M.Ĭ.: Pigmentoutvorennya u Dunaliella salina Teod. v umovakh
        riznoĭ solonosti seredovishcha. [Pigment formation in Dunaliella salina Teod.
        under various water salinity.] - Ukr. bot. Zh. 41 (6): 65-70, 1984. [In Ukr,
        ab: E, R.]

14196 - RADIN, J.W.: Stomatal responses to water stress and to abscisic acid in phos-
        phorus-deficient cotton plants. - Plant Physiol. 76: 392-394, 1984.

14197 - RADIN, J.W., EIDENBOCK, M.P.: Hydraulic conductance as a factor limiting leaf
        expansion of phosphorus-deficient cotton plants. - Plant Physiol. 75: 372-
        377, 1984.

*14198 - RAI, R., PRASAD, V.: Studies on compatibility of nitrogen fixation by light-
        -temperature-adapted Rhizobium strains and Vigna radiata genotypes at two
        moisture levels in calcareous soil. - J. agr. Sci. 101. 377-381, 1983.

*14199 - RAJ, K.P.S., CHINOY, J.J.: Effect of drought on sulfhydryl content of barley
        grown under two photoperiods. - Agrochimica 24: 9-13, 1980.

*14200 - RAJ, K.P.S., CHINOY, J.J.: Effect of drought on sugar content of barley grown
        under two photoperiods. - Agrochimica 24: 108-112, 1980.

*14201 - RAKHMANINA, K.P., MOLOTKOVSKIĬ, Yu.I.: Vodnyĭ rezhim ėdifikatorov chal'noĭ
        rastitel'nosti Yuzhnogo Tadzhikistana. [Water regime of the edificators of
        the chalny vegetation in the South Tadjikistan.] - Ėkologiya 1983 (6): 9-16,
        1983. [In R, ab: E.]

14202 - RAMBAL, S.: Water balance and pattern of root water uptake by a Quercus coc-
        cifera L. evergreen scrub. - Oecologia 62: 18-25, 1984.

*14203 - RAMBAL, S., BERGER, A., PARISOT, J.M.: Evapotranspiration reelle, extraction
        racinaire, regime hydrique et production de cultures de luzerne. - In: Iso-
        tope and Radiation Techniques in Soil Physics and Irrigation Studies 1983.
        Pp. 291-299. International Atomic Energy Agency, Vienna 1983.

*14204 - RAMSAY, A.J., BALL, K.T.: Estimation of algae in New Tealand pasture soil
        and litter by culturing and by chlorophyll a extraction. - N. Zeal. J. Sci.
        26: 493-503, 1983.

14205 - RAMSEUR, E.L., QUISENBERRY, V.L., WALLACE, S.U., PALMER, J.H.: Yield and
        yield components of 'Braxton' soybeans as influenced by irrigation and intra-
        row spacing. - Agron. J. 76: 442-446, 1984.

14206 - RAO, A.N., SEETAMBARAM, Y., DAS, V.S.R.: Growth and productivity of three
        sorghum cultivars under low light stress. - Ind. J. exp. Biol. 22: 260-266,
        1984.

14207 - **RAO, P., AGARWAL, S.K.:** Diurnal variation in leaf water potential, stomatal
conductance, and irradiance of winter crop under different moisture levels.
- Biol. Plant. 26: 1-4, 1984.

*14208 - **RAO, P.V., VENKATACHARI, A., REDDY, K.A.:** Consumptive use of oilseed crops
in winter season. - Oilseeds J. 6: 28-33, 1976.

14209 - **RASMUSSEN, K.R., RASMUSSEN, S.:** The summer water balance in a Danish oak
stand. - Nordic Hydrol. 15: 213-222, 1984.

14210 - **RATHERT, G.:** Sucrose and starch content of plant parts as a possible indicator for salt tolerance of crops. - Aust. J. Plant Physiol. 11: 491-495,
1984.

*14211 - **RATKOVIČ, S., BAČIČ, G.:** 323-water exchange in *Nitella cells*: A PMR study
in the presence of paramagnetic Mn^{2+} ions. - Bioelectrochem. Bioenerg. 7:
405-412, 1980. Also in: J. Electroanal. Chem. 116: 405-412, 1980.

*14212 - **RATKOVIČ, S., DENIČ, M., LAHAJNAR, G.:** Kinetics of water imbibition by seed:
Why normal and opaque-2 maize kernels differ in their hydration properties?
- Period. biol. 84: 180-182, 1982.

*14213 - **RATKOVIČ, S., DENIČ, M., LAHAJNAR, G., ZUPANČIČ, I.:** Biological systems with
low water content: NMR approach to the state of water in plant seed. - In:
FRANKS, F., MATHIAS, S.F. (ed.): Biophysics of Water. Pp. 312-314. John Wiley
& Sons, Chichester - New York - Brisbane - Toronto - Singapore 1982.

14214 - **RAUPACH, M.R., LEGG, B.J.:** The uses and limitations of flux-gradient relationships in micrometeorology. - Agr. Water Manage. 8: 119-131, 1984. Also
in: SHARMA, M.L. (ed.): Evapotranspiration from Plant Communities. Pp. 119-
131. Elsevier Science Science Publishers, Amsterdam - Oxford - New York -
Tokyo 1984.

14215 - **RAVEN, J.A.:** Energetics and Transport in Aquatic Plants. - A.R. Liss. Inc..
New York 1984.

14216 - **RAVEN, J.A., GRIFFITHS, H., ALLEN, S.:** N source, transpiration rate and
stomatal aperture in *Ricinus*. - In: CRAM, W.J., JANÁČEK, K., RYBOVÁ, R.,
SIGLER, K. (ed.): Membrane Transport in Plants. Pp. 161-162. Academia, Praha
1984.

14217 - **RAWSON, H.M., MUNNS, R.:** Leaf expansion in sunflower as influenced by salinity and short-term changes in carbon fixation. - Plant Cell Environ. 7: 207-
213, 1984.

*14218 - **RAY, N., KHADDAR, V.K.:** Formation of adventious and floating roots in cotton
under waterlogged conditions. - Curr. Sci. 52: 826-828, 1983.

14219 - **RAZNOPOLOV, O.N., TERKULOVA, L.P., KRASAVTSEV, O.A.:** Morozostoĭkost' tsvet-
kovykh pochek kostochkovykh i sostoyanie v nikh vody pri otritsatel'nykh
temperaturakh. [Frost-hardiness of flower buds in stone fruits and state
of water in them at temperatures below zero.] - Fiziol. Biokhim. kul't.
Rast. 16: 161-166, 1984. [In R, ab: E.]

*14220 - **REDDY, K.A., VENKATACHARI, A.:** Actual evapotranspiration from wheat crop and
its relationship with estimates based on empirical formulae. - Időjárás
(Budapest) 82: 225-229, 1978.

*14221 - **REDDY, P.R., ASNANI, V.L.:** Effect of opaque-2 gene on kernel moisture content in maize. - Ind. J. Genet. Plant Breed. 43: 203-207, 1983.

*14222 - **REDDY, S.J.:** Agroclimatic classification of the semi-arid tropics. I. A
method for the computation of classificatory variables. - Agr. Meteorol. 30:
185-200, 1983.

14223 - REDDY, S.J.: Agroclimatic classification of the semi-arid tropics. III.
        Characteristics of variables relevant to crop production potential. - Agr.
        Meteorol. 30: 269-292, 1984.

14224 - REDDY, S.J., AMORIM NETO, M. da S.: A method for the estimation of potential
        evapotranspiration and for open pan evaporation over Brazil. - Pesq. agropec.
        Bras. 19: 247-267, 1984.

*14225 - REDDY, V.M, TANNER, J.W.: The effects of irrigation, inoculants and ferti-
        lizer nitrogen on peanuts (Arachis hypogaea L.). I. Nitrogen fixation. -
        Peanut Sci. 7: 114-119, 1980.

14226 - REDL, H.: Das Wasserpotential von Blättern der Rebsorten Grüner Veltliner und
        Rheinriesling und mögliche Beziehungen zur Stiellähme der Weintrauben. -
        Bodenkultur 35: 127-138, 1984.

14227 - REDL, H.: Der Einfluss der Erziehungshöhe auf das Blattwasserpotential bei
        der Rebsorte Grüner Veltliner. - Mitt. Klosterneuburg 34: 47-50, 1984.

*14228 - REED, R.H.: The osmotic responses of Polysiphonia lanosa (L.) Tandy from
        marine and estuarine sites: evidence for incomplete recovery of turgor. - J.
        exp. mar. Biol. Ecol. 68: 169-193, 1983.

14229 - REED, R.H., CHUDEK, J.A., FOSTER, R., STEWART, W.D.P.: Osmotic adjustment in
        cyanobacteria from hypersaline environments. - Arch. Microbiol. 138: 333-
        337, 1984.

14230 - REED, R.H., RICHARDSON, D.L., WARR, S.R.C., STEWART, W.D.P.: Carbohydrate
        accumulation and osmotic stress in cyanobacteria. - J. gen. Microbiol. 130:
        1-4, 1984.

14231 - REICH, P.B.: Oscillations in stomatal conductance of hybrid poplar leaves in
        the light and dark. - Physiol. Plant. 61: 541-548, 1984.

14232 - REICH, P.B.: Relationships between leaf age, irradiance, leaf conductance,
        CO_2 exchange, and water use efficiency in hybrid poplar. - Photosynthetica
        18: 445-453, 1984.

14233 - REICH, P.B.: Leaf stomatal density and diffusive conductance in three amphi-
        stomatous hybrid poplar cultivars. - New Phytol. 98: 231-239, 1984.

14234 - REICH, P.B.: Loss of stomatal function in ageing hybrid poplar leaves. - Ann.
        Bot. 53: 691-698, 1984.

14235 - REICH, P.B., BORCHERT, R.: Water stress and tree phenology in a tropical dry
        forest in the lowlands of Costa Rica. - J. Ecol. 72: 61-74, 1984.

14236 - REICH, P.B., LASSOIE, J.P.: Effects of low level O_3 exposure on leaf diffu-
        sive conductance and water-use efficiency in hybrid poplar. - Plant Cell
        Environ. 7: 661-668, 1984.

14237 - REID, D.M., BRADFORD, K.J.: Effect of flooding on hormone relations. - In:
        KOZLOWSKI, T.T. (ed.): Flooding and Plant Growth. Pp. 195-219. Academic
        Press, Orlando - San Diego - San Francisco - New York - London - Toronto -
        Montreal - Sydney - Tokyo - São Paulo 1984.

14238 - REID, J.B., HASHIM, O., GALLAGHER, J.N.: Relations between available and
        extractable soil water and evapotranspiration from a bean crop. - Agr. Water
        Manage. 9: 193-209, 1984.

14239 - REID, R.J., JEFFERIES, R.L., PITMAN, M.G.: Ion and osmotic adjustment in
        Lamprothamnium.   In: CRAM, W.J., JANÁČEK, K., RYBOVÁ, R., SIGLER, K. (ed.):
        Membrane Transport in Plants. Pp. 117-118. Academia, Praha 1984.

14240 - REINERS, W.A., HOLLINGER, D.Y., LANG, G.E.: Temperature and evapotranspiration gradients of the White Mountains, New Hampshire, U.S.A. - Arctic alpine
Res. 16: 31-36, 1984.

14241 - REINHOLD, L., SEIDEN, A., VOLOKITA, M.: Is modulation of the rate of proton
pumping a key event in osmoregulation? - Plant Physiol. 75: 846-849, 1984.

14242 - RENARD, C., KARAMAGA, P.: Étude des relations hydriques chez *Coffea arabica*
L. III. Evolution de la conductance stomatique et des composantes du potentiel hydrique chez deux cultivars soumis à la sécheresse en conditions
contrôlées. - Café Cacao Thé 28: 155-164, 1984.

*14243 - RENGER, M., STREBEL, O.: Wasserverbrauch und Ertrag von Pflanzenbeständen. -
Kali-Briefe 15: 135-143, 1980.

*14244 - REUTHER, G.: The effect of water stress on photosynthesis and transpiration
of *Vitis vinifera* under different ecological conditions. - In: METZNER, H.
(ed.): Photosynthesis and Plant Productivity. Pp. 78-82. Wissenschaftliche
Verlagsgesellschaft, Stuttgart 1983.

14245 - REYNOLDS, D.N.: Alpine annual plants: phenology, germination, photosynthesis,
and growth of three Rocky Mountain species. - Ecology 65: 759-766, 1984.

14246 - REYNOLDS, J.F., KEMP, P.R., CUNNINGHAM, G.L.: Photosynthetic responses of
saltgrass (*Distichlis spicata*) to irradiance, temperature and salinity growth
treatments: A modeling synthesis. - Photosynthetica 18: 100-110, 1984.

*14247 - REZNICEK, S.A., SVOBODA, J.: Tundra communities along a microenvironmental
gradient at Coral Harbour, Southampton Island, N.W.T. - Natur. Can. (Quebec)
109: 583-595, 1982.

14248 - RHEINBABEN, W., von, TROLLDENIER, G.: Influence of plant growth on denitrif-
ication in relation to soil moisture and potassium nutrition. - Z. Pflanzenernähr. Bodenk. 147: 730-738, 1984.

14249 - RHOADES, J.D.: Principles and methods of monitoring soil salinity. - In:
SHAINBERG, I., SHALHEVET, J. (ed.): Soil Salinity under Irrigation. Processes
and Management. Pp. 130-142. Springer-Verlag, Berlin - Heidelberg - New York
- Tokyo 1984.

14250 - RHOADS, F.M., STANLEY, R.L., Jr.: Yield and nutrient utilization efficiency
of irrigated corn. - Agron. J. 76: 219-223, 1984.

*14251 - RICHARDSON, D.H.S., NIEBOER, E.: Ecophysiological responses of lichens to
sulphur dioxide. - J. Hattori bot. Lab. 54: 331-351, 1983.

14252 - RICHTER, W.: Zur Abhängigkeit des Sommergerstenkornertrages auf einem
schwachlehmigen Sandboden vom Niederschlagsangebot im Hauptwasserbedarfs-
zeitraum. - Arch. Acker- Pflanzenbau Bodenk. 28: 163-168, 1984.

14253 - RIDGE, R.W., LONERAGAN, W.A., BELL, D.T., COLQUHOUN, I.J., KUO, J.: Comparative studies in selected species of *Eucalyptus* used in rehabilitation of the
northern Jarrah forest, Western Australia. II. Wood and leaf anatomy. -
Aust. J. Bot. 32: 375-386, 1984.

14254 - RIOU, C.: Experimental study of potential evapotranspiration (PET) in Central
Africa. - J. Hydrol. 72: 275-288, 1984.

14255 - RITCHIE, G.A., SHULA, R.G.: Seasonal changes of tissue-water relations in
shoots and root systems of Douglas-fir seedlings. - Forest Sci. 30: 538-548,
1984.

*14256 - RITCHIE, J.T.: Soil water availability. - Plant Soil 58: 327-338, 1981.

*14257 - RIVIER, L., LÉONARD, J.-F., COTTIER, J.-P.: Rapid effect of osmotic stress on
the content and exodiffusion of abscisic acid in *Zea mays* roots. - Plant Sci.
Lett. 31: 133-137, 1983.

*14258 - RIZZO, V., Di BARI, V., LOSAVIO, N.: I consumi idrici per evapotraspirazione
del mais da granella in coltura principale nell'ambiente del Tavoliere
pugliese. [Water consumptive use by evapotranspiration from grain maize as
first crop in the apulian "Tavoliere" region.] - Riv. Agron. 14: 263-274,
1980. [In Ital, ab: E.]

14259 - ROBERTS, A.H.C., THOMSON, N.A.: Seasonal distribution of pasture production
in New Zealand XVIII. South Taranaki. - J. exp. Agr. 12: 83-92, 1984.

14260 - ROBERTS, A.H.C., THOMSON, N.A.: Seasonal distribution of pasture production
in New Zealand XVIV. Central Taranaki. - J. exp. Agr. 12: 93-101, 1984.

14261 - ROBERTS, B.R., DOMIR, S.C.: Effect of plant water stress on regrowth of
American sycamore seedlings injected with maleic hydrazide. - Forest Ecol.
Manage. 7: 291-296, 1983/1984.

14262 - ROBERTS, J., WALLACE, J.S., PITMAN, R.M.: Factors affecting stomatal conductance of bracken below a forest canopy. - J. appl. Ecol. 21: 643-655, 1984.

14263 - ROBERTS, S.W., BOWMAN, W.D.: Osmotic and turgor relations in selected
chaparral shrub species. - In: MARGARIS, N.S., ARIANOUSTOU-FARAGGITAKI, M.,
OECHEL, W.C.: Being Alive on Land. Pp. 77-84. Dr W. Junk Publishers, The
Hague - Boston - Lancaster 1984.

14264 - ROBERTSON, J.A., CHAPMAN, G.W., WILSON, R.L., Jr., RUSSELL, R.B.: Effect of
moisture content of oil type sunflower seed on fungal growth and seed quality
during storage. - J. Amer. Oil Chem. Soc. 61: 768-771, 1984.

14265 - ROBICHAUX, R.H., PEARCY, R.W.: Evolution of C_3 and C_4 plants along an environmental moisture gradient: patterns of photosynthetic differentiation in
Hawaiian *Scaevola* and *Euphorbia* species. - Amer. J. Bot. 71: 121-129, 1984.

14266 - ROCHESTER, E.W., BACKMAN, P.A., McGUIRE, J.A., CURTIS, L.M., STARLING, J.,
IVEY, H.: Irrigation schedules for peanut production. - Alabama agr. exp.
Sta. Bull. 556: 3-28, 1984.

14267 - RØEGGEN, O.: Effect of temperature and moisture on germination in spinach
(*Spinacia oleracea* L.). - Scientia Hort. 24: 221-229, 1984.

14268 - ROGERS, H.H., SIONIT, N., CURE, J.D., SMITH, J.M., BINGHAM, G.E.: Influence
of elevated carbon dioxide on water relations of soybeans. - Plant Physiol.
74: 233-238, 1984.

*14269 - ROGERS, H.H., THOMAS, J.F., BINGHAM, G.E.: Response of agronomic and forest
species to elevated atmospheric carbon dioxide. - Science 220: 428-429, 1983.

14270 - ROOD, S.B., MAJOR, D.J.: Influence of plant density, nitrogen, water supply
and pod or leaf removal on growth of oilseed rape. - Field Crops Res. 8:
323-331, 1984.

14271 - ROSE, C.W.: Modelling evapotranspiration: an approach to heterogenous commu-
nities. - Agr. Water Manage 8: 203-221, 1984. Also in: SHARMA, M.L. (ed.):
Evapotranspiration from Plant Communities. Pp. 203-221. Elsevier Science
Publishers, Amsterdam - Oxford - New York - Tokyo 1984.

14272 - ROSE, C.W., SHARMA, M.L.: Summary and recommendations of the workshop on
"evapotranspiration from plant communities". - Agr. Water Manage. 8: 325-342,
1984. Also in: SHARMA, M.L. (ed.): Evapotranspiration from Plant Communities.
Pp. 203-221. Elsevier Science Publishers, Amsterdam - Oxford - New York -
Tokyo 1984.

*14273 - **ROSEN, P.M., GOOD, G.L., STEPONKUS, P.L.:** Desiccation injury and direct freezing injury to evergreen azaleas: a comparison of cultivars. - J. Amer. Soc. hort. Sci. 108: 28-31, 1983.

*14274 - **ROSENBERG, N.J.:** The increasing CO_2 concentration in the atmosphere and its implication on agricultural productivity II. Effects through CO_2-induced climatic change. - Climatic Change 4: 239-254, 1982.

14275 - **ROSENOW, D.T., QUISENBERRY, J.E., WENDT, C.W., CLARK, L.E.:** Drought tolerant sorghum and cotton germplasm. - Agr. Water Manage. 7: 207-222, 1983. Also In: STONE, J.F., WILLIS, W.O. (ed.): Plant Production and Management under Drought Conditions. Pp. 207-222. Elsevier Science Publishers, Amsterdam - Oxford - New York - Tokyo 1983.

14276 - **ROTH, D., KACHEL, K., RICHTER, W.:** Untersuchungen zum Niederschlagsbedarf der Zuckerrüben in Abhangigkeit von Witterung, Standort und Ertragsniveau. - Arch. Acker- Pflanzenbau Bodenk.28: 677-687, 1984.

14277 - **ROTH, I.:** Stratification of Tropical Forests as Seen in Leaf Structure. (Tasks for Vegetation Science Volume 6.) - Dr W. Junk Publishers, The Hague - Boston - Lancaster 1984.

14278 - **ROUNDY, B.A.:** Estimation of water potential components of saline soils of Great basin rangelands. - Soil Sci. Soc. Amer. J. 48: 645-650, 1984.

14279 - **ROUNDY, B.A., EVANS, R.A., YOUNG, J.A.:** Surface soil and seedbed ecology in salt-desert plant communities. - In: TIEDEMANN, A.R., McARTHUR, E.D., STUTZ, H.C., STEVENS, R., JOHNSON, K.L. (ed.): Proceedings - Symposium on the Biology of Atriplex and Related Chenopods. Pp. 66-74. Intermountain Forest and Range Experiment Station, Ogden 1984.

*14280 - **ROW, K.S., VENKATESWARLU, B.:** Influence of varied moisture regimes at different growth phases on yield components in rice (*Oryza sativa* L.) - Ind. J. Plant Physiol. 26: 126-132, 1983.

*14281 - **ROW, K.S., VENKATESWARLU, B.:** Diurnal variation in evapotranspiration and transpiration at different levels of moisture supply in rice. - Ind. J. Plant Physiol. 26: 214-219, 1983.

14282 - **ROY, J.:** Growth of *Dactylis glomerata* L. and *Bromus erectus* Huds. in natural habitats and along light and water gradients. - In: MARGARIS, N.S., ARIA-NOUSTOU-FARAGGITAKI, M., OECHEL, W.C. (ed.): Being Alive on Land. Pp. 167-172. Dr W. Junk Publishers, The Hague - Boston - Lancaster 1984.

*14283 - **RÜHLE, W., WEINZETTEL, A., WILD, A.:** Isolation of photochemically active chloroplasts from wheat grown at low and high light intensities and with different nitrogen supply. - In: METZNER, H. (ed.): Photosynthesis and Plant Productivity. Pp. 210-215. Wissenschaftliche Verlagsgesellschaft, Stuttgart 1983.

14284 - **RUMBAUGH, M.D., ASAY, K.H., JOHNSON, D.A.:** Influence of drought stress on genetic variances of alfalfa and wheatgrass seedlings. - Crop Sci. 24: 297-303, 1984.

14285 - **RUMNEY, R.P., HARRIS, C.E., DHANOA, M.S.:** The effect of rewetting on some drying characteristics of Italian ryegrass leaves in simulated hay-making conditions. - Agr. Forest Meteorol. 32: 133-144, 1984.

14286 - **RUNNING, S.W.:** Documentation and preliminary validation of H2OTRANS and DAYTRANS, two models for predicting transpiration and water stress in western coniferous forests. - USDA Forest Serv. RM Res. Pap. RM-252: 1-45, 1984.

14287 - **RUNNING, S.W.:** Microclimate control of forest productivity: analysis by computer simulation of annual photosynthesis/transpiration balance in different environments. - Agr. Forest Meteorol. 32: 267-288, 1984.

*14288 - **RUNNING, S.W., KNIGHT, D.H., FAHEY, T.J.:** Description and application of H2OTRANS a stand level hydrologic model for western coniferous forests. - In: LAUENROTH, W.K., SKOGERBOE, G.V., FLUG, M. (ed.): Analysis of Ecological Systems: State-of-the-Art in Ecological Modelling. Pp. 489-496. Elsevier Science Publishers, Amsterdam 1983.

14289 - **RUSSO, D.:** Statistical analysis of crop yield-soil water relationships in heterogeneous soil under trickle irrigation. - Soil Sci. Soc. Amer. J. 48: 1402-1410, 1984.

14290 - **RUSSO, D.:** Spatial variability considerations in salinity management. - In: SHAINBERG, I., SHALHEVET, J. (ed.): Soil Salinity under Irrigation. Processes and Management. Pp. 198-216. Springer-Verlag, Berlin - Heidelberg - New York - Tokyo 1984.

*14291 - **RUWALI, K.N., SIROHI, G.S., TOMAR, O.P.S.:** Physiological nature of hybrid vigour in bajra hybrid-1 (*Pennisetum typhoides* (Burm S & H) in relation to its parents. II. Yield analysis of bajra hybrid-1 and its parents under rainfed and irrigated conditions. - Ind. J. Plant Physiol. 26: 45-54, 1983.

14292 - **SABALE, A.B., BHOSALE, L.J.:** C_3-C_4 photosynthesis in *Bouganvillea* cv. Mary Palmer. - Photosynthetica 18: 84-89, 1984.

*14293 - **SABATIER, G., VARTANIAN, N.:** Cinétique de la rhizogenèse adaptive à la sécheresse. Relation avec l'évolution des paramètres hydriques et morphologiques chez le *Sinapis alba*, à deux niveaux d'énergie lumineuse. - Physiol. Plant. 59: 501-507, 1983.

14294 - **SADEGHI, A.M., HANCOCK, G.D., WAITE, W.P., SCOTT, H.D., RAND, J.A.:** Microwave measurements of moisture distributions in the upper soil profile. - Water Resour. Res. 20: 927-934, 1984.

14295 - **SAIKIA, D.N., DEY, S.K.:** Leaf-water potential and stomatal resistance of tea leaf as influenced by soil moisture and potash application. - Ind. J. agr. Sci. 54: 727-732, 1984.

*14296 - **SAIKIA, D.N., DEY, S.K.:** Water consumption by young tea (*Camellia sinensis* L.) under potassium manuring. - Two Bud 25: 36-41, 1978.

14297 - **SAINI, H.S., SEDGLEY, M., ASPINALL, D.:** Developmental anatomy in wheat of male sterility induced by heat stress, water deficit or abscisic acid. - Aust. J. Plant Physiol. 11: 243-253, 1984.

*14298 - **SAINT-CLAIR, P.M.:** Etude de quelques aspects de la résistance à la sécheresse. - Turrialba 29: 139-146, 1979.

14299 - **SAIRAM, R.K., DUBE, S.D.:** Effect of moisture stress on proline accumulation in wheat in relation to drought tolerance. - Ind. J. agr. Sci. 146-147, 1984.

14300 - **SAITO, S.M.T., MONTANHEIRO, M.N.S., VICTÓRIA, R.L., REICHARDT, K.:** The effects of N fertilizer and soil moisture on the nodulation and growth of *Phaseolus vulgaris*. - J. agr. Sci. 103: 87-93, 1984.

14301 - **SAKAI, W.S., SANFORD, W.G.:** A developmental study of silicification in the abaxial epidermal cells of sugarcane leaf blades using scanning electron microscopy and energy dispersive X-ray analysis. - Amer. J. Bot. 71: 1315-1322, 1984.

*14302 - **SALEM, S.A., CAESAR, K., BARTELS, M.:** Field studies on the irrigation levels and cultivars in faba bean (*Vicia faba* L.) in Northern Egypt. - Z. Acker- Pflanzenbau 152: 325-335, 1983.

14303 - **SALIH, A.M.A., SENDIL, U.:** Evapotranspiration under extremely arid climates. - J. Irrig. Drain Eng. 110: 289-303, 1984.

14304 - **SALIM, M., PITMAN, M.G.:** Pressure-induced water and solute flow through plant roots. - J. exp. Bot. 35: 869-881, 1984.

14305 - **SALIM, M., PITMAN, M.G.:** Water and solute flow through mung bean roots under applied pressure. - Physiol. Plant. 61: 263-270, 1984.

14306 - **SALLEO, S., LO GULLO, M.A., SIRACUSANO, L.:** Distribution of vessel ends in stems of some diffuse- and ring-porous trees: the nodal regions as 'safety zones' of the water conducting system. - Ann. Bot. 54: 543-552, 1984.

14307 - **SAMEJIMA, T., NISHIKATA, Y., YANO, T.:** Moisture permeability of cured tobacco epidermis. - Agr. biol. Chem. 48: 439-443, 1984.

14308 - **SAMET, J.S., CORTES, P.M., SINCLAIR, T.R.:** Diurnal and seasonal changes in abscisic acid of field-grown soybeans subjected to drought stress. - Field Crops Res. 8: 49-59, 1984.

*14309 - **SAMMIS, T.W., GREGORY, E.J., KALLSEN, C.E.:** Estimating evapotranspiration with water-production functions or the Blaney-Criddle method. - Trans. ASAE 25: 1656-1661, 1982.

*14310 - **SAMMLER, P., EGEL, G., SCHMIDT, A., BERGMANN, H.:** Der Einfluss von Wasser- stress auf die generative Entwicklung und die Abscission reproduktiver Organe von *Vicia faba* L. - Arch. Acker- Pflanzenbau Bodenk. 26: 227-236, 1982.

*14311 - **SANCHEZ CONDE, M.P., AZUARA, P.:** Evaluacion del efecto de soluciones nutri- tivas equilibras de distintas presiones osmoticas sobre la planta de lechuga. [An evaluation of the effect of osmotic pressures in equilibrated nutritive solutions on the lettuce plant.] - Agrochimica 24: 176-183, 1980. [In Ital, ab: E.]

14312 - **SANDERS, T.H., BLANKENSHIP, P.D., COLE, R.J., HILL, R.A.:** Effect of soil temperature and drought on peanut pod and stem temperatures relative to *Aspergillus flavus* invasion and aflatoxin contamination. - Mycopathologia 86: 51-54, 1984.

14313 - **SANDERSON, J.B., IVANY, J.A., WHITE, R.P.:** Effect of time of desiccation on seed potato yield and size distribution. - Amer. Potato J. 61: 691-696, 1984.

14314 - **SANDS, R., KRIEDEMANN, P.E., COTTERILL, P.P.:** Water relations and photosyn- thesis in three families of radiata pine seedlings known to differ in their response to weed control. - Forest Ecol. Manage. 9: 173-184, 1984.

14315 - **SANDS, R., NAMBIAR, E.K.S.:** Water relations of *Pinus radiata* in competition with weeds. - Can. J. Forest Res. 14: 233-237, 1984.

14316 - **SANFORD, J.O., HAIRSTON, J.E.:** Effects of N fertilization on yield, growth, and extraction of water by wheat following soybeans and grain sorghum. - Agron. J. 76: 623-627, 1984.

14317 - **SANTARIUS, K.A.:** The role of the chloroplast coupling factor in the inacti- vation of thylakoid membranes at low temperatures. - Physiol. Plant. 61: 591- 598, 1984.

14318 - **SANTOS, I., SALEMA, R.:** Effects of hydration conditions on the stroma inclu- sion of some CAM chloroplasts. - Plant Cell Environ 7: 541-544, 1984.

*14319 - **SARADA DEVI, C., RAO, G.R.:** Influence of salinity on stomatal behaviour in groundnut. - Ind. J. Plant Physiol. 23: 174-180, 1980.

14320 - **SASAHARA, T.:** Varietal variations in leaf anatomy as related to photosynthesis in soybean (*Glycine Max* (L.) Merr.). - Jap. J. Breed. 34: 295-303, 1984.

14321 - **SASAKI, A.:** [Varietal differences in response to excess soil moisture at internode elongation stage in barley.] - Jap. J. Breed. 34: 79-86, 1984. [In Jap.]

*14322 - **SATO, T., SAKAI, A.:** Cold tolerance of gametophytes and sporophytes of some cool temperate ferns native to Hokkaido. - Can. J. Bot. 59: 604,608, 1981.

*14323 - **SAUGIER, B.:** Plant growth and its limitations in crops and natural communities. - In: MOONEY, H.A., GORDON, M. (ed.): Disturbance and Ecosystems Components of Response. Pp. 159-174. Springer-Verlag, Berlin - Heidelberg - New York 1983.

14324 - **SAVAGE, J.V.:** Modification of a Stevens type-F recorder for simultaneous recording of rainfall and water level. - Trans. ASAE 27: 443-444, 1984.

14325 - **SAVAGE, M.J., CASS, A.:** Measurement errors in field calibration of *in situ* leaf psychrometers. - Crop Sci. 24: 371-372, 1984.

14326 - **SAVAGE, M.J., CASS, A.:** Psychrometric field measurement of water potential changes following leaf excision. - Plant Physiol. 74: 96-98, 1984.

*14327 - **SAVAGE, M.J., CASS, A., De JAGER, J.M.:** Statistical assessment of some errors in thermocouple hygrometric water potential measurement. - Agr. Meteorol. 30: 83-97, 1983.

14328 - **SAVAGE, M.J., CASS, A., WIEBE, H.H.:** Effect of excision on leaf water potential. - J. exp. Bot. 35: 204-208, 1984.

14329 - **SAVAGE, M.J., WIEBE, H.H., CASS, A.:** Effect of cuticular abrasion on thermocouple psychrometric *in situ* measurement of leaf water potential. - J. exp. Bot. 35: 36-42, 1984.

14330 - **SAWADA, S., HAYAKAWA, T.:** Effect of Bordeaux mixture on net photosynthetic rate and stomatal and intracellular resistances in apple leaves. - Photosynthetica 18: 69-73, 1984.

*14331 - **SAWADA, S., IGARASHI, T., MIYACHI, S.:** Effects of phosphate nutrition on photosynthesis, starch and total phosphorus levels in single rooted leaf of dwarf bean. - Photosynthetica 17: 484-490, 1983.

14332 - **SAWADA, S., SUGAI, M.:** Responses of transpiration and CO_2 exchange characteristics to soil moisture stress in four *Plantago* species. - Photosynthetica 18: 34-42, 1984.

*14333 - **SAWADA, S., SUGAI, M., HIMORI, H.:** Water status and physical properties of soil and vegetation at habitats of some *Plantago* species. - Jap. J. Ecol. 33: 149-160, 1983.

14334 - **SCHÄFER, W., GÜNTHER, K.:** Schätzung des Wasserverbrauchs von beregneten Pflanzenbeständen mit Hilfe eines flachen Verdunstungskessels. - Arch. Acker-Pflanzenbau Bodenk. 28: 451-456, 1984.

14335 - **SCHAFFER, B., BUBENHEIM, D.L., BARDEN, J.A.:** Net gas exchange by leaves of intact and excised apple shoots as influenced by vapor pressure gradient. - HortScience 19: 556-557, 1984.

14336 - **SCHAT, H.**: A comparative ecophysiological study on the effects of waterlogging and submergence on dune slack plants: growth, survival and mineral nutrition in sand culture experiments. - Oecologia 62: 279-286, 1984.

14337 - **SCHLEIFF, U., SCHAFFER, G.**: The effect of decreasing soil osmotic and soil matric water potential in the rhizosphere of a loamy and a sandy soil on the water uptake rate of wheat roots. - Z. Acker- Pflanzenbau 153: 373-384, 1984.

14338 - **SCHLESINGER, W.H., JONES, C.S.**: The comparative importance of overland runoff and mean annual rainfall to shrub communities of the Mojave desert. - Bot. Gaz. 145: 116-124, 1984.

14339 - **SCHLOTT, G., MALICKY, G.**: Biomasse und Phosphorgehalt der Makrophyten in der NO-Bucht des Lunzer Untersees (Austria) in Abhängigkeit von nährstoffreichen zuflüssen und vom Sediment. - Arch. Hydrobiol. 101: 265-277, 1984.

14340 - **SCHMIDT, R.E., SNYDER, V.**: Effects of N, temperature, and moisture stress on the growth and physiology of creeping bentgrass and response to chelated iron. - Agron. J. 76: 590-594, 1984.

14341 - **SCHMORL, G., JAUERT, R.**: Effektivität von Stickstoffdüngung und Beregnung zu Ackergras (Welsches Weidelgras) auf verschiedenen Standorten. - Arch. Acker-Pflanzenbau Bodenk. 28: 485-489, 1984.

14342 - **SCHNABL, H., KOTTMEIER, C.**: Determination of malate levels during the swelling of vacuoles isolated from guard-cell protoplasts. - Planta 161: 27-31, 1984.

14343 - **SCHNEIDER, R.W.**: Transient changes in hydraulic resistance caused in corn roots by *Fusarium moniliforme*. - Phytopathology 74: 1230-1233, 1984.

14344 - **SCHOCH, P.G., JACQUES, R., LECHARNY, A., SIBI, M.**: Dependence of the stomatal index on environmental factors during stomatal differentiation in leaves of *Vigna sinensis* L. II. Effect of different light quality. - J. exp. Bot. 35: 1405-1409, 1984.

14345 - **SCHOETTLE, A.W., LEOPOLD, A.C.**: Solute leakage from artificially aged soybean seeds after imbibition. - Crop Sci. 24: 835-838, 1984.

14346 - **SCHOFIELD, N.J.**: A simulation model predicting winter interception losses from reforestation stands in south-west W.A. - Aust. Forest Res. 14: 105-127, 1984.

*14347 - **SCHONBECK, M.W., NORTON, T.A.**: Drought-hardening in the upper-shore seaweeds *Fucus spiralis* and *Pelvetia canaliculata*. - J. Ecol. 67: 687-696, 1979.

14348 - **SCHOPFER, P., PLACHY, C.**: Control of seed germination by abscisic acid. II. Effect on embryo water uptake in *Brassica napus* L. - Plant Physiol. 76: 155-160, 1984.

*14349 - **SCHUBER, M., KLUGE, M.**: *In situ* studies on crassulacean acid metabolism in *Sedum acre* L. and *Sedum mite* Gil. - Oecologia 50: 82-87, 1981.

*14350 - **SCHULZE, E.-D.**: Root-shoot interactions and plant life forms. - Neth. J. agr. Sci. 31: 291-303, 1983.

*14351 - **SCHULZE, E.-D.**: The effect of light, nutrients, age and plant water status on the operational range of CO_2 assimilation and leaf conductance. - In: METZNER, H. (ed.): Photosynthesis and Plant Productivity. Pp. 176-177. Wissenschaftliche Verlagsgesellschaft, Stuttgart 1983.

14352 - **SCHULZE, E.-D., BLOOM, A.J.**: Relationship between mineral nitrogen influx and transpiration in radish and tomato. - Plant Physiol. 76: 827-828, 1984.

14353 - SCHULZE, E.-D., EHLERINGER, J.R.: The effect of nitrogen supply on growth and water-use efficiency of xylem-tapping mistletoes. - Planta 162: 268-275, 1984.

14354 - SCHULZE, E.-D., TURNER, N.C., GLATZEL, G.: Carbon, water and nutrient relations of two mistletoes and their hosts: A hypothesis. - Plant Cell Environ. 7: 293-299, 1984.

14355 - SCHWAB, K.B., HEBER, U.: Thylakoid membrane stability in drought-tolerant and drought-sensitive plants. - Planta 161: 37-45, 1984.

*14356 - SCHWAEGERLE, K.E.: A method for maintaining constant soil moisture availability for potted plants. - Soil Sci. Soc. Amer. J. 47: 608-610, 1983.

14357 - SCHWARTZ, A., ZEIGER, E.: Metabolic energy for stomatal opening. Roles of photophosphorylation and oxidative phosphorylation. - Planta 161: 129-136, 1984.

14358 - SCHWARZ, M., GALE, J.: Growth response to salinity at high levels of carbon dioxide. - J. exp. Bot. 35: 193-196, 1984.

14359 - SEDDIGH, M., JOLLIFF, G.D.: Physiological responses of field-grown soybean leaves to increased reproductive load induced by elevated night temperatures. Crop Sci. 24: 952-957, 1984.

14360 - SEILER, J.R., JOHNSON, J.D.: Growth and acetylene reduction of black alder seedlings in response to water stress. - Can. J. Forest Res. 14: 477-480, 1984.

14361 - SEILER, J.R., JOHNSON, J.D.: Abnormal needle morphology in loblolly pine induced by low humidity. - HortScience 19: 521-522, 1984.

*14362 - SEKHON, N., SINGH, N.T.: Plant water status and salt uptake in wheat, barley and gram at two moisture and salt levels. - Agrochimica 27: 386-396, 1983.

14363 - SEMENIUK, G.: Common leafspot of alfalfa: ascospore germination and disease development in relation to moisture and temperature. - Phytopathol. Z. 110: 281-289, 1984.

14364 - SEMENIUK, G.: Common leafspot of alfalfa: ascospore discharge and plant infection in the field. - Phytopathol. Z. 110: 290-300, 1984.

*14365 - SEMIKHATOVA, O.A., VOZNESENSKIĬ, V.L., FILIPPOVA, L.A.: Pamyati Olega Vyacheslavovicha Zalenskogo (4 IX 1915 - 12 XII 1982). [In memoriam of Oleg Vyacheslavovich Zalenskiĭ (4 September 1915 - 12 December 1982).] - Bot. Zh. 68: 1270-1277, 1983. [In R.]

14366 - SEN, H., JANA, P.K., MAITY, S.P.: Effect of soil-moisture tensions on evapotranspiration and moisture-extraction pattern of tossa jute. - Ind. J. agr. Sci. 54: 482-484, 1984.

14367 - SENARATNA, T., McKERSIE, B.D., STINSON, R.H.: Association between membrane phase properties and dehydration injury in soybean axes. - Plant Physiol. 76: 759-762, 1984.

*14368 - SENNERBY-FORSSE, L., SIRÉN, G., LESTANDER, T.: Results from the first preliminary test with short rotation willow clones. - Tek. Rapp. Proj. Energiskogsodling (ESO) 30: 1-35, 1983.

14369 - SERGEEVA, D.S., SIL'CHENKO, V.M.: Ustoĭchivost' koriandra k nizkim temperaturam. [Resistance of coriander to low temperatures.] - Fiziol. Biokhim. kul't. Rast. 16: 52-55, 1984. [In R, ab: E.]

14370 - **SEROPIAN, C., PLANCHON, C.:** Physiological responses of six bread wheat and durum wheat genotypes to water stress. - Euphytica 33: 757-767, 1984.

14371 - **ŠESTÁK, Z., BENEŠOVÁ, H., ZIMA, J., POSPÍŠILOVÁ, J., KUTÍK, J.:** Effects of age and water potential of leaves on photochemical activities of immobilized chloroplasts. - In: SYBESMA, C. (ed.): Advances in Photosynthesis Research. Volume IV. Pp. 407-410. Martinus Nijhoff / Dr W. Junk Publishers, The Hague - Boston - Lancaster 1984.

*14372 - **SEVERI, A., BARONI FORNASIERO, R.:** Effects of metabolic inhibitors and ionophores on the stomatal apparatus in *Spirodela oligorrhiza* (Kurz.) Hegelm. and in *Lemna minor* L. - Z. Pflanzenphysiol. 113: 39-46, 1983.

14373 - **SEYMOUR, V.A., HSIAO, T.C.:** A soil surface psychrometer for measuring humidity and studying evaporation. - Agr. Forest Meteorol. 32: 61-70, 1984.

14374 - **SHACKEL, K.A.:** Theoretical and experimental errors for *in situ* measurements of plant water potential. - Plant Physiol. 75: 766-772, 1984.

*14375 - **SHACKEL, K.A., FOSTER, K.W., HALL, A.E.:** Genotypic differences in leaf osmotic potential among grain sorghum cultivars grown under irrigation and drought. - Crop Sci. 22: 1121-1125, 1982.

14376 - **SHACKEL, K.A., HALL, A.E.:** Effect of intercropping on the water relations of sorghum and cowpea. - Field Crops Res. 8: 381-387, 1984.

14377 - **SHAINBERG, I., SHALHEVET, J. (ed.):** Soil Salinity under Irrigation. Processes and Management. (Ecological Studies, Volume 51.). Springer-Verlag, Berlin - Heidelberg - New York - Tokyo 1984.

14378 - **SHALHEVET, J.:** Management of irrigation with brackish water. - In: SHAINBERG, I., SHALHEVET, J. (ed.): Soil Salinity under Irrigation. Processes and Management. Pp. 298-318. Springer-Verlag, Berlin - Heidelberg - New York - Tokyo 1984.

*14379 - **SHANNON, M.C.:** Differences in salt tolerance within 'Empire' lettuce. - J. Amer. Soc. hort. Sci. 105; 944-947, 1980.

14380 - **SHANNON, M.C., McCREIGHT, J.D.:** Salt tolerance of lettuce introductions. - HortScience 19: 673-675, 1984.

14381 - **SHARAF, N.S., AL-MUSA, A.M., NAZER, I.K.:** Control of tomato yellow leaf curl virus in Jordan. I. Effects of two irrigation regimes alone or in combination with pirimiphos-methyl on whitefly (*Bemisia tabaci* Genn.) population and the incidence of tomato yellow leaf curl virus. - Z. Pflanzenkrankheiten Pflanzenschutz 91: 404-410, 1984.

*14382 - **SHARENKOVA, Kh.A., SELIVANKINA, S.Yu., ROMANKO, E.G., TVORUS, E.K., LOBANOVA, T.A., GENKEL', P.A., BASKANOV, Yu.A., KULAEVA, O.N.:** Aktivatsiya kartolinom RNK-polimerazy v list'yakh yachmenya pri deTstvii zasukhi. [Activation by cartolin of RNA-polymerase in barley leaves as affected by drought.] - Fiziol. Rast. 30: 1042-1044, 1983. [In R.]

14383 - **SHARKEY, T.D.:** Transpiration-induced changes in the photosynthetic capacity of leaves. - Planta 160: 143-150, 1984.

*14384 - **SHARMA, B.R., CHAUDHARY, T.N.:** Wheat root growth, grain yield and water uptake as influenced by soil water regime and depth of nitrogen placement in a loamy sand soil. - Agr. Water Manage. 6: 365-373, 1983.

14385 - **SHARMA, B.R., CHAUDHARY, T.N.:** Nitrogen uptake by wheat in relation to water regime and depth of N placement in a loamy sand soil: measurement and model simulation. - J. Soil Sci. 35: 141-148, 1984.

14386 - **SHARMA, M.L. (ed.):** Evapotranspiration from Plant Communities. (Developments in Agricultural and Managed-Forest Ecology 13.). - Elsevier Science Publishers, Amsterdam - Oxford - New York - Tokyo 1984.

14387 - **SHARMA, M.L.:** Evapotranspiration from a *Eucalyptus* community. - Agr. Water Manage. 8: 41-56, 1984.

*14388 - **SHARMA, R.A., VERMA, G.P., GUPTA, R.K., KATRE, R.K.:** Moisture depletion pattern and use by unirrigated wheat and safflower grown on vertisol as influenced by cultural practices. - Z. Acker Pflanzenbau 151: 267-274, 1982.

14389 - **SHARMA, S.K., GARG, O.P.:** Nitrate reductase, nitrite reductase activities and leaf diffusion resistance in wheat under water deficit. - Biol. Plant. 26: 151-153, 1984.

14390 - **SHARP, R.E., MATTHEWS, M.A., BOYER, J.S.:** Kok effect and the quantum yield of photosynthesis. Light partially inhibits dark respiration. - Plant Physiol. 75: 95-101, 1984.

14391 - **SHEREMET'EV, S.N.:** Vodnyĭ rezhim *Pistacia vera (Anacardiaceae)* v Badkhyze. [Water relations of *Pistacia vera (Anacardiaceae)* from Badkhyz.] - Bot. Zh. 69: 149-157, 1984. [In R, ab: E.]

14392 - **SHERIFF, D.W.:** Phases in water uptake during leaf rehydration, experiments and a heuristic model. - Ann. Bot. 53: 865-873, 1984.

14393 - **SHERIFF, D.W.:** Epidermal transpiration and stomatal responses to humidity: Some hypotheses explored. - Plant Cell Environ. 7: 669-677, 1984.

14394 - **SHERIFF, D.W., LUDLOW, M.M.:** Physiological reactions to an imposed drought by *Macroptilium atropurpureum* and *Cenchrus ciliaris* in a mixed sward. - Aust. J. Plant Physiol. 11: 23-34, 1984.

14395 - **SHERIFF, D.W., WHITEHEAD, D.:** Photosynthesis and wood structure in *Pinus radiata* D. Don during dehydration and immediately after rewatering. - Plant Cell Environ. 7: 53-62, 1984.

14396 - **SHERRATT, D.J., WHEATER, H.S.:** The use of surface resistance-soil moisture relationships in soil water budget models. - Agr. Forest Meteorol. 31: 143-157, 1984.

*14397 - **SHEVCHENKO, N.N.:** Aerozol'noe uvlazhnenie sel'skokhozyaĭstvennykh kul'tur. [Aerosol irrigation of crops.] - In:SHMAT'KO, I.G. (ed.): Vodnyĭ Rezhim Rasteniĭ v Svyazi s Deĭstviem Faktorov Sredy. Pp. 175-187, 198. Naukova Dumka, Kiev 1983. [In R.]

*14398 - **SHEVCHENKO, N.N.:** Optimizatsiya vodnogo i pitatel'nogo rezhimov torfyanoĭ pochvy. [Optimization of water relations and nutrition in peat soils.] - In: SHMAT'KO, I.G. (ed.): Vodnyĭ Rezhim Rasteniĭ v Svyazi s Deĭstviem Faktorov Sredy. Pp. 141-174, 197-198. Naukova Dumka, Kiev 1983. [In R.]

14399 - **SHIBABE, S., YODA, K.:** Transport of calcium, germanium and rubidium ions in rice seedlings in deuterium oxide. - Radioisotope 33: 606-610, 1984.

14400 - **SHIBABE, S., YODA, K.:** Hydrogen isotope effect on transport of potassium ion in rice seedlings equilibrated with deuterium oxide. - Radioisotopes 33: 675-679, 1984.

14401 - **SHIBABE, S., YODA, K.:** The effect of different concentrations of deuterium oxide on rice seedling shoot and root tissue water. - Environ. exp. Bot. 24: 369-375, 1984.

14402 - **SHIH, S.F.:** Data requirement for evapotranspiration estimation. - J. Irrig. Drain. Eng. 110: 263-274, 1984.

14403 - SHIH, S.F.: Impact of water cutback on sugarcane production in the Florida
Everglades. - Water Resour. Bull. 20: 187-191, 1984.

14404 - SHIH, S.F., RAHI, G.S.: Evapotranspiration of lettuce in relation to water
table depth. - Trans. ASAE 27: 1074-1080, 1984.

14405 - SHIH, S.F., SNYDER, G.H.: Evapotranspiration and water use efficiency of taro.
- Trans. ASAE 27: 1745-1748, 1984.

14406 - SHIRALIPOUR, A., WEST, S.H.: Inhibition of specific protein synthesis in
maize seedlings during water stress. - Soil Crop Sci. Soc. Florida Proc. 43:
102-106, 1984.

14407 - SHIRALIPOUR, A., WEST, S.H.: Uptake and preferential synthesis of proline in
tissue sections of maize seedlings during water stress. - Soil Crop Sci. Soc.
Florida Proc. 43: 107-110, 1984.

14408 - SHITOLE, M.G., JOSHI, G.V.: Effect of sodium chloride on the balance between
C_3 and C_4 carbon fixation pathways and growth. - Photosynthetica 18: 377-384,
1984.

*14409 - SHMAT'KO, I.G. (ed.): Vodnyĭ Rezhim Rasteniĭ v Svyazi s Deĭstviem Faktorov
Sredy. [Plant Water Relations in Dependence on Environmental Factors.] -
Naukova Dumka, Kiev 1983. [In R.]

*14410 - SHMAT'KO, I.G., GRIGORYUK, I.A.: Osobennosti vodnogo rezhima sortov pshe-
nitsy v zasushlivykh usloviyakh. [Peculiarities of water regime in wheat
cultivars under drought conditions.] - In: SHMAT'KO, I.G. (ed.): Vodnyĭ
Rezhim Rasteniĭ v Svyazi s Deĭstviem Faktorov Sredy. Pp. 40-63, 192-193.
Naukova Dumka, Kiev 1983. [In R.]

*14411 - SHMAT'KO, I.G., PETRENKO, N.I.: Vodoobmen organov yarovoĭ pshenitsy pri
nedostatochoĭ vodoobespechennosti i deĭstvii povyshennoĭ temperatury. [Water
relation of spring wheat organs under water stress and high temperature.] -
In: SHMAT'KO, I.G. (ed): Vodnyĭ Rezhim Rasteniĭ v Svyazi s Deĭstviem Faktorov Sredy. Pp. 22-39, 190-192. Naukova Dumka, Kiev 1983. [In R.]

*14412 - SHOMER-ILAN, A., SAMISH, Y.B., KIPNIS, T., ELMER, D., WAISEL, Y.: Effects of
salinity, N-nutrition and humidity on photosynthesis and protein metabolism
of *Chloris gayana* Kunth. - Plant Soil 53: 477-486, 1979.

*14413 - SHONE, M.G.T., GALE, J.: Effect of sodium chloride stress and nitrogen source
on respiration, growth and photosynthesis in lucerne (*Medicago sativa* L.). -
J. exp. Bot. 34: 1117-1125, 1983.

14414 - SHTEFYRTSĖ, A.A., KUSHNIRENKO, M.D.: Prezhdevremennoe opadanie list'ev yablo-
ni gruppy Golden v zavisimosti ot usloviĭ vneshneĭ sredy. [Premature defoliation of the Golden group apple-tree depending on the environmental conditions.] - Fiziol. Biokhim. kul't. Rast. 16: 65-72, 1984. [In R, ab: E.]

*14415 - SHUKLA, J., MINTZ, Y.: Influence of land-surface evapotranspiration on the
earth's climate. - Science 215: 1498-1500, 1982.

*14416 - SHUTTLEWORTH, W.J.: Evaporation. - Inst. Hydrol. (Wallingford), Rep. 56: 1-
61, 1979.

14417 - SHUTTLEWORTH, J.W., GASH, J.H.C., LLOYD, C.R., MOORE, C.J., ROBERTS, J.,
MARQUES FILHO, A. de O., FISCH, G., SILVA FILHO, V. de P., GÔES RIBEIRO, M.
de N., MOLION, L.C.B., De ABREU SÁ, L.D., NOBRE, J.C.A., CABRAL, O.M.R.,
PATEL, S.R., De MORAES, J.C.: Eddy correlation measurements of energy parti-
tion for Amazonian forest. - Quart. J. roy. meteorol. Soc. 110: 1143-1162, 1984
1984.

14418 - **SHVALEVA, A.L., GUSEV, N.A., CHEREZOV, S.N.:** IK-spektry list'ev pshenitsy v svyazi s parametrami ikh vodoobmena pri raznykh usloviyakh vodoobespecheniya rasteniĭ. [IR-spectra of wheat leaf blades in view of parameters of their water exchange under different conditions of water supply in plants.] - Fiziol. Biokhim. kul't. Rast. 16: 471-476, 1984. [In R, ab: E.]

*14419 - **SHVEDOVA, O.E., SHMAT'KO, I.G.:** Parametry vodnogo rezhima pshenitsy v ontogeneze v svyazi s ėkologicheskimi faktorami i sortovymi osobennostyami. [Parameters of water regime during wheat ontogeny in connection with ecological factors and cultivar peculiarities.] - In: SHMAT'KO, I.G. (ed.): Vodnyĭ Rezhim Rasteniĭ v Svyazi s Deĭstviem Faktorov Sredy. Pp. 64-86, 193-194. Naukova Dumka, Kiev 1983. [In R.]

*14420 - **SIEVERDING, E.:** Influence of soil water regimes on VA mycorrhiza. II. Effect of soil temperature and water regime on growth, nutrient uptake, and water utilization of *Eupatorium odoratum* L. - Z. Acker- Pflanzenbau 152: 56-67, 1983.

14421 - **SIEVERDING, E.:** Influence of soil water regimes on VA mycorrhiza. III. Comparison of three mycorrhizal fungi and their influence on transpiration. - Z. Acker- Pflanzenbau 153: 52-61, 1984.

*14422 - **SILCOCK, R.G., WILSON, D.:** The effects of leaf orientation on photosynthesis, transpiration and diffusive conductance of leaves of contrasting *Festuca* species. - New Phytol. 90: 27-36, 1982.

*14423 - **SILCOCK, R.G., WILSON, D.:** Effect of ambient humidity and leaf disposition on leaf diffusive conductance of *Festuca arundinacea* cv. S170. - New Phytol. 90: 201-208, 1982.

14424 - **SILK, W.K.:** Quantitative descriptions of development. - Annu. Rev. Plant Physiol. 35: 479-518, 1984.

14425 - **SILVA, P.R.F., da, FLECK, N.G., HECKLER, J.C.:** Desfolhamento artificial durante a formação do botão floral do girassol. [Artificial defoliation during budding stage in sunflower.] - Pesq. agropec. Bras. 19: 149-156, 1984. [In Port, ab: E.]

14426 - **SILVEIRA, P.M., da, STEINMETZ, S., GUIMARÃES, C.M., AIDAR, H., CARVALHO, J.R. P., de:** Lâminas de água e turnos de rega na cultura do feijoeiro de inverno. [Frequency and levels of irrigation in dry beans during winter season.] - Pesq. agropec. Bras. 19: 219-223, 1984. [In Port, ab: E.]

14427 - **SILVOLA, J., AALTONEN, H.:** Water content and photosynthesis in the peat mosses *Sphagnum fuscum* and *S. angustifolium*. - Ann. bot. Fenn. 21: 1-6, 1984.

14428 - **ŠIMKO, J., KEČKEMÉTHY, A.:** Sledovanie rastu nadzemnej a podzemnej hmoty lucerny siatej v rôznych ekologických podmienkach. [The growth of the aboveground and underground matter of alfalfa under different ecological conditions.] - Rost. Výroba (Praha) 30: 139-144, 1984. [In Slov, ab: R, E.]

14429 - **SIMON, E.W.:** Early events in germination. - In: MURRAY, D.R. (ed.): Seed Physiology. Volume 2. Germination and Reserve Mobilization. Pp. 77-115. Academic Press, Sydney - Orlando - San Diego - Petaluma - New York - London - Toronto - Montreal - Tokyo 1984.

14430 - **SIMON, E.W.:** Respiration and membrane reorganisation during imbibition. - In: PALMER, J.M. (ed.): The Physiology and Biochemistry of Plant Respiration. Pp. 17-31. Cambridge University Press, Cambridge - London - New York - New Rochelle - Melbourne - Sydney 1984.

14431 - ŠIMON, J.: Výnosy a jakost zavlažované cukrovky (Beta vulgaris subsp. altissima Döll. var. saccharifera) na lehkých půdách při různé agrotechnice. [The yields and quality of irrigated sugar-beet (Beta vulgaris subsp. altissima Döll. var. saccharifera) grown with different cultural practices on light-textured soils.] - Rost. Výroba (Praha) 30: 969-976, 1984. [In Czech, ab: R, E.]

14432 - SIMPSON, D.G.: Filmforming antitranspirants: Their effects on root growth capacity, storability, moisture stress avoidance, and field performance of containerized conifer seedlings. - Forest Chron. 60: 335-339, 1984.

*14433 - SINCLAIR, R.: Water relations of tropical epiphytes. I. Relationships between stomatal resistance, relative water content and the components of water potential. - J. exp. Bot. 34: 1664-1675, 1983.

*14434 - SINCLAIR, R.: Water relations of tropical epiphytes. II. Performance during droughting. - J. exp. Bot. 34: 1664-1675, 1983.

14435 - SINCLAIR, R.: Water relations of tropical epiphytes. III. Evidence for Crassulacean acid metabolism. - J. exp. Bot. 35: 1-7, 1984.

14436 - SINCLAIR, T.R., TANNER, C.B., BENNETT, J.M.: Water-use efficiency in crop production. - BioScience 34: 36-40, 1984.

*14437 - SINGH, B., NARAIN, P.: Effect of the salinity of irrigation water on wheat yield and soil properties. - Ind. J. agr. Sci. 50: 422-427, 1980.

14438 - SINGH, B.N., SRIVASTAVA, S.P., HAZARIKA, U.K.: Effect of continuous flow of irrigation water on growth and yield of rice grown at high altitudes. - Ind. J. agr. Sci. 54: 201-203, 1984.

14439 - SINGH, J., BLACKWELL, B.A., MILLER, R.W., BEWLEY, J.D.: Membrane organization of the desiccation-tolerant moss Tortura ruralis in dehydrated states. - Plant Physiol. 75: 1075-1079, 1984.

14440 - SINGH, N.T., AGGARWAL, G.C., BRAR, G.S.: Effect of soil-moisture stress on heat-unit requirement of wheat at maturity. - Ind. J. agr. Sci. 54: 442-444, 1984.

14441 - SINGH, P.N., PRASAD, R., SALIM, M., SHARGA, A.: An improved system of subjecting plants to water stress. - Biol. Plant. 26: 16-21, 1984.

*14442 - SINGH, S., SINGH, N.P., SINGH, M.: Growth, yield and water use by pea as influenced by irrigation and phosphorus application. - Ind. J. Agron. 28: 376-382, 1983.

*14443 - SINGH, U.P., SINGH, R.B.: The effect of soil texture, soil mixture, soil moisture and depth of soil on carpogenic germination of Sclerotinia sclerotiorum. - Z. Pflanzenkrankheiten Pflanzenschutz 90: 662-669, 1983.

14444 - SINGH, Y.V., BHANDARI, R.C.: Effects of irrigation frequency and amount of water on the yield of cauliflower in an arid region. - Ind. J. agr. Sci. 54: 581-586, 1984.

*14445 - SINGHVI, N.R., SHARMA, K.D.: Stomatal response to riboflavin and folic acid. - Acta bot. Ind. 11: 256-257, 1983.

14446 - SINGLETON, P.W., BOHLOOL, B.B.: Effect of salinity on nodule formation by soybean. - Plant Physiol. 74: 72-76, 1984.

14447 - SIONIT, N., ROGERS, H.H., BINGHAM, G.E., STRAIN, B.R.: Photosynthesis and stomatal conductance with CO_2-enrichment of container- and field-grown soybeans. Agron. J. 76: 447-451, 1984.

*14448 - SIPOS, S., NAGY, J.: A lucernatermés növelésének lehetőségei réti agyagtalajon.
[Possibility of increasing lucerne yield on meadow clay soil.] - Növenyter-
melés 30: 425-435, 1981. [In Hung, ab: E.]

*14449 - SIVANAPPAN, R.K., RAMA RAO, G.: Estimation of evapotranspiration and crop
coefficient (K) for high yielding variety of paddy for Coimbatore region. -
Riso 29: 107-110, 1980.

14450 - ŠKOPEK, V.: Method and procedure of ascertaining the water-regime of a geo-
ecological locality. - Ekológia 3: 305-316, 1984.

*14451 - SLUKHAĬ, S.I.: Znachenie urovnya vodoobezpecheniya i azotnogo pitaniya v pro-
duktivnosti i kachestve ozimogo yachmenya. [Importance ot water supply and
nitrogen nutrition in productivity and quality of winter barley.] - In:
SHMAT'KO, I.G. (ed.): Vodnyĭ Rezhim Rasteniĭ v Svyazi s Deĭstviem Faktorov
Sredy. Pp. 122-140, 196-197. Naukova Dumka, Kiev 1983. [In R.]

*14452 - SLUKHAI, S.Ĭ., PETRENKO, N.I.: Vodnyĭ rezhim kukuruzy pri razlichnom urovne
kaliĭnogo pitaniya. [Water regime of maize under different level of potassium
nutrition.] - In: SHMAT'KO, I.G. (ed.): Vodnyĭ Rezhim Rasteniĭ v Svyazi s
Deĭstviem Faktorov Sredy. Pp. 107-121, 196. Naukova Dumka, Kiev 1983. [In
R.]

14453 - SMAJSTRLA, A.G., CLARK, G.A., SHIH, S.F., ZAZUETA, F.S., HARRISON, D.S.:
Characteristics of potential evapotranspiration in Florida. - Soil Crop Sci.
Soc. Florida Proc. 43: 40-46, 1984.

14454 - SMITH, C.J., DELAUNE, R.D.: Effect of sediment moisture on carbon dioxide
exchange in *Spartina alterniflora*. - Plant Soil 79: 291-293, 1984.

*14455 - SMITH, C.M., SATOH, K., FORK, D.C.: Interruption of photosynthetic processes
of the red alga *Porphyra perforata* by salinity. - Carnegie Inst. Washington
Year Book 82: 68-72, 1983.

*14456 - SMITH, E.A., BAILEY, P.H., INGRAM, G.W.: Prediction of the field moisture con-
tent of mature barley and wheat by commonly used drying equations. - J. agr.
Eng. Res. 26: 171-178, 1981.

14457 - SMITH, J.A.C., STEUDLE, E., LÜTTGE, U.: Day-night changes in cell turgor pressure during Crassulacean acid metabolism (CAM) measured with the pressure
probe. - In: CRAM, W.J., JANÁČEK, K., RYBOVÁ, R., SIGLER, K. (ed.): Membrane
Transport in Plants. Pp. 119-120. Academia, Praha 1984.

14458 - SMITH, M.A.L., PALTA, J.P., McCOWN, B.H.: The measurement of isotonicity and
maintenance of osmotic balance in plant protoplast manipulations. - Plant Sci.
Lett. 33: 249-258, 1984.

14459 - SMITH, R.L., SCHANK, S.C., LITTELL, R.C.: The influence of shading on associ-
ative N_2 fixation. - Plant Soil 80: 43-52, 1984.

14460 - SMITH, W.K., YOUNG, D.R., CARTER, G.A., HADLEY, J.L., McNAUGHTON, G.M.: Autumn
stomatal closure in six conifer species of the Central Rocky Mountains. -
Oecologia 63: 237-242, 1984.

14461 - SMIT-SPINKS, B., SWANSON, B.T., MARKHART, A.H., III: Changes in water relations, water flux, and root exudate abscisic acid content with cold acclimation of *Pinus sylvestris* L. - Aust. J. Plant Physiol. 11: 431-441, 1984.

*14462 - SMITTLE, D.A.: Response of lima bean (*Phaseolus lunatus* L.) to irrigation,
nitrogen fertilization, and seed grading. - J. Amer. Soc. hort. Sci. 104:
176-178, 1979.

14463 - SMOLANDER, H., LAPPI, J.: The interactive effect of water stress and temperature on the CO_2 response of photosynthesis in *Salix*. - Silva Fenn. 18: 133-
139, 1984.

14464 - SNAITH, P.J., MANSFIELD, T.A.: Studies of the inhibition of stomatal opening
by naphth-1-ylacetic acid and abscisic acid. - J. exp. Bot. 35: 1410-1418,
1984.

14465 - SOARES, J.M., BERNARDO, S., BRITO, R.A.L., FERREIRA, P.A.: Irrigação por
sulco com e sem utilização da água de escoamento superficial. [Furrow irrigation with and without use of runoff water.] - Pesq. agropec. Bras. 19: 59-
66, 1984. [In Port, ab: E.]

14466 - SOBRADO, M.A., RAWSON, H.M.: Leaf expansion as related to plant water availability in a wild and a cultivated sunflower. - Physiol. Plant. 60: 561-566,
1984.

14467 - SOJKA, R.E., CAMP, C.R., PARSONS, J.E., KARLEN, D.L.: Measurement variability
in soybean water status and soil-nutrient extraction in a row spacing study
in the U.S. Southeastern Coastal Plain. - Commun. Soil Sci. Plant Anal. 15:
1111-1134, 1984.

14468 - SOLÁROVÁ, J., POSPÍŠILOVÁ, J.: The effect of water stress during ontogeny of
primary bean leaves on the light-induced stomatal opening. - Biol. Plant. 26:
56-61, 1984.

14469 - SOLOMON, K.H.: Yield related interpretations of irrigation uniformity and efficiency measures. - Irrig. Sci. 5: 161-172, 1984.

14470 - SOMERVILLE, C.R., SOMERVILLE, S.C.: Les photosynthèses des plantes. - Recher-
che 15: 490-501, 1984.

14471 - SOUZA, J.G., de, BELTRÃO, N.E. de M., SILVA, J.V., da: Supressão da floração
na assimilação crescimento e nutração mineral do algodoeiro. [Casting off
flower buds on assimilation, growth and mineral nutrition of cotton.] -
Pesq. agropec. Bras. 19: 1327-1333, 1984. [In Port, ab: E.]

14472 - SOUZA, J.G., de, GILES, J.A., NETO, M.B., VIEIRA da SILVA,J.: Selection for
water stress tolerance in Upland cotton in the northeast of Brazil. - Trop.
Agr. 61: 2-4, 1984.

14473 - SPENCE, R.D., SHARPE, P.J.H., POWELL, R.D.: The role of the epidermal cells
in the carbon dioxide response of stomata of *Vicia faba* L. - New Phytol. 97:
145-154, 1984.

14474 - SPENCE, R.D., SHARPE, P.J.H., POWELL, R.D., WU, H.: Response of guard cells
to temperature at different concentrations of carbon dioxide in *Vicia faba* L.
- New Phytol. 97: 129-144, 1984.

14475 - SPEPANYAN, Dzh.P.: Vliyanie poliva i udobreniya na urozhaĭnost' kartofelya.
[Effect of irrigation and fertilization on potato productivity.] - In: Uve-
lichenie Urozhaĭnosti Zernovykh, Kormovykh Kul'tur i Kartofelya. Pp. 34-38.
Armyanskiĭ nauchno-issledovatel'skiĭ Institut Zemledeliya, Êchmiadzin 1984.
[In R.]

14476 - SPIERS, A.G., HOPCROFT, D.H.: Influence of leaf age, leaf surface and frequency of stomata on the susceptibility of poplar cultivars to *Marssonina
brunnea*. - Europ. J. Forest Pathol. 14: 270-282, 1984.

*14477 - SPIERTZ, J.H.J.: Ear development, assimilate supply and grain growth of wheat.
- In: PENNING de VRIES, F.W.T., VAN LAAR, H.H.: (ed.): Simulation of Plant
Growth and Crop Production. Pp. 136-144. PUDOC, Wageningen 1982.

*14478 - SPITTLEHOUSE, D.L., BLACK, T.A.: Measuring and modelling forest evapotranspiration. - Can. J. chem. Eng. 59: 173-180, 1981.

14479 - SQUIRE, G.R., GREGORY, P.J., MONTEITH, J.L., RUSSELL, M.B., SINGH, P.: Control of water use by pearl millet (Pennisetum typhoides). - Exp. Agr. 20: 135-149, 1984.

*14480 - SRIVASTAVA, K., RANDHAWA, A.K.: Eco-physiological exploitation of Triticale seeds with pre-sowing treatments to develop hardiness against moisture stress. - Acta Agron. Acad. Sci. Hung. 32: 362-370, 1983.

14481 - STADELMANN, E.J.: The derivation of the cell wall elasticity function from the cell turgor potential. - J. exp. Bot. 35: 859-868, 1984.

14482 - STADELMANN, E.J., LEE-STADELMANN, O.Y., BUSHNELL, W.R.: Cell wall - membrane interactions and passive permeability. - In: CRAM, W.J., JANÁČEK, K., RYBOVÁ, R., SIGLER, K. (ed.): Membrane Transport in Plants. Pp. 90-95. Academia, Praha 1984.

14483 - STAMP, P., THIRAPORN, R., GEISLER, G.: Leaf anatomy of maize lines from different latitudes at sub- and supraoptimal temperatures. - J. exp. Bot. 35: 384-388, 1984.

*14484 - STANKOVA, P.G.: V"zmozhnosti za otsenka na z"rneno-zhitnite kulturi po sukhoustoĭchivost v zavisimost ot vodozad"rzhashchata sposobnost na listata. [Possibility for evaluating the water-stress resistance of grain crops by the water holding capacity of the leaves.] - Fiziol. Rast. (Sofia) 9 (3): 53-58, 1983. [In Bulg, ab: R, E.]

14485 - STANLEY, C.D., HARBAUGH, B.K.: Estimating daily water use for potted chrysanthemum using pan evaporation and plant height. - HortScience 9: 287-288, 1984.

14486 - STAPPER, M.F., LYDA, S.D., JORDAN, W.R.: Temperature x water potential interactions on growth and sclerotial germination of Phymatotrichum omnivorum. - Phytopathology 74: 509-513, 1984.

14487 - STARK, C.: Ertragsvermögen und Kalium-Natrium-Relationen von drei glykophytischen Gräsern unter permanentem Salzstress. - Arch. Acker- Pflanzenbau Bodenk. 28: 735-739, 1984.

*14488 - STARK, J.C., JARRELL, W.M., LETEY, J.: Evaluation of irrigation-nitrogen management practices for celery using continuous-variable irrigation. - Soil Sci. Soc. Amer. J. 47: 95-98, 1983.

14489 - STASSART, J.M., BOGEMANS, J.: The effect of salinity on transport and distribution of ions in barley. - In: CRAM, W.J., JANÁČEK, K., RYBOVÁ, R., SIGLER, K. (ed.): Membrane Transport in Plants. Pp. 510-511. Academia, Praha 1984.

*14490 - STAVAREK, S.J., RAINS, D.W.: Mechanisms for salinity tolerance in plants. - Iowa State J. Res. 57: 457-476, 1983.

14491 - STEPONKUS, P.L.: Role of the plasma membrane in freezing injury and cold acclimation.- Annu. Rev. Plant Physiol. 35: 543-584, 1984.

14492 - STEUDLE, E., TYERMAN, S.D.: Determination of solute permeability of cells with the pressure probe. - In: CRAM, W.J., JANÁČEK, K., RYBOVÁ, R., SIGLER, K. (ed.): Membrane Transport in Plants. Pp. 121-122. Academia, Praha 1984.

14493 - STEUDLE, E., ZIMMERMANN, U.: Water relations of plant cells: further development of the pressure probe and of techniques for measuring pressure-dependent transport. - In: CRAM, W.J., JANÁČEK, K., RYBOVÁ, R., SIGLER, K. (ed.): Membrane Transport in Plants. Pp. 73-82. Academia, Praha 1984.

*14494 - STEWART, A.C., LARKUM, A.W.D.: Photosynthetic electron transport in thylakoid preparations from two marine red algae (*Rhodophyta*). - Biochem. J. 210: 583-589, 1983.

*14495 - STEWART, B.A., MUSICK, J.T.: Conjunctive use of rainfall and irrigation in semiarid regions. - Adv. Irrig. 1: 1-24, 1982.

14496 - STEWART, H.T.L., FLINN, D.W.: Establishment and early growth of trees irrigated with wastewater at four sites in Victoria, Australia. - Forest Ecol. Manage. 8: 243-256, 1984.

14497 - STEWART, J.B.: Measurement and prediction of evaporation from forested and agricultural catchments. - Agr. Water Manage. 8: 1-28, 1984. Also in: SHARMA, M.L. (ed.): Evapotranspiration from Plant Communities. Pp. 1-28. Elsevier Science Publishers, Amsterdam - Oxford - New York - Tokyo 1984.

*14498 - STIGTER, C.J.: On reference crop evaporation in Tanzania. - East Afr. agr. Forest. J. 44: 112-118, 1978.

14499 - STOCK, H.-G.: Untersuchungen zur Ermittlung des optimalen Beregnungsregimes für Körnerfuttererbsen. - Arch. Acker- Pflanzenbau Bodenk. 28: 477-483, 1984.

14500 - STOKER, R., CARTER, K.E.: Effect of irrigation and nitrogen on yield and quality of oilseed rape. - N. Zeal J. exp. Agr. 12: 219-224, 1984.

14501 - STOLZY, L.H., SOJKA, R.E.: Effects of flooding on plant disease. - In: KOZLOWSKI, T.T. (ed.): Flooding and Plant Growth. Pp. 221-264. Academic Press, Orlando - San Diego - San Francisco - New York - London - Toronto - Montreal - Sydney - Tokyo - São Paulo 1984.

14502 - STONE, J.F., WILLIS, W.O. (ed.): Plant Production and Management under Drought Conditions. (Developments in Agricultural and Managed-Forest Ecology 12.) - Elsevier Science Publishers, Amsterdam - Oxford - New York - Tokyo 1984.

14503 - STONE, L.F., LIBARDI, P.L., REICHARDT, K.: Deficiência hídrica, vermiculita e cultivares. I. Efeito na produtividade do arroz. [Water stress, vermiculite and cultivars. I. Effect on rice yield.] - Pesq. agropec. Bras. 19: 695-707, 1984. [In Port, ab: E.]

14504 - STONE, L.F., LIBARDI, P.L., REICHARDT, K.: Deficiência hídrica, vermiculita e cultivares. II. Efeito na utilização do nitrogênio pelo arroz. [Water stress, vermiculite and cultivars. II. Effect on nitrogen utilization by rice. rice.] - Pesq. agropec. Bras. 19: 1403-1416, 1984. [In Port, ab: E.]

*14505 - STOYANOV, Zh.V.: Razseĭvane na energiyata, neto i bruto fotosinteza v razlichni oblasti na spekt'ra. [Dispersion of energy, net and gross photosynthesis in various regions of the spectrum.] - In: Fiziologiya na Rasteniyata. Volume 5. Pp. 677-681. B"lg. Akad. Nauk, Sofiya 1980. [In Bulg, ab: R, E.]

*14506 - STRANG, J.G., LOMBARD, P.B., WESTWOOD, M.N., WEISER, C.J.: Effect of duration and rate of freezing and tissue hydration on 'Barlett' pear buds, flowers, and small fruit. - J. Amer. Soc. hort. Sci. 105: 102-107, 1980.

*14507 - STROOSNIJDER, L.: Simulation of the soil water balance. - In: PENNING de VRIES, F.W.T., VAN LAAR, H.H. (ed.): Simulation of Plant Growth and Crop Production. Pp. 175-193. PUDOC, Wageningen 1982.

14508 - STUART, B.L., HARRISON, S.K., ABERNATHY, J.R., KRIEG, D.R., WENDT, C.W.: The response of cotton (*Gossypium hirsutum*) water relations to smooth pigweed (*Amaranthus hybridus*) competition. - Weed Sci. 32: 126-132, 1984.

14509 - STUART, B.L., ISBELL, V.R., WENDT, C.W., ABERNATHY, J.R.: Modification of cotton water relations and growth with mepiquat chloride. - Agron. J. 76: 651-655, 1984.

14510 - SUBBAIAH, C.C.: A polyethylene glycol based medium for *in vitro* germination of cashew pollen. - Can. J. Bot. 62: 2473-2475, 1984.

14511 - SUEHIRO, K., HOZUMI, K., SHINOZAKI, K.: Growth of three species of *Bidens* under different levels of soil moisture content. - Bot. Mag. (Tokyo) 97: 163-170, 1984.

*14512 - SULLIVAN, C.Y.: Genetic variability in physiological mechanisms of drought resistance. - Iowa State J. Res. 57: 423-439, 1983.

*14513 - SUNDQVIST, C., RYBERG, H.: A long-wavelength absorbing form of protochlorophyll and crystaloids in the inner seed coat of *Luffa cylindrica*. - Physiol. Plant. 59: 42-45, 1983.

14514 - SUTTER, E.G., HUTZELL, M.: Use of humidity tents and antitranspirants in the acclimatization of tissue-cultured plants to the greenhouse. - Scientia Hort. 23: 303-312, 1984.

14515 - SUTTON, J.C., ROWELL, P.M., JAMES, T.D.W.: Effects of leaf wax, wetness duration and temperature on infection of onion leaves by *Botrytis squamosa*. - Phytoprotection 65: 65-68, 1984.

14516 - SVARADZH, K., SHISHCHENKO, S.V., KOZLOVA, G.I., ANDREEVA, I.N., ZHIZNEVSKAYA, G.Ya.: Deǐstvie vodnogo defitsita na simbioticheskuyu azotfiksatsiyu u soi. [Effect of water deficit on symbiotic nitrogen fixation in soybean.] - Fiziol. Rast. 31: 833-840, 1984. [In R, ab: E.]

14517 - SVESHNIKOVA, V.M.: O vodnom rezhime rasteniǐ ostrova Vrangelya. [On the water regime of the plants of the Wrangel Island.] - Bot. Zh. 69: 1167-1172, 1984. [In R, ab: E.]

*14518 - SWANEY, D.P., MISHOE, J.W., JONES, J.W., BOGGESS, W.G.: Using crop models for management: impact of weather characteristics on irrigation decisions in soybeans. - Trans. ASAE 26: 1808-1814, 1983.

14519 - SWEATT, M.R., DAVIES, F.T., Jr.: Mycorrhizae, water relations, growth, and nutrient uptake of geranium grown under moderately high phosphorus regimes. - J. Amer. Soc. hort. Sci. 109: 210-213, 1984.

14520 - SWIETLIK, D., BUNCE, J.A., MILLER, S.S.: Effect of foliar application of mineral nutrients on stomatal aperture and photosynthesis in apple seedlings. - J. Amer. Soc. hort. Sci. 109: 306-312, 1984.

14521 - SYMONS, M., MALKIN, S., KORENSTEIN, R.: External electric effects on photosynthetic membrane vesicles. Kinetic characterization of two electrophotoluminescence phases in hypotonically swollen chloroplasts. - Biochim. biophys. Acta 767: 223-230, 1984.

14522 - SYVERTSEN, J.P.: Light acclimation in citrus leaves. II. CO_2 assimilation and light, water, and nitrogen use efficiency. - J. Amer. Soc. hort. Sci. 109: 812-817, 1984.

14523 - SYVERTSEN, J.P., SMITH, M.L., Jr.: Light acclimation in citrus leaves. I. Changes in physical characteristics, chlorophyll, and nitrogen content. - J. Amer. Soc. hort. Sci. 109: 807-812, 1984.

*14524 - SZLOVÁK, S.: The effect of increasing nitrogen doses upon dry matter production, transpiration and water utilization of maize plants. - Acta bot. Acad. Sci. Hung. 29: 293-306, 1983.

14525 - TAIZ, L.: Plant cell expansion: regulation of cell wall mechanical properties. - Annu. Rev. Plant Physiol. 35: 585-657, 1984.

14526 - **TAJCHMAN, S.J.:** Distribution of the radiative index of dryness and forest site quality in a mountainous watershed. - Can. J. Forest Res. 14: 717-721, 1984.

14527 - **TAKAHASHI, N.:** Seed germination and seedling growth. - In: TSUNODA, S., TAKAHASHI, N. (ed.): Biology of Rice. Pp. 71-88. Japan Scientific Societies Press, Tokyo and Elsevier, Amsterdam - Oxford - New York - Tokyo 1984.

*14528 - **TAL, M., SHANNON, M.C.:** Effects of dehydration and high temperature on the stability of leaf membranes of *Lycopersicon esculentum*, *L. cheesmanii*, *L. peruvianum* and *Solanum pennellii*. - Z. Pflanzenphysiol. 112: 411-416, 1983.

*14529 - **TALANOVA, V.V., TITOV, A.F., DROZDOV, S.N., AKIMOVA, T.V.:** Effect of temperature on the thermoresistance and respiration of tomato leaves (*Lycopersicon esculentum* Mill.). - Biochem. Physiol. Pflanz. 178: 601-605, 1983.

14530 - **TANG, Z.C., KOZLOWSKI, T.T.:** Ethylene production and morphological adaptation of woody plants to flooding. - Can. J. Bot. 62: 1659-1664, 1984.

14531 - **TANG, Z.C., KOZLOWSKI, T.T.:** Water relations, ethylene production, and morphological adaptation of *Fraxinus pensylvanica* seedlings to flooding. - Plant Soil 77: 183-192, 1984.

14532 - **TANG, Z.-C., WANG, Y.-Q., WU, Y.-H., WANG, H.-C.:** [Growth and physiological responses of sorghum seedlings to hypertonic culture solution and their stress-tolerance.] - Acta phytophysiol. Sin. 10: 37-45, 1984. [In Chin, ab: E.]

*14533 - **TANIYAMA, T., MIZUNO, T.:** [Studies on injurious effects of air pollutants on crop plants XVIII Relationship between complex air pollutants of low concentration and growth of rice plant at early growth stage.] - Rep. environ. Sci. Mie Univ. 8: 91-97, 1983. [In Jap, ab: E.]

*14534 - **TANIYAMA, T., MIZUNO, T.:** Studies on injurious effects of air pollutants on crop plants XIX On the estimation of atmospheric ozone concentration by rubber cracking method during long period and its improved method. - Rep. environ. Sci. Mie Univ. 8: 99-105, 1983.

14535 - **TANJI, A., BOULET, C., HAMMOUMI, M.:** Contribution à l'étude de la biologie de *Solanum elaeagnifolium* Cav. (Solanacées), adventice des cultures dans le périmetre du Tadla (Maroc). - Weed Res. 24: 401-409, 1984.

14536 - **TANZARELLA, O.A., DE PACE, C., FILIPPETTI, A.:** Stomatal frequency and size in *Vicia faba* L. - Crop Sci. 24: 1070-1076, 1984.

*14537 - **TARGON, P.G.:** Izmenenie soderzhaniya vody i vodouderzhivayushchei sposobnosti tkanei u drevesnykh introducentov v osenne-zimnii period. [Changes in water content and water retention capacity of leaf tissues of woody introducers in autumn-winter period.] - In: Ekologiya i Fiziologiya Rastenii Vodnykh i Nazemnykh Biotsenozov. Pp. 79-85. Shtiintsa, Kishinev 1983. [In R.]

*14538 - **TAYLOR, A.A., DE-FELICE, J., HAVILL, D.C.:** Nitrogen metabolism in *Poterium sanguisorba* during water stress. - New Phytol. 90: 19-25, 1982.

14539 - **TENHUNEN, J.D., LANGE, O.L., GEBEL, J., BEYSCHLAG, W., WEBER, J.A.:** Changes in photosynthetic capacity, carboxylation efficiency, and CO_2 compensation point associated with midday stomatal closure and midday depression of net CO_2 exchange of leaves of *Quercus suber*. - Planta 162: 193-203, 1984.

*14540 - **TERAMURA, A.H.:** Effects of ultraviolet-B radiation on the growth and yield of crop plants. - Physiol. Plant. 58: 415-427, 1983.

14541 - TERAMURA, A.H., FORSETH, I.N., LYDON, J.: Effects of ultraviolet-B radiation on plants during mild water stress. IV. The insensitivity of soybean internal water relations to ultraviolet-B radiation. - Physiol. Plant. 62: 384-389, 1984.

14542 - TERAMURA, A.H., PERRY, M.C., LYDON, J., McINTOSH, M.S., SUMMERS, E.G.: Effects of ultraviolet-B radiation on plants during mild water stress. III. Effects on photosynthetic recovery and growth in soybean. - Physiol. Plant. 60: 484-492, 1984.

*14543 - TERASAWA, T., ASANO, H., HIROSE, S.: [Ecological studies on environmental adaptation in weeds. 3. The effect of soil moisture on growth and seed productive structure of large crabgrass and common purslane.] - Weed Res. (Japan) 26: 14-18, 1981. [In Jap, ab: E.]

*14544 - TERASAWA, T., HIROSE, S., ASANO, H.: [Ecological studies on environmental adaptation in weeds. 5. Effects of the cultural conditions on germination in common purslane.] - Weed Res. (Japan) 27:.245-250, 1982. [In Jap, ab: E.]

14545 - TERJUNG, W.H., HAYES, J.T., O'ROURKE, P.A., TODHUNTER, P.E.: Yield responses of crops to changes in environment and management practices: model sensitivity analysis. I. Maize. - Int. J. Biometeorol. 28: 261-278, 1984.

14546 - TERJUNG, W.H., HAYES, J.T., O'ROURKE, P.A., TODHUNTER, P.E.: Yield responses of crops to changes in environment and management practices: model sensitivity analysis. II. Rice, wheat, and potato. - Int. J. Biometeorol. 28: 279-292, 1984.

14547 - TERJUNG, W.H., JI, H-Y., HAYES, J.T., O'ROURKE, P.A., TODHUNTER, P.E.: Actual and potential yield for rainfed and irrigated wheat in China. - Agr. Forest Meteorol. 31: 1-23, 1984.

14548 - TERJUNG, W.H., JI, H-Y., HAYES, J.T., O'ROURKE, P.A., TODHUNTER, P.E.: Actual and potential yield for rainfed and irrigated maize in China. - Int. J. Biometeorol. 28: 115-135, 1984.

14549 - TERJUNG, W.H., JI, H-Y., HAYES, J.T., O'ROURKE, P.A., TODHUNTER, P.E.: Crop water requirements for rainfed and irrigated wheat in China and Korea. - Agr. Water Manage. 8: 411-427, 1984.

14550 - TERJUNG, W.H., LIVERMAN, D.M., HAYES, J.T.: Climatic change and water requirements for grain corn in the North American Great Plains. - Climatic Change 6: 193-220, 1984.

*14551 - TERRY, N.: Limiting factors in photosynthesis. IV. Iron stress-mediated changes in light-harvesting and electron transport capacity and its effects on photosynthesis *in vivo*. - Plant Physiol. 71: 855-860, 1983.

14552 - TERRY, N., FARQUHAR, G.D.: Photochemical capacity and photosynthesis. - In: PEARSON, C.J. (ed.): Control of Crop Productivity. Pp. 43-56. Academic Press, Sydney - Orlando - San Diego - Petaluma - New York - London - Toronto - Montreal - Tokyo 1984.

*14553 - TERRY, N., WALDRON, L.J., TAYLOR, S.E.: Environmental influences on leaf expansion. - In: DALE, J.E., MILTHORPE, F.L. (ed.): The Growth and Functioning of Leaves. Pp. 179-205. Cambridge University Press, Cambridge - London - New York - New Rochelle - Melbourne - Sydney 1983.

*14554 - TERSKOV, I.A. (ed.): Voprosy Upravleniya Biosintezom Nizshikh Rastenii. [Problems of Control of Biosynthesis in Lower Plants.] - Nauka, Sibirskoe Otdelenie, Novosibirsk 1982. [In R.]

*14555 - **TESHA, A.J., KUMAR, D.:** Effects of soil moisture, potassium and nitrogen on mineral absorption and growth of *Coffea arabica* L. - Turrialba 29: 213-218, 1979.

14556 - **TESKEY, R.O., GRIER, C.C., HINCKLEY, T.M.:** Change in photosynthesis and water relations with age and season in *Abies amabilis*. - Can. J. Forest Res. 14: 77-84, 1984.

14557 - **TESKEY, R.O., HINCKLEY, T.M., GRIER, C.C.:** Temperature-induced change in the water relations of *Abies amabilis* (Dougl.) Forbes. - Plant Physiol. 74: 77-80, 1984.

*14558 - **THOMAS, A.W., BRUCE, R.R., CURTIS, A.A.:** Prediction of rainfall excess and soil water flux variable storms on layered field soils. - Trans. ASAE 25: 1589-1596, 1982.

14559 - **THOMAS, T.H.:** Changes in endogenous cytokinins of celery (*Apium graveolens* L.) seeds following an osmotic priming or growth regulator soak treatment. - Plant Growth Regulation 2: 135-141, 1984.

14560 - **THOMPSON, J.R., MUELLER, P.W., FLÜCKIGER, W., RUTTER, A.J.:** The effect of dust on photosynthesis and its significance for roadside plants. - Environ. Pollut., Ser. A. 34: 171-190, 1984.

*14561 - **THORPE, N.:** The role of phosphoenolpyruvate carboxylase in the guard cell of *Commelina cyanea*. - Plant Sci. Lett. 30: 331-338, 1983.

14562 - **THORPE, N.:** Stomatal physiology. In: PEARSON, C.J. (ed.): Control of Crop Productivity. Pp. 33-40. Academic Press, Sydney - Orlando - San Diego - Petaluma - New York - London - Toronto - Montreal - Tokyo 1984.

14563 - **THORPE, N., MILTHORPE, F.L.:** Transport of metabolites between the mesophyll and epidermis of *Commelina cyanea* R. Br. - Aust. J. Plant Physiol. 11: 59-68, 1984.

14564 - **TICHÁ, I.:** Quantitative Veränderungen der Kenngrösse Nettophotosyntheserate mit Rücksicht auf Blattontogenese und -insertion. - Wiss. Z. Humboldt-Univ., math.-naturwiss. Reihe 33: 300-302, 1984.

*14565 - **TIETZ, A., DINGKUHN, M., LUDEWIG, M., DÖRFFLING, K.:** Effect of abscisic acid on transport and storage of assimilates. - In: METZNER, H. (ed.): Photosynthesis and Plant Productivity. Pp. 182-185. Wissenschaftliche Verlagsgesellschaft, Stuttgart 1983.

14566 - **TIMMERMANN, B.N., STEELINK, C., LOEWUS, F.A. (ed.):** Phytochemical Adaptations to Stress. (Recent Advances in Phytochemistry, Volume 18. Proceedings of the Twenty-third Annual Meeting of the Phytochemical Society of North America, Tucson, Arizona, July, 1983.) - Plenum Press, New York - London 1984.

*14567 - **TING, I.P., BURK, J.H.:** Aspects of carbon metabolism in *Welwitschia*. - Plant Sci. Lett. 32: 279-285, 1983.

14568 - **TITUS, J.E., WAGNER, D.J.:** Carbon balance for two *Sphagnum* mosses: water balance resolves a physiological paradox. - Ecology 65: 1765-1774, 1984.

*14569 - **TKACHUK, E.S., SHMAT'KO, I.G.:** Vodopotreblenie ozimoĭ pshenitsy v raznykh tsenoticheskikh usloviyakh. [Water consumption of winter wheat in different communities.] - In: SHMAT'KO, I.G. (ed.): Vodnyĭ Rezhim Rasteniĭ v Svyazi s Deĭstviem Faktorov Sredy. Pp. 87-106, 195. Naukova Dumka, Kiev 1983. [In R.]

14570 - **TOLLEY, L.C., STRAIN, B.R.:** Effects of CO_2 enrichment and water stress on growth of *Liquidambar styraciflua* and *Pinus taeda* seedlings. - Can. J. Bot. 62: 2135-2139, 1984.

*14571 - **TOMATI, U., VERI, G., GALLI, E.:** Effect of water status on photosynthesis and nitrate reductase activity in maize plants. - Riv. Agron. 12: 119-122, 1978.

*14572 - **TOMBESI, L.:** Elementi di Climatologia Agraria e Valutazione della Produttività Ambientale. [Elements of Agricultural Climatology and Assesment of Environmental Production.] - Istituto Sperimentale per la Nutrizione delle Piante, Roma 1982. [In Ital.]

14573 - **TOMMERUP, I.C.:** Effect of soil water potential on spore germination by vesicular-arbuscular mycorrhizal fungi. - Trans. Brit. mycol. Soc. 83: 193-202, 1984.

14574 - **TOMOS, A.D., LEIGH, R.A., SHAW, C.A., WYN JONES, R.G.:** A comparison of methods for measuring turgor pressures and osmotic pressures of cells of red beet storage tissue. - J. exp. Bot. 35: 1675-1683, 1984.

14575 - **TOMPSETT, P.B.:** Desiccation studies in relation to the storage of *Araucaria* seed. - Ann. appl. Biol. 105: 581-586, 1984.

14576 - **TOMPSETT, P.B.:** The effect of moisture content and temperature on the seed storage life of *Araucaria columnaris*. - Seed Sci. Technol. 12: 801-816, 1984.

14577 - **TOOMING, Kh.G., TAMMETS, T.Kh.:** Svyaz' udel'noĭ poverkhnostnoĭ plotnosti list'ev nekotorykh vidov rasteniĭ s radiatsieĭ prisposobleniya i rezhimom FAR. [Relationship of specific leaf weight to irradiation density of adaptation and PhAR regime in some plant species.] - Fiziol. Rast. 31: 258-265, 1984. [In R, ab: E.]

14578 - **TORELLO, W.A., SYMINGTON, A.G.:** Screening of turfgrass species and cultivars for NaCl tolerance. - Plant Soil 82: 155-161, 1984.

14579 - **TORIYAMA, K., INOUE, K.:** [The effect of micrometeorological elements on sterility due to cool injury in rice plants: An application of simulation model for the prediction of canopy climate.] - Jap. J. Crop Sci. 53: 387-395, 1984. [In Jap, ab: E.]

*14580 - **TORRES, A., DIAZ, B.:** Influencia de la humedad del suelo y la fertilización sobre el regimen hídrico de las posturas de citrus. Déficit hídrico y contenido de agua en las hojas. [Influence of soil moisture and fertilization upon water regime of citrus seedlings. Water deficit and content in leaves.] - Cultivos trop. (La Habana) 5: 551-563, 1983. [In Span, ab: E.]

14581 - **TRAN, V.N., CAVANAGH, A.K.:** Structural aspects of dormancy. - In: MURRAY, D. R. (ed.): Seed Physiology. Volume 2. Germination and Reserve Mobilization. Pp. 1-44. Academic Press, Sydney - Orlando - San Diego - Petaluma - New York - London - Toronto - Montreal - Tokyo 1984.

14582 - **TRAPANI, N., GENTINETTA, E.:** Screening of maize genotypes using drought tolerance tests. - Maydica 29:.89-100, 1984.

14583 - **TRAPANI, N., MOTTO, M.:** Combining ability for drought tolerance tests in maize populations. - Maydica 29: 325-334, 1984.

14584 - **TREICHEL, S., BRINCKMANN, E., SCHEITLER, B., WILLERT, D.J., von:** Occurrence and changes of proline content in plants in the southern Namib Desert in relations to increasing and decreasing drought. - Planta 162: 236-242, 1984.

14585 - **TRESHOW, M.:** Epilogue: a biochemical overview. - In: KOZIOŁ, M.J., WHATLEY, F.R. (ed.):.Gaseous Air Pollutants and Plant Metabolism. Pp. 425-437. Butterworths, London - Boston - Durban - Singapore - Sydney - Toronto - Wellington 1984.

*14586 - **TRIPPI, V., THIMANN, K.V.:** The exudation of solutes during senescence of oat leaves. - Physiol. Plant. 58: 21-28, 1983.

14587 - **TROMP, J.**:Diurnal fruit shrinkage in apple as affected by leaf water potential and vapour pressure deficit of the air. - Scientia Hort. 22: 81-87, 1984.

*14588 - **TSEL'NIKER, Yu.A.**: Funktsional'naya i strukturnaya organizatsiya fotosinteti- cheskogo apparata lesnykh rasteniĭ. [Functional and structural organization of the photosynthetic apparatus of woody plants.] - In: Ékologo-fiziologi- cheskie Issledovaniya Fotosinteza i Vodnogo Rezhima Rasteniĭ v Polevykh Uslo- viyakh. Pp. 5-15, 160. Sibirskoe Otdelenie, Akademiya Nauk SSSR, Irkutsk 1983. [In R.]

14589 - **TSENOV, A., PETROVA, D.**: Metodi za otsenka na selektsionnite materiali ot zimnite zhitni i z"rneno-bobovite kulturi k"m stresovi v"zdejstviya. [Methods of evaluating stress effects on winter cereals and grain-leguminous crops.] - Rasteniev. Nauki 21 (6): 77-86, 1984. [In Bulg, ab: R, E.]

*14590 - **TSENOVA, M.P.**: Fotosintetichno fosforilirane pri izolirani khloroplasti v za- visimost ot kontsentratsiyata na bikarbonatniya anion. [Photosynthetic phos- phorylation of isolated chloroplasts as a function of bicarbonate concentra- tion.] - Fiziol. Rast. (Sofiya) 4 (3): 12-17, 1978. [In Bulg, ab: E, R.]

14591 - **TSUKAHARA, H., KOZLOWSKI, T.T.**: Effect of flooding on *Larix leptolepis* seed- lings. - J. Jap. Forest Soc. 66: 333-336, 1984.

14592 - **TSUNODA, S.**: Adjustment of photosynthetic structures in three steps of rice evolution. - In: TSUNODA, S., TAKAHASHI, N. (ed.): Biology of Rice. Pp. 89- 115. Japan Scientific Societies Press, Tokyo and Elsevier, Amsterdam - Oxford - New York - Tokyo 1984.

14593 - **TSUNODA, S., TAKAHASHI, N. (ed.)**: Biology of Rice. (Developments in Crop Science Volume 7.). - Japan Scientific Societies Press, Tokyo and Elsevier, Amsterdam - Oxford - New York - Tokyo 1984.

14594 - **TU, J.C., McDONNELL, M.M.**: Effect of rhizobial inoculation and nodulation on yield and winter survival of some alfalfa cultivars. - Can. J. Plant Sci. 64: 151-159, 1984.

14595 - **TUBA, Z.**: Changes of carotenoids in various drought adapted species during subsequent dry and wet periods. - Acta bot. Hung. 30: 217-228, 1984.

14596 - **TUBA, Z.**: Rearrangement of photosynthetic pigment composition in C_4, C_3 and CAM species during drought and recovery. - J. Plant Physiol. 115: 331-338, 1984.

14597 - **TUBA, Z.**: Adaptations of the photosynthetic pigment system to ecological conditions with respect to water in different terrestrial plants. - In: MARGARIS, N.S., ARIANOUSTOU-FARAGGITAKI, M., OECHEL, W.C.: Being Alive on Land. Pp. 85-90. Dr W. Junk Publishers, The Hague - Boston - Lancaster 1984.

14598 - **TUCHMAN, M.L., STOERMER, E.F., CARNEY, H.J.**: Effect of increased salinity on the diatom assemblage of Fonds Lake, Michigan. - Hydrobiologia 109: 179-188, 1984.

14599 - **TURNER, A.K., CHANMEESRI, N.**: Shallow flow of water through non-submerged vegetation. - Agr. Water Manage. 8: 375-385, 1984.

14600 - **TURNER, N.C., SCHULZE, E.-D., GOLLAN, T.**: The responses of stomata and leaf gas exchange to vapour pressure deficits and soil water content. I. Species comparisons at high soil water contents. - Oecologia 63: 338-342, 1984.

14601 - **TURNER, N.C., SINGH, D.P.**: Responses of adaxial and abaxial stomata to light and water deficits in sunflower and sorghum. - New Phytol. 96: 187-195, 1984.

14602 - TURNER, N.C., SPURWAY, R.A., SCHULZE, E.-D.: Comparison of water potentials measured by *in situ* psychrometry and pressure chamber in morphologically different species. - Plant Physiol. 74: 316-319, 1984.

14603 - TYERMAN, S.D., HATCHER, A.I., WEST, R.J., LARKUM, A.W.D.: *Posidonia australis* growing in altered salinities: leaf growth, regulation of turgor and the development of osmotic gradients. - Aust. J. Plant Physiol. 11: 35-47, 1984.

14604 - TYERMAN, S.D., STEUDLE, E.: Determination of solute permeability in *Chara* internodes by a turgor minimum method. Effects of external pH. - Plant Physiol. 74: 464-468, 1984.

*14605 - TYMMS, M.J., GAFF, D.F., HALLAM, N.D.: Protein synthesis in the desiccation tolerant angiosperm *Xerophyta villosa* during dehydration. - J. exp. Bot. 33: 332-343, 1982.

*14606 - TYREE, M.T.: Maple sap uptake, exudation, and pressure changes correlated with freezing exotherms and thawing endotherms. - Plant Physiol. 73: 277-285, 1983.

14607 - TYREE, M.T., DIXON, M.A., THOMPSON, R.G.: Ultrasonic acoustic emissions from the sapwood of *Thuja occidentalis* measured inside a pressure bomb. - Plant Physiol. 1046-1049, 1984.

*14608 - UDEH, C.N., BUSCH, J.R.: Optimal irrigation management using probabilistic hydrologic and irrigation efficiency parameters. - Trans. ASAE 25: 954-960, 1982.

14609 - ULRICH, B.: Waldsterben durch saure Niederschläge. - Umschau 1984 (11): 348-355, 1984.

14610 - UNSWORTH, M.H.: Evaporation from forests in cloud enhances the effects of acid deposition. - Nature 312: 262-263, 1984.

14611 - UNSWORTH, M.H., HEAGLE, A.S., HECK, W.W.: Gas exchange in open-top field chambers-I. Measurement and analysis of atmospheric resistances to gas exchange. - Atmos. Environ. 18: 373-380, 1984.

14612 - UNSWORTH, M.H., HEAGLE, A.S., HECK, W.W.: Gas exchange in open-top field chambers-II. Resistances to ozone uptake by soybeans. - Atmos. Environ. 18: 381-385, 1984.

*14613 - UPADHYAYA, H.D., SMITHSON, J.B., HAWARE, M.P., KUMAR, J.: Resistance to wilt in chickpea. II. Further evidence for two genes for resistance to race 1. - Euphytica 32: 749-755, 1983.

14614 - UPCHURCH, G.R., Jr.: Cuticle evolution in early Cretaceous angiosperms from the Potomac Group of Virginia and Maryland. - Ann. Missouri bot. Garden 71: 522-550, 1984.

14615 - USUDA, H., EDWARDS, G.E.: Is photosynthesis during the induction period in maize limited by the availability of intercellular carbon dioxide? - Plant Sci. Lett. 37: 41-45, 1984.

14616 - VÁCLAVÍK, J.: Photosynthetic CO_2 uptake by *Zea mays* leaves as influenced by unilateral irradiation of adaxial and abaxial leaf surfaces. - Biol. Plant. 26: 206-214, 1984.

14617 - VALANNE, N.: Photosynthesis and photosynthetic products in mosses. - In: DYER, A.F., DUCKETT, J.G. (ed.): Experimental Biology of Bryophytes. Pp. 257-273. Academic Press, London 1984.

14618 - **VAN ANDEL, J., NELISSEN, H.J.M., WATTEL, E., VAN VALEN, T.A., WASSENAAR, A.T.:** Theil's inequality index applied to quantify population variation of plants with regard to dry matter allocation. - Acta bot. Neerl. 33: 161-175, 1984.

14619 - **VAN BAVEL, C.H.M., LASCANO, R.J., STROOSNIJDER, L.:** Test and analysis of a model of water use by sorghum. - Soil Sci. 137: 443-456, 1984.

14620 - **Van der PLAS, L.H.W., WAGNER, M.J.:** Influence of osmotic stress on the respiration of potato tuber callus. - Physiol. Plant. 62: 398-403, 1984.

14621 - **VAN GENUCHTEN, M.T., HOFFMAN, G.J.:** Analysis of crop salt tolerance data. - In: SHAINBERG, I., SHALHEVET, J. (ed.): Soil Salinity under Irrigation. Processes and Management. Pp. 258-271. Springer-Verlag, Berlin - Heidelberg - New York - Tokyo 1984.

*14622 - **VAN KEULEN, H.:** Crop production under semi-arid conditions, as determined by moisture availability. - In: PENNING de VRIES, F.W.T., VAN LAAR, H.H. (ed.): Simulation of Plant Growth and Crop Production. Pp. 159-174. PUDOC, Wageningen 1982.

*14623 - **VAN KEULEN, H.:** Crop production under semi-arid conditions, as determined by nitrogen and moisture availability. - In: PENNING de VRIES, F.W.T., VAN LAAR, H.H. (ed.): Simulation of Plant Growth and Crop Production. Pp. 234-249. PUDOC, Wageningen 1982.

14624 - **VAN KEULEN, H., De WIT, C.T.:** To what extent can agricultural production be expanded? - Options IAMZ-84/1: 55-76, 1984.

14625 - **VAN NOORDWIJK, M., DE WILLIGEN, P.:** Mathematical models on diffusion of oxygen to and within plant roots, with special emphasis on effects of soil-root contact. II. Applications. - Plant Soil 77: 233-241, 1984.

14626 - **VAN STEVENING, R.F.M.:** Regulatory role of abscisic acid in nutrient transport and osmoregulation. - In: CRAM, W.J., JANÁČEK, K., RYBOVÁ, R., SIGLER, K. (ed.): Membrane Transport in Plants. Pp. 453-458. Academia, Praha 1984.

14627 - **VAN VOLKENBURGH, E., CLELAND, R.E.:** Control of leaf growth by changes in cell wall properties. - What's New Plant Physiol. 15: 25-28, 1984.

14628 - **VAPAAVUORI, E.M., KORPILAHTI, E., NURMI, A.H.:** Photosynthetic rate in willow leaves during water stress and changes in the chloroplast ultrastructure, with special reference to crystal inclusions. - J. exp. Bot. 35: 306-321, 1984.

*14629 - **VARADAN, K.M., RAMAMOORTHY, B.:** Nutritional criterion in wheat as a factor of their drought resistance. - Ind. J. Agron. 28: 185-186, 1983.

14630 - **VARTANIAN, N.:** Un modèle de processus adaptatif à la sécheresse: aspects phylétiques, génétiques et physiologiques. - Bull. Soc. Bot. France, Actual. bot. 131: 59-67, 1984.

14631 - **VARTANIAN, N., LEMÉE, G.:** La notion d'adaptation à la sécheresse. - Bull. Soc Soc. Bot. France, Actual. bot. 131: 7-15, 1984.

14632 - **VARTANIAN, N., LESQUOY, E.:** Modélisation de la rhizogénèse de sécheresse. - Acta oecol. - Oecol. gen. 5: 3-19, 1984.

14633 - **VASANDER, H.:** Effect of forest amelioration on diversity in an ombrotrophic bog. - Ann. bot. Fenn. 21: 7-15, 1984.

14634 - **VASIČ, G.:** Uticaj navodnjavanja na vodni režim černozema Zemunskog polja i prinos kukuruza. [Effect of irrigation on chernozem water regime in Zemun polje and maize yield.] - Archiv poljopr. Nauke 45: 65-95, 1984. [In Croat, ab: E.]

14635 - VAVASSEUR, A., GARREC, J.-P., LAFFRAY, D.: Electronic microprobe study of
        variations in potassium concentration of stomatal complex of two *Commelina-
        ceae*. - Physiol. Vég. 22: 841-849, 1984.

14636 - VEERANJANEYULU, K., DAS, V.S.R.: Purple pigmentation in leaves of some tropi-
        cal weed species. - Biol. Plant 26: 215-220, 1984.

*14637 - VENEZIAN SCARASCIA, M.E., LOSAVIO, N.: Riposta morfo-fisiologica di una
        coltura di mais alla carenza idrica in coincidenza di alcuni stadi fenologici.
        [Morpho-physiological response of corn plants to soil drought conditions
        during some phenological stages.] - Riv. Agron. 14: 194-209, 1980. [In
        Ital, ab: E.]

*14638 - VENKATACHARI, A., REDDY, K.A.: Relationship of evapotranspiration with pan
        evaporation and evaluation of crop coefficient. - Acta agron. Acad. Sci.
        Hung. 27: 107-110, 1978:

*14639 - VENKATACHARI, A., REDDY, K.A., RAO, R.S., REDDY, S.S., RAO, D.M.: Consumptive
        use and economics of different crops of winter season in multiple cropping
        system. - In: Proceedings from Second World Congress., International Water
        Resources Association. Volume I. Pp. 301-307. New Delhi 1975.

14640 - VENKATARAMANA, S., SHUNMUGASUNDARAM, S., NAIDU, K.M.: Growth behaviour of
        field grown sugarcane varieties in relation to environmental parameters and
        soil moisture stress. - Agr. Forest Meteorol. 31: 251-260, 1984.

*14641 - VENKATESWARA RAO, K., MADHAVA RAO, K.V.: Influence of potassium nutrition on
        stomatal behaviour, transpiration rate and leaf water potential of pigeon pea
        (*Cajanus cajan* (L.) Mill sp.) in sand culture. - Proc. Ind. Acad. Sci., Plant
        Sci. 92: 323-330, 1983.

14642 - VENKATESWARLU, M.S., NAIDU, K.J., REDDI, G.H.S.: Production potential of
        'Kalyani' fingermillet under different irrigation frequencies. - Ind. J. agr.
        Sci. 54: 51-54, 1984.

14643 - VERESOGLOU, D.S., FITTER, A.H.: Spatial and temporal patterns of growth and
        nutrient uptake of five co-existing grasses. - J. Ecol. 72: 259-272, 1984.

*14644 - VERMA, S.K., RAY, N., KHADDAR, V.K.: Effect of soil alkalinity on growth and
        yield of dwarf rice in stagnated and drained soil moisture conditions. -
        Curr. Agr. 7: 45-50, 1983.

14645 - VERTUCCI, C.W., LEOPOLD, A.C.: Bound water in soybean seed and its relation
        to respiration and imbibitional damage. - Plant Physiol. 75: 114-117, 1984.

14646 - VESELKOV, B.M., TIKHOV, P.V.: Svyaz' transporta vody po ksileme s intensiv-
        nost'yu transpiratsii u sosny obyknovennoĭ. [Relationship between water
        transport in the xylem and transpiration rate in pine.] - Fiziol. Rast. 31:
        1099-1106, 1984. [In R, ab: E.]

*14647 - VETÖ, F.: On the role of "vicinal" water in membrane permeability. - Acta
        Biochim. Biophys. Acad. Sci. Hung. 18: 23, 1983.

*14648 - VEVERKA, K.: Vliv půdní vlhkosti na vzcházivost obalovaného osiva cukrovky.
        [The effect of soil moisture content on the emergence rate of pelleted sugar-
        -beet seeds.] - Rost. Výroba (Praha) 29: 937-942, 1983. [In Czech, ab: R,
        E.]

14649 - VEVERKA, K.: Vliv vysoké vlhkosti během fáze předcházející klíčení na násled-
        ný průběh klíčení osiva cukrovky (*Beta vulgaris* subsp. *altissima* Döll. var.
        saccharifera). [The effect of high moisture in the pre-germination stage on
        the subsequent germination of the seeds of sugar-beet (*Beta vulgaris* subsp.
        altissima Döll. var. *saccharifera*).] - Rost. Výroba (Praha) 30: 953-960,
        1984. [In Czech, ab: R, E.]

*14650 - **VICHERKOVÁ, M., PÁTKOVÁ, J.**: Adsorption of vaporized terpenes on soil affecting rye grain germination rate and seedling growth. - Scripta Fac. Sci. nat. Univ. Purkynianae Brunensis (Brno) 12: 382-392, 1982.

14651 - **VIDAL, A., POGNONEC, J.-C.**: Effet de l'alimentation en eau sur quelques caractères morphologiques et anatomiques des feuilles de soja (*Glycine max* (L.) Merrill). - Agronomie 4: 967-975, 1984.

14652 - **VIEIRA da SILVA, J.**: Applications des études fondamentales à l'amélioration de plantes résistantes à la sécheresse. - Bull. Soc. Bot. France, Actual. bot. 131: 51-57, 1984.

14653 - **VILLALOBOS-RODRIGUEZ, E., SHIBLES, R., GREEN, D.E.**: Response of stem termination types of soybean to supplemental irrigation. - Iowa State J. Res. 59: 45-52, 1984.

14654 - **VLAKHOVA, M., V"LEV, V., ANDONOVA, P., K"DREV, T.**: Efektivno izpolzuvane na polivnite ploshchi chrez optimizirane na toreneto pri otglezhdane na tsarevitsa za silazh p"rva kultura. [Effective exploitation of irrigated areas by optimizing fertilizer application to silage maize grown as first crop.] - Rasteniev. Nauki 21 (5): 133-139, 1984. [In Bulg, ab: R, E.]

14655 - **VODYANIK, A.S., VODYANIK, T.M.**: O metabolizme azota v nadzemnykh organakh gorokha pri neblagopriyatnykh usloviyakh uvlazhneniya. [On nitrogen metabolism in pea shoots under unfavourable moisture conditions.] - Sel'skokhoz. Biol. 1984 (10): 12-15, 1984. [In R, ab: E.]

14656 - **VODYANIK, A.S., VODYANIK, T.M.**: Rost i fiziologicheskaya aktivnost' kornevoĭ sistemy gorokha v usloviyakh razlichnogo rezhima uvlazhneniya. [Growth and physiological activity of pea root system under different soil moisture.] - Sel'skokhoz. Biol. 1984 (1): 33-35, 1984. [In R, ab: E.]

14657 - **VOETBERG, G., STEWART, C.R.**: Steady state proline levels in salt-shocked barley leaves. - Plant Physiol. 76: 567-570, 1984.

14658 - **VOGEL, S.**: The lateral thermal conductivity of leaves. - Can. J. Bot. 62: 741-744, 1984.

*14659 - **VOGELLEHNER, D.**: Phylogenie des sekundären Wasserleitsystems. - Ber. Deutsch. bot. Ges. 96: 365-374, 1983.

14660 - **Von BERNUTH, R.D., MARTIN, D.L., GILLEY, J.R., WATTS, D.G.**: Irrigation system capacities for corn production in Nebraska. - Trans. ASAE 27: 419-424, 428, 1984.

14661 - **VOSKRESENSKAYA, N.P., KUMAKOV, A.V., DROZDOVA, I.S.**: Fotosinteticheskiĭ gazoobmen CO_2 lista yachmenya i ego vozrastnye izmeneniya u rasteniĭ, vyrashchivaemykh na svetu razlichnogo spektral'nogo sostava. [Photosynthetic CO_2 exchange in barley leaf as related to the age of plants growing under various spectral composition of light.] - Fiziol. Rast. 31: 233-240, 1984. [In R, ab: E.]

14662 - **VOZNESENSKIĬ, V.L.**: Izmenenie vodnogo defitsita i intensivnost' fotosinteza. [Variations in water deficit and the rate of photosynthesis.] - Fiziol. Rast. 31: 598-601, 1984. [In R.]

*14663 - **VOZNESENSKIĬ, V.L., LEDYAĬKINA, N.A., YUDINA, O.S., NABIEV, M.M.**: Osobennosti gazoobmena rasteniĭ zharkoĭ pustyni Karakum. [Peculiarities of the gas exchange of plants from the hot desert Karakum.] - In: Ékologo-fiziologicheskie issledovaniya Fotosinteza i Vodnogo Rezhima Rasteniĭ v Polevykh Usloviyakh. Pp. 51-56, 162. Sibirskoe Otdelenie, Akademiya Nauk SSSR, Irkutsk 1983. [In R.]

*14664 - **V"RBANOV, M.**: Vliyanie na toreneto i vlazhnostta na pochvata v"rkhu rastezha, formata na listata i produktivnostta na rasteniyata pri izpolzuvane na metoda na lineĩnata regresiya za opredelyane na listna ploshch na 2x, 3x i 4x formi zakharno tsveklo. [The effect of fertilizer application and soil moisture on growth, leaf shape and plant productivity by applying the method of linear regression in determining the leaf area of 2x, 3x and 4x sugar beet forms.] - Rasteniev. Nauki 20 (8): 42-48, 1983. [In Bulg, ab: R, E.]

14665 - **WABER, J.**: Effect of heavy water (99 + %D_2O) on beet root permeability. - Environ. exp. Bot. 24: 253-257, 1984.

14666 - **WAGENET, R.J.**: Salt and water movement in the soil profile. - In: SHAINBERG, I., SHALHEVET, J. (ed.): Soil Salinity under Irrigation. Processes and Management. Pp. 100-114. Springer-Verlag, Berlin - Heidelberg - New York - Tokyo 1984.

14667 - **WAGNER, D.J., TITUS, J.E.**: Comparative desiccation tolerance of two *Sphagnum* mosses. - Oecologia 62: 182-187, 1984.

*14668 - **WAGNER, G.**: Higher plant vacuoles and tonoplasts. - In: HALL, J.L., MOORE, A. L. (ed.): Isolation of Membranes and Organelles from Plant Cells. Pp. 83-118. Academic Press, London - New York - Paris - San Diego - San Francisco - Saõ Paulo - Sydney - Tokyo - Toronto 1983.

*14669 - **WAHAB, K., IRUTHAYARAJ, M.R.**: Effect of irrigation and systems of planting on growth, yield and water use of *Sorghum* varieties. - Sorghum News Lett. 22: 34-36, 1979.

*14670 - **WAHAB, K., IRUTHAYARAJ, M.R.**: Nutrient uptake in *Sorghum* as influenced by irrigation regimes and systems of planting. - Sorghum News Lett. 29: 134-135, 1979.

*14671 - **WAHAB, K., SINGH, K.N.**: Effect of irrigation applied at different critical growth stages on growth characters and yield of hulled and hull-less barley. - Ind. J. Agron. 28: 412-417, 1983.

*14672 - **WAHAB, K., SINGH, K.N.**: Effect of irrigation on nutritional uptake and protein content in two types of barley. - Ind. J. Agron. 28: 418-424, 1983.

14673 - **WAHAB, K., SINGH, K.N.**: Effect of irrigation on the yield of two types of barley. - Z. Acker- Pflanzenbau 153: 175-180, 1984.

14674 - **WAINWRIGHT, S.J.**: Adaptation of plants to flooding with salt water. - In: KOZLOWSKI, T.T. (ed.): Flooding and Plant Growth. Pp. 295-343. Academic Press, Orlando - San Diego - San Francisco - New York - London - Toronto - Montreal - Sydney - Tokyo - São Paulo 1984.

14675 - **WALKER, K.A., KEYS, A.J., GIVAN, C.V.**: Effect of L-methionine sulphoximide on the products of photosynthesis in wheat (*Triticum aestivum*) leaves. - J. exp. Bot. 35: 1800-1810, 1984.

*14676 - **WALKER, P.N., THORNE, M.D., BENHAM, E.C., SIPP, S.K.**: Yield response of corn and soybeans to irrigation and drainage on a claypan soil. - Trans. ASAE 25: 1617-1621, 1982.

14677 - **WALKER, S., OOSTERHUIS, D.M., WIEBE, H.H.**: Ratio of cut surface area to leaf sample volume for water potential measurements by thermocouple psychrometers. - Plant Physiol. 75: 228-230, 1984.

·14678 - **WALLACE, J.S., LLOYD, C.R., ROBERTS, J., SHUTTLEWORTH, W.J.**: A comparison of methods for estimating aerodynamic resistance of heather (*Calluna vulgaris* (L.) Hull) in the field. - Agr. Forest Meteorol. 32: 289-305, 1984.

14679 - **WALLACE, L.L., McNAUGHTON, S.J., COUGHENOUR, M.B.:** Compensatory photosynthetic responses of three African graminoids to different fertilization, watering, and clipping regimes. - Bot. Gaz. 145: 151-156, 1984.

14680 - **WALLENDER, W.W., KELLER, J.:** Foliar fluoride accumulation under sprinkle irrigation. - Trans. ASAE 27: 449-455, 1984.

14681 - **WALTER, H., BRECKLE, S.-W.:** Spezielle Ökologie der Tropischen und Subtropischen Zonen. (Ökologie der Erde, Band 2.) - Gustav Fischer Verlag, Stuttgart 1984.

14682 - **WALTERS, G.A., BARTHOLOMEW, D.P.:** *Acacia koa* leaves and phyllodes: gas exchange, morphological, anatomical, and biochemical characteristics. - Bot. Gaz. 145: 351-357, 1984.

*14683 - **WAMPLE, R.L., CULVER, E.B.:** The influence of paclobutrozol, a new growth regulator, on sunflowers. - J. Amer. Soc. hort. Sci. 108: 122-125, 1983.

14684 - **WAMPLE, R.L., THORNTON, R.K.:** Differences in the response of sunflower (*Helianthus annuus*) subjected to flooding and drought stress. - Physiol. Plant. 61: 611-616, 1984.

14685 - **WANG, T.L., DONKIN, M.E., MARTIN, E.S.:** The physiology of a wilty pea: abscisic acid production under water stress. - J. exp. Bot. 35: 1222-1232, 1984.

14686 - **WANJURA, D.F., KELLY, C.A., WENDT, C.W., HATFIELD, J.L.:** Canopy temperature and water stress of cotton crops with complete and partial ground cover. - Irrig. Sci. 5: 37-46, 1984.

*14687 - **WARING, R.H.:** Land of the giant conifers. - Natur. hist. 91: 55-62, 1982.

14688 - **WEAVER, M.L., NG, H., BURKE, D.W., SILBERNAGEL, M.J., FOSTER, K., TIMM, H.:** Effect of soil moisture tension on pod retention and seed yield of beans. - HortScience 19: 567-569, 1984.

14689 - **WEBB, E.K.:** Evaluation of evapotranspiration and canopy resistance: an alternative combination approach. - Agr. Water Manage. 8: 151-166, 1984. Also in: SHARMA, M.L. (ed.): Evapotranspiration from Plant Communities. Pp. 151-166. Elsevier Science Publishers, Amsterdam - Oxford - New York - Tokyo 1984.

14690 - **WEBSTER, C.P., DOWDELL, R.J.:** Effect of drought and irrigation on the fate of nitrogen applied to cut permanent grass swards in lysimeters: leaching losses. - J. Sci. Food Agr. 35: 1105-1111, 1984.

*14691 - **WĘGLARZ, Z.:** Wpływ czynników agrotechnicznych na przechodzenie kminku zwyczajnego (*Carum carvi* L.) z fazy wegetatywnej w generatywną. II. Wpływ nawożenia i wilgotności gleby na rozwój i plonowanie kminku zwyczajnego. [Effect of agricultural agents on transition of *Carum carvi* L. from vegetative to generative phase. II. Effect of fertilization and soil moisture on the development and cropping of *Carum carvi* L.]- Herba Polonica 29: 21-26, 1983. [In Pol, ab: E, R.]

14692 - **WEIHE, J.:** Benetzung und Interzeption von Buchen- und Fichtenbeständen. IV. Die Verteilung des Regens unter Fichtenkronen. - Allg. Forst. Jagdzeit. 155: 241-252, 1984.

14693 - **WEIMBERG, R., LERNER, H.R., POLJAKOFF-MAYBER, A.:** Changes in growth and water-soluble solute concentrations in *Sorghum bicolor* stressed with sodium and potassium salt. - Physiol. Plant. 62: 472-480, 1984.

14694 - **WEIR, A.H., BRAGG, P.L., PORTER, J.R., RAYNER, J.H.:** A winter wheat crop simulation model without water or nutrient limitations. - J. agr. Sci. 102: 371-382, 1984.

*14695 - **WEISS, A., HIPPS, L.E., BLAD, B.L., STEADMAN, J.R.:** Comparison of within-canopy microclimate and white mold disease (*Sclerotinia sclerotiorum*) development in dry edible beans as influenced by canopy structure and irrigation. - Agr. Meteorol. 22: 11-21, 1980.

14696 - **WEISSENBÖCK, G., SCHNABL, H., SACHS, G., ELBERT, C., HELLER, F.-O.:** Flavonol content of guard cell and mesophyll cell protoplasts isolated from *Vicia faba* leaves. - Physiol. Plant. 62: 356-362, 1984.

14697 - **WENDLER, S., ZIMMERMANN, U.:** Pressure clamp studies on algal cells. - In: CRAM, CRAM, W.J., JANÁČEK, K., RYBOVÁ, R., SIGLER, K. (ed.): Membrane Transport in Plants. Pp. 123-124. Academia, Praha 1984.

14698 - **WENDLER, S., ZIMMERMANN, U., BENTRUP, F.-W.:** Turgor mediated processes in *Acetabularia mediterranea*. - In: CRAM, W.J., JANÁČEK, K., RYBOVÁ, R., SIGLER, K. (ed.): Membrane Transport in Plants. Pp. 125-126. Academia, Praha 1984.

14699 - **WENDT, C.W., ISBELL, V.R., STUART, B.L., ABERNATHY, J.R.:** Effects of 1,1-dimethylpiperidinium chloride on growth and water relations of cotton in a semiarid environment. - In: ORY, R.L., RITTIG, F.R. (ed.): Bioregulators. Chemistry and Uses. Pp. 205-213. American Chemical Society, Washington 1984.

14700 - **WEST, D.W., TAYLOR, J.A.:** Response of six grape cultivars to the combined effects of high salinity and rootzone waterlogging. - J. Amer. Soc. hort. Sci. 109: 844-851, 1984.

14701 - **WESTGATE, M.E., BOYER, J.S.:** Transpiration- and growth-induced water potentials in maize. - Plant Physiol. 74: 882-889, 1984.

14702 - **WHITE, E.G.:** A multispecies simulation model of grassland producers and consumers I. Validation. - Ecol. Model. 24: 137-157, 1984.

*14703 - **WHITE, E.M.:** Growth of spring wheat treated with a constricting force on upper roots. - Proc. South Dakota Acad. Sci. 59: 70-79, 1980.

*14704 - **WHITE, J.M., STRANDBERG, J.O.:** Physical factors affecting carrot growth: water saturation of soil. - J. Amer. Soc. hort. Sci. 104: 414-416, 1979.

14705 - **WHITEHEAD, D., JARVIS, P.G., WARING, R.H.:** Stomatal conductance, transpiration, and resistance to water uptake in a *Pinus sylvestris* spacing experiment. - Can. J. Forest Res. 14: 692-700, 1984.

*14706 - **WHITEHEAD, D.C.:** The influence of frequent defoliation and of drought on nitrogen and sulphur in the roots of perennial ryegrass and white clover. - Ann. Bot. 52: 931-934, 1983.

14707 - **WIEBE, H.H.:** Water condensation on Peltier-cooled thermocouple psychrometers: a photographic study. - Agron. J. 76: 166-168, 1984.

14708 - **WIEBE, H.H., GREER, R.L., VAN ALFEN, N.K.:** Frequency and grouping of vessel endings in alfalfa (*Medicago sativa*) shoots. - New Phytol. 97: 583-590, 1984.

*14709 - **WIEGAND, C.L., NIXON, P.R., JACKSON, R.D.:** Drought detection and quantification by reflectance and thermal responses. - Agr. Water Manage. 7: 303-321, 1983. Also in: STONE, J.F., WILLIS, W.O. (ed.): Plant Production and Management under Drought Conditions. Pp. 303-321. Elsevier Science Publishers, Amsterdam - Oxford - New York - Tokyo 1983.

14710 - **WIENCKE, C., KNOTH, A., STELZER, R.:** Changes of protoplasmic and vacuolar volume during osmotic adaptation and ion compartmentation in *Porphyra umbilicalis*. - Hydrobiologia 116/117: 481-484, 1984.

14711 - **WIESNER, B., HAGEDORN, R., HOFFMANN, P., MEINL, G.:** Effects of intermittent light on physiological parameters of wheat seedlings. - Arch. Züchtungsforsch. 14: 359-366, 1984.

*14712 - **WILHELM, C., WILD, A.:** Growth and photosynthesis of *Nanochlorum eucaryotum,* a new and extremely small eucaryotic green alga. - Z. Naturforsch. 37: 115- 119, 1982,

14713 - **WILLE, A.C., LUCAS, W.J.:** Ultrastructural and histochemical studies on guard cells. - Planta 160: 129-142, 1984.

14714 - **WILLERT, D.J. von, BRINCKMANN, E., ELLER, B.M., SCHEITLER, B.:** Water loss and malate fluctuations during the day for plants in the southern Namib desert - Oecologia 61: 393-397, 1984.

14715 - **WILLIAMS, P.M., DE MALLORCA, M.S.:** Effect of osmotically induced leaf moisture stress on nodulation and nitrogenase activity of *Glycine max.* - Plant Soil 80: 267-283, 1984.

14716 - **WILLIAMS, R., ZANGVIL, A., KARNIELI, A.:** A portable evaporimeter for rapid measurement of the evaporation rate of water. - Agr. Forest Meteorol. 32: 217-224, 1984.

*14717 - **WILLIAMS, R.J.:** Frost desiccation: an osmotic model. - In: OLIEN, C.R., SMITH, M.N. (ed.): Analysis and Improvement of Plant Cold Hardiness. Pp. 90-113. CRC Press, Inc., Boca Raton 1980.

*14718 - **WILLMER, C.M.:** Stomata. - Longman Inc., London - New York 1983.

14719 - **WILSON, J.:** Microscopic features of wind damage to leaves of *Acer pseudoplatanus* L. - Ann. Bot. 53: 73-82, 1984.

*14720 - **WILSON, J.R., BROWN, R.H.:** Nitrogen response of *Panicum* species differing in CO_2 fixation pathways. I. Growth analysis and carbohydrate accumulation. - Crop Sci. 23: 1148-1153, 1983.

*14721 - **WILSON, J.R., MUCHOW, R.C.:** Effect of water stress on dry matter digestibility and concentration of nitrogen and phosphorus in seven tropical grain legumes. - J. Aust. Inst. agr. Sci. 1983: 167-169, 1983.

14722 - **WINTER, E.:** Water-use efficiency of *Trifolium alexandrinum* exposed to salt stress under different environmental conditions. - Flora 175: 1-13, 1984.

*14723 - **WITT, M., BARFIELD, B.J.:** Environmental stress and plant productivity. - In: RECHCIGL, M., Jr. (ed.): CRC Handbook of Agricultural Productivity. Volume 1. Plant Productivity. Pp. 347-374. CRC Press, Inc., Boca Raton 1982.

14724 - **WIUM-ANDERSEN, S., BORUM, J.:** Biomass variation and autotrophic production of an epiphyte-macrophyte community in a coastal Danish area: I. Eelgrass (*Zostera marina* L.) biomass and net production. - Ophelia 23: 33-46, 1984.

*14725 - **WOLF, D.D., PARRISH, D.J.:** Short-term growth responses of tall fescue to changes in soil water potential and to defoliation. - Crop Sci. 22: 996-999, 1982.

14726 - **WOLSWINKEL, P., AMMERLAAN, A.:** Turgor-sensitive sucrose and amino acid transport into developing seeds of *Pisum sativum*. Effect of a high sucrose or mannitol concentration in experiments with empty ovules. - Physiol. Plant. 61: 172-182, 1984.

*14727 - **WOLSWINKEL, P., AMMERLAAN, A., KUYVENHOVEN, H.:** Effect of KCN and p-chloromercuribenzenesulfonic acid on the release of sucrose and 2-amino(1-^{14}C)isobutyric acid by the seed coat of *Pisum sativum*. - Physiol. Plant. 59: 375-386, 1983.

14728 - **WOMACK, C.L.:** Reduction in photosynthetic and transpiration rates of alfalfa caused by potato leafhopper (*Homoptera: Cicadellidae*) infestations. - J. econ. Entomol. 77: 508-513, 1984.

14729 - **WONG, D.H., BARBETTI, M.J., SIVASITHAMPARAM, K.:** Effects of soil temperature and moisture on the pathogenicity of fungi associated with root rot of subterranean clover. - Aust. J. agr. Res. 35: 675-684, 1984.

14730 - **WOODRUFF, J.R., LIGON, J.T., SMITH, B.R.:** Water table depth interaction with nitrogen rates on subirrigated corn. - Agron. J. 76: 280-283, 1984.

14731 - **WOOLACOTT, D., AYRES, P.G.:** Effects of plant age and water stress on production of conidia by *Erysiphe graminis* f. sp. *hordei* examined by non-destructive sampling. - Trans. Brit. mycol Soc. 82: 449-454, 1984.

*14732 - **WOOLHOUSE, H.W., JENKINS, G.I.:** Physiological responses, metabolic changes and regulation during leaf senescence. - In: DALE, J.E., MILTHORPE, F.L. (ed.): (ed.): The Growth and Functioning of Leaves. Pp. 449-487. Cambridge University Press, Cambridge - London - New York - New Rochelle - Melbourne - Sydney 1983.

14733 - **WRONSKI, E.:** A model of canopy drying. - Agr. Water Manage. 8: 243-262, 1984. Also in: SHARMA, M.L. (ed.): Evapotranspiration from Plant Communities. Pp. 243-262. Elsevier Science Publishers, Amsterdam - Oxford - New York - Tokyo 1984.

14734 - **WU, M.-T., WALLNER, S.J., WADDELL, J.W.:** Heat stress responses in cultured plant cells. Effect of culture handling and age. - Plant Physiol. 74: 944-946, 1984.

14735 - **WYN JONES, R.G.:** Phytochemical aspects of osmotic adaptation. - In: TIMMERMANN, B.N., STEELINK, C., LOEWUS, F.A. (ed.): Phytochemical Adaptations to Stress. Pp. 55-78. Plenum Press, New York - London 1984.

14736 - **XU, D.-Q., LI, D.-Y., SHEN, Y.-G., LIANG, G.-A.:** [On midday depression of photosynthesis of wheat leaf under field conditions.] - Acta phytophysiol. Sin. 10: 269-276, 1984. [In Chin, ab: E.]

14737 - **YADAVA, R.B.R., PATIL, B.D.:** Screening of cowpea (*Vigna unguiculata* L.) varieties for drought tolerance. - Z. Pflanzenzücht. 93: 259-262, 1984.

14738 - **YAMAMOTO, A., TAKEDA, G., NAKAJIMA, T., YAMAZAKI, K.:** [Studies on the variability of stomatal density in soybean cultivars. I. Influence of leaf sides and leaf positions on the differences among cultivars.] - Jap. J. Crop Sci. 53: 463-471, 1984. [In Jap, ab: E.]

14739 - **YAMAMOTO, T.:** Non-destructive *in situ* measurements of pear fruit moisture during growth. - Scientia Hort. 22: 97-103, 1984.

*14740 - **YAMAMOTO, T., WATANABE, S.:** Studies on leaf burn of pear trees. XI. The effects of changes in soil moisture tensions, rainfalls and foliar sprays of water on leaf burn of bartlett pears. - Bull. Yamagata Univ. (Agr. Sci.) 8: 643-655, 1980.

14741 - **YAMBAO, E.B., O'TOOLE, J.C.:** Effects of nitrogen nutrition and root medium water potential on growth, nitrogen uptake and osmotic adjustment of rice. - Physiol. Plant. 60: 507-515, 1984.

14742 - **YANG, C.-Y., WENG, J.-H., CHEN, C.-Y.:** [Studies on photosynthesis and dry matter production of soybean (II) the photosynthetic characteristics of winter crop soybean.] - J. agr. Assoc. China 126: 34-43, 1984. [In Chin, ab: E.]

14743 - **YANG, J.-F., CHEN, B.-S., WANG, H.-C.:** [Influence of high temperature and low humidity on the fatty acid compositions of membrane lipids in wheat.] - Acta bot. Sin. 26: 386-391, 1984. [In Chin, ab: E.]

14744 - **YAO, N., LI, J.:** The relationship between the resistance of intertidal marine benthic algae against osmotic shock and their content of soluble carbohydrates. - Hydrobiologia 116: 485-487, 1984.

14745 - **YAZAR, A.:** Evaporation and drift losses from sprinkler irrigation systems under various operating conditions. - Agr. Water Manage. 8: 439-449, 1984.

*14746 - **YELENOSKY, G.:** Water-stress induced cold hardening of young citrus trees. - J. Amer. Soc. hort. Sci. 104: 270-273, 1979.

14747 - **YEO, A.R., FLOWERS, T.J.:** Nonosmotic effects of polyethylene glycols upon sodium transport and sodium-potassium selectivity by rice roots. - Plant Physiol. 75: 298-303, 1984.

14748 - **YIANOULIS, P., TYREE, M.T.:** A model to investigate the effect of evaporative cooling on the pattern of evaporation in sub-stomatal cavities. - Ann. Bot. 53: 189-206, 1984.

14749 - **YOSHIDA, S.:** Studies on freezing injury of plant cells. I. Relation between thermotropic properties of isolated plasma membrane vesicles and freezing injury. - Plant Physiol. 75: 38-42, 1984.

14750 - **YOSHIDA, S., UEMURA, M.:** Protein and lipid compositions of isolated plasma membranes from orchard grass (*Dactylis glomerata* L.) and changes during cold acclimation. - Plant Physiol. 75: 31-37, 1984.

*14751 - **YOUNG, F.L., WYSE, D.L., JONES, R.J.:** Effect of irrigation on quackgrass (*Agropyron repens*) interference in soybeans (*Glycine max*). - Weed Sci. 31: 720-727, 1983.

14752 - **ZACK, C.D., LOY, J.B.:** Comparative effects of gibberellic acid and N^6-benzyladenine on dry matter partitioning and osmotic and water potentials in seedling organs of dwarf watermelon. - J. Plant Growth Regulation 3: 65-73, 1984.

14753 - **ZAGDAŃSKA, B.:** Influence of water stress upon photosynthetic carbon metabolism in wheat. - J. Plant Physiol. 116: 153-160, 1984.

14754 - **ZAGDAŃSKA, B.:** Effect of water stress on CO_2 exchange in flag leaves of spring wheat. - Acta Physiol. Plant. 6: 215-221, 1984.

*14755 - **ZAROGIANNIS, V.:** Einfluss der·Beregnung und der Bestandesdichte auf den Befall von Maiszünsler und Maisbeulebrand bei Körnemais. - Bodenkultur 33: 146-154, 1982.

*14756 - **ZAVADSKAYA, I.G., ANTROPOVA, T.A.:** Tsitofiziologicheskie reaktsii na obezvozhivanie ustoĭchivogo i chustvitel'nogo k zasukhe sortov yachmenya *Hordeum sativum* (*Poaceae*). [Cytophysiological reactions on dehydration in drought resistant and drought sensitive varieties of barley *Hordeum sativum* (*Poaceae*),] - Bot. Zh. 68: 625-631, 1983. [In R.]

14757 - **ZAVADSKAYA, I.G., ANTROPOVA, T.A.:** Vliyanie obezvozhivaniya list'ev ustoĭchivykh i chuvstvitel'nykh k zasukhe rasteniĭ na pervichnuyu i obshchuyu teploustoĭchivost' kletok. [The effect of dehydration of leaves of drought--resistant and drought-sensitive plants on primary and general thermostability of cells.] - Bot. Zh. 69: 229-235, 1984. [In R.]

*14758 - **ZEIDE, B.:** Ranking of forest growth factors. - Environ. exp. Bot. 20: 421-427, 1980.

14759 - **ZEIGER, E.:** Blue light and stomatal function. - In: SENGER, H. (ed.): Blue
        Light Effects in Biological Systems. Pp. 484-494. Springer-Verlag, Berlin -
        Heidelberg - New York - Tokyo 1984.

14760 - **ŻELAWSKI, W., SZLENK, W.:** A concept of vegetative growth and a dynamic model
        of dry matter accumulation in plants [on an example of Scots pine (*Pinus
        sylvestris* L.)seedlings]. - Wiss. Z. Humboldt-Univ., math.-naturwiss. Reihe
        33: 359-361, 1984.

14761 - **ZEMÁNEK, M.:** Sklizňový index genotypů ozimé pšenice v modelově rozdílných
        podmínkách zásobení vodou. [The harvest index of winter wheat genotypes
        under different conditions of water supply.] - Rost. Výroba (Praha) 30:
        1231-1244, 1984. [In Czech, abL R, E.]

14762 - **ZHEKOV, Zh., DANAILOVA, S.:** Vliyanie na vodniya defitsit na listata v"rkhu
        fotokhimichnata aktivnost na khloroplasti ot tsarevitsa. [Influence of water
        deficit in leaves on chloroplast photochemical activity in maize.] -Nauch.
        Tr. Plovdiv. Univ."Paisii Khilendarski" - Biol. 20 (4): 153-159, 1982. [In
        Bulg, ab: E, R.]

14763 - **ZHIVKOV, Zh.V.:** Optimizirane na polivniya rezhim na zakharno tsveklo pri
        nedostig na voda za napoyavane. [Optimizing the irrigation regime of sugar beet
        beet in case of water deficit.] - Rasteniev. Nauki 21 (7): 72-78, 1984.
        [In Bulg, ab: R, E.]

14764 - **ZIEGLER, H.:** Gedanken zu einer Biophysik der Pflanze. - Wiss. Z. Humboldt-
        -Univ., math.-naturwiss. Reihe 33: 331-334, 1984.

14765 - **ZUR, B., JONES, J.W.:** Diurnal changes in the instantaneous water use effi-
        ciency of a soybean crop. - Agr. Forest Meteorol. 33: 41-51, 1984.

AUTHORS' INDEX

(Cumulative Index to volumes 6 - 10)

Authors' names are presented in the form in which they appear in the respective publication. The names from papers published in Cyrillic character are transcribed as shown in Instructions for Use. Alternative spelling and form of the name of the same author are usually cross-indexed.

A

AMOR, R.L. 13399
AMORIM NETO, M.da S. 12953, 14224
AMORY, A.M. 9699
AMOUR, G.S. 13247
AMTHOR, J.S. 11231
AMUNDSON, R.G. 8167, 12385, 13307
ANAGNOSTOPOULOS, G.D. 8862, 9700
ANAND REDDY, K. 11541
ANDALES, S.C. 6640
ANDEL, J., van see VAN ANDEL, J.
ANDEREGG, B.N. 8168
ANDEREGG, J.C. 7408
ANDERSEN, A.S. 7675, 7676, 7687
ANDERSEN, K. 8169
ANDERSEN, P.C. 9701, 12954, 12955,
12956, 12957, 12958, 12959
ANDERSEN, R.L. 6810, 8310
ANDERSEN, S. 8169
ANDERSON, C.A. 12960
ANDERSON, C.K. 9702
ANDERSON, D.J. 8122
ANDERSON, J.A. 11232
ANDERSON, J.E. 6641, 7413, 9306,
9703
ANDERSON, L.C. 10633, 13891
ANDERSON, L.E. 12385, 12386, 12387
ANDERSON, R.C. 6614, 12210
ANDERSON, W.K. 6642, 6643, 12961
ANDERSSON, Γ. 8170
ANDERSSON, L.-Å. 8171
ANDONOV, D. 8172, 8173, 9704,
9705, 11233, 12962 see also
ANDONOV, D.N. 9706
ANDONOVA, P. 7320, 11234, 14654
ANDONOVA, T. 11234
ANDRADE LIMA, D., de 8872
ANDRÉ, M. 7425, 9707, 10457,
11014
ANDREEV, I.M. 11059
ANDREEVA, I.N. 14516
ANDREEVA, T.F. 9550
ANDREI, R. 7271
ANDREWS, C.J. 8174, 11653, 13170,
14162
ANDREWS, K.L. 8175
ANDREWS. T.J. 11466
ANGELIS, J.D. de see De ANGELIS, J.D.
ANGSTRÖM, J. 9682
ANGUILLESI, M.C. 6644
ANGUS, D.E. 11235, 12963
ANGUS, J.F. 6645, 11236, 11237
ANIKEENKO, A.P. 6929
ANISIMOV, A.V. 13315
ANISIMOVA, K.I. 11366
ANNAMMA, Y. 7648
ANNAN, A.P. 7980
ANSARI, R. 10566
ANTAL, E. 8166
ANTIPOV, N.I. 6646, 9708, 9709,
11238

ANTLFINGER, A.E. 6647, 11239
ANTONIELLI, M. 10932
ANTONOV, A.V. 8454
ANTONOV, I. 12964
ANTOSZEWSKI, R. 11240, 11967, 12029
ANTROPOVA, T.A. 11175, 14756, 14757
AONO, H. 8176
APARICIO-TEJO, P. 9710, 10812 see
also
APARICIO-TEJO, P.M. 6648, 6649,
11241, 12965, 13022
APEL, P. 6650, 8177, 12966
APELBAUM, A. 8178
APOSTOLAKOS, P. 13394
APPAJI RAO, N. 13678, 13679,
13680, 13681, 13682
APPEL, D.N. 12967
APPELBAUM, S. 6651
APPLEBY, A.P. 8979
APPLEBY, R.F. 11242, 11243
AQUINO, A.R.L., de see De AQUINO, A.R.L.
AQUINO, R.F. 13892
ARAD (MALIS), S. 11244
ARAGON, E.L. 9711
ARAKI, T. 11773, 13555
ARAUJO, J.P., de 11245
ARDITTI, J. 6652, 12235, 12665
AREKAL, G.D. 7513
ARES, J.O. 12968
ARGEL, P.J. 11246, 11247
ARGUELLO, J.A. 8283
ARIANOUSTOU-FARRAGITAKI, M. 13882
ARIAS, I. 7453
ARIF, L.H. 10016
ARKIN, G.F. 11565
ARKLEY, R.J. 12969
ARMITAGE, A.M. 9712, 11248, 11249,
12970
ARMOND, P. 8116, 9668, 11179
ARMOND, P.A. 6731, 7076, 7673,
8589
ARMSTRONG, W. 10813
ARNAUDET, L. 9713
ARNAUDO, D. 9586, 11091
ARNDT, U. 13698
ARNEKE, W.-W. 10490
ARNOLD, K.E. 8922
ARNOTT, H.J. 12822
ARNOUX, M. 9586, 9587, 11091
ARONSSON, A. 8179, 8722
ARORA, B.B. 8180
ARTECA, R.N. 11250
ARTYKOV, K. 6653
AS, H., van see VAN AS, H.
ASADA, S. 6654
ASANO, H. 14543, 14544
ASAY, K.H. 8761, 10279, 14284
ASBELL, C.W. 7993
ASFARY, A.F. 11251
ASHCROFT, G.L. 7111

BASHAW, E.C.   12328
BASHE, D.   13015
BASKANOV, Yu.A.   14382
BASKIN, C.C.   9736, 9737
BASKIN, J.M.   9736, 9737
BASSETT, E.N.   9738
BASSMAN, J.H.   9739
BASSTANIE, L.   9722, 11078
BASTIANPILLAI, V.A.   11295
BASU, R.N.   6688, 6689, 10441,
12466, 12483
BATAILLÉ, A.   13333
BATAL, K.M.   8217
BATANOUNY, K.H.   8218, 8219, 8220
BATCHELOR, C.H.   9607, 11296, 11712,
13016
BATCHELOR, J.T.   13017
BATES, L.M.   8221, 9740, 9741,
13831
BATSON, D.B.   9790, 9791
BÄTZ, G.   8305
BAUCHER, G.   12975
BAUDER, J.W.   6690, 8222
BAUER, A.   11297, 13018, 13380
BAUER, H.   6691, 7305, 8895, 10225,
10388, 11298, 13019
BAUER, M.E.   9742, 11207
BAUER, U.   6691, 12302, 13020
BAUER, V.   11299
BAUMEISTER, W.   8223
BAUSHER, M.G.   9498
BAUWE, H.   12966
BAVEL, C.H.M., van see VAN BAVEL,
C.H.M.
BAYASGALAN, L.   13624
BAYER, D.E.   11067
BAYFIELD, N.G.   10670
BAZALIĬ, G.G.   10620
BAZALIĬ, V.V.   10620
BAZOS, L.J.   12755
BAZZAZ, F.A.   6823, 8981, 9844,
9991, 11656, 13021, 14137
B"CHVAROV, N.   10350
BEADLE, C.L.   8224
BEADLE, N.B.   13575
BEALE, O.W.   7632
BEALL, P.T.   8225
BEAMES, D.   13823
BEAN, J.N.   12236
BEARD, J.B.   8067, 10277, 11846
BEARD, J.S.   11300
BEARDSELL, D.V.   6692
BEBAWI, F.F.   9743
BECANA, M.   13022
BECHSTÄDT, O.   7920
BECK, D.E.   13936
BECK, E.   13023, 13024
BECKER, M.   11301
BECKERS, F.   8123
BECWAR, M.R.   8226, 11302, 11303,
11304, 11305, 11306, 11672, 12809

BEDELL, T.E.   8552
BEDENKO, V.P.   8227
BEDFORD, C.L.   7875
BEDUNAH, D.   6693, 8228
BEER, S.   6694
BEESE, F.   6695, 8012
BEEVER, R.E.   11307
BEFFAGNA, N.   14136
BEGG, J.E.   6696, 9558
BÉGUIN, C.   6697
BEGUN, S.   7405
BEHAIRY, A.K.A.   13025
BEHBOUDIAN, M.H.   6698
BEHL, R.   13026
BEHRENDT, S.   8842
BEHRNS, G.T.   6931
BEĬKAL, M.N.   11972
BEINEKE, W.E.   6699
BEJENARU, A.   12933
BEKER, M.J.   12903
BEL, A.J.E., van see VAN BEL, A.J.E.
BELAN, F.   6700
BELAYA, G.A.   6701, 6702
BELBACHIR, O.   9744
BELCHEVA, L.   13027
BELEITES, F.   8513
BELESKY, D.P.   9745, 13028, 13029,
13030
BELFORD, R.K.   6703, 6815, 6816,
6817, 8229, 13031
BELIČ, B.   12153
BELIČ, J.   8230
BEL'KOVICH, T.M.   13315
BELL, A.A.   9746
BELL, C.J.   8231, 8232, 8233, 9747
BELL, D.H.   7070
BELL, D.T.   8579, 13178, 14253
BELL, J.N.B.   9719, 12983
BELL, K.L.   6704, 6705
BELL, K.R.   8234
BELLES, W.S.   7887
BELMANS, C.   7011, 9748, 10229,
13032
BELOKOBYL'SKIĬ, I.M.   8235
BELOT, Y.   6706, 13033
BELOW, F.E.   9880, 13203
BELT, G.H.   6707
BELTRAO, N.N. de M.   14471
BELTZ, C.K.   9425
BEN-AMOTZ, A.   6708, 6709, 6710,
9749, 9750, 11308
BEN ASHER, J.   9751
BENECKE, P.   6711, 8012, 8236,
8237
BENECKE, U.   6712, 7936, 8238
BENEŠOVÁ. H.   12544, 13034, 14371
BENGTSON, C.   8239, 13458
BENHAM, E.C.   14676
BEN-HAYYIM, G.   11309
BENHAM, E.C.   11098
BEN-HERUT, Z.   12017

DAVIES, W.J. 6912, 7578, 7579,
8413, 8951, 9123, 9618, 9951,
10782, 11242, 11243, 11340, 11516,
11517, 12155, 12772, 12773, 13065,
13066, 13877
DAVIS, C.B. 9581
DAVIS, D.C. 6640
DAVIS, D.D. 7058, 8249, 8250,
8411, 9007, 9450, 9776, 13945
DAVIS, D.W. 12212
DAVIS, E.A. 6913
DAVIS, J. 13954, 13955
DAVIS, J.L. 7980
DAVIS, K.R. 10719, 13581, 13583
DAVIS, M.A. 9952
DAVIS, R.G. 8767, 13643 ·
DAVIS, R.W. 12811
DAVIS, T.D. 8414
DAVISON, A. 9953
DAVY de VIRVILLE, J. 13228
DAVYDOVA, G.V. 12757
DAWES, C.J. 8415, 9952
DAWSON, J.H. 8058
DAWSON, P. 9954, 13229
DAWSON, R.F. 11518
DAWSON, R.N. 6767
DAY, J.M. 10470, 12458
DAY, J.W., Jr. 13952
DAY, W. 7595, 7596, 8416,
8417, 8901, 9148, 11519, 11995,
12287
DAYANAND 10888
DAYNARD, T.B. 13267, 13799
DAYTON, A.D. 7931
DE, R. 8585, 8586, 11520,
11521, 11522, 11688
De ABREU SA, L.D. 14417
DE ALBUQUERQUE, J.A.S. 9955
De ALBUQUERQUE, M.M. see
ALBUQUERQUE, M.M., de
DE ALBUQUERQUE, T.C.S. 9955
DEAN, B.B. 13089
DEAN, C.A. 13230
DEAN, T.J. 11414, 11523
De ANDRADE LIMA, D. see ANDRADE
LIMA, D., de
DEANE-DRUMMOND, C.E. 11524, 11690
De ANGELIS, J.D. 11525, 11526, 11527
DEANS, J.D. 10402
DE AQUINO, A.R.L. 7919
De ARAÚJO, J.P. see ARAÚJO, J.P., de
DEARMAN, A.S. 6724, 8384, 9773,
9774
DEARMAN, J. 9773
DEATON, D.E. 7696
De AZEVEDO, J.A. 13326
DeBELL, D.S. 13231, 13923
DEBERGH, P. 8418
De BOER, A.H. 8868, 13232, 13233,
13234, 13235

DEBRECZENI, I. 9956
De BRUYN, L.P. 9957
DEBRY, R. 11176
DeBUSK, T.A. 11528
De CADENA, G. 13312, 13313
DÉCAMPS, O. 6914
De CARVALHO, J.R.P. see CARVALHO,
J.R.P., de
DE CARVALHO E SILVA, W.L. 9958
De CASTRO, M.E. 12454
DECKER, W.L. 11529
DECKERS, J.C. 13355
De CORMIS, L. see CORMIS, L., de
De DATTA, S.K. 9711, 10237, 13362,
13593, 13594
De FARIA, C.M.B. see FARIA, C.M.B., de
DE-FELICE, J. 14538
DE FRANÇA, G.E. 12777
DEGANI, H. 9959
DEGEN, B. 10294
De GROOTH, B.G. 8419
DEHAN, K. 7258
DEHGAN, B. 8420
DEIBERT, E.J. 6631, 9307, 13104
De JABRUN, P.L.M. see JABRUN, P.L.M., de
DEJAEGERE, R. 8421
De JAGER, J.M. 8295, 10825, 10826,
11395, 14327
De JONG, T.M. 6915, 8422, 8423,
9960, 9961, 11530, 12730, 13236,
13237, 13820
DEKKER, L.W. 9748
DeKOCK, P.C. 9962
DE KONING, A. 13238
De KOUCHKOVSKY, Y. see KOUCHKOVSKY,
Y., de
DEKOV, I. 6916, 9963, 14140
DEKOV, I.Ch. 13958
DELACHIAVE, M.E.A. 12449
DELA CRUZ, A. 10351
DE LA GUARDIA, M.D. 8424
DELANE, R. 9964, 10546
DELAUNE, R.D. 14454
DELECOLLE, R. 10496
DELBRIDGE, S.G. 13568
DELEGHER, V. 8421
De LEÓN, F. 7346
DELMAS, J. 10977
DELMAS, R.E. 7667
De LÓPEZ, M.C.P. 11710
De LOTH, C. 11421
Del RÍO ESCURRA, G.A. 10924
DELUCIA, E.H. 13239
De MAGALHÃES, A.A. see MAGALHÃES,
A.A., de
De MALACH, Y. 10656
DE MALLÓRCA, M.S. 14715
DEMARLY, Y. 9987
DEMARTY, M. 13240
De MENEZES, E.M. see MENEZES, E.M., de

EREMIN, G.V. 10094, 12363
ERICKSON, P.I. 6987
ERICSON, M.C. 13321
ERIE, L.J. 8303, 10022
ERISMANN, K.H. 7033, 7034, 8540
ERKAN, Z. 6988
ERMAKOV, I.P. 12221
ERNST, W. 6989
ERREDE, L.A. 11602, 11603, 11604
ERWEE, M.G. 13322
ERWIN, J.E. 13524
ESASHI, Y. 13622, 13623
ESAULENKO, G.P. 10689
ESCHBACH, J.M. 11605
ESHEL, A. 6694, 11501
ESIKOV, A.D. 13323
ESPELIE, K.E. 10023
ESPINOZA, W. 10024, 10025, 10026,
10110, 11606, 13324, 13325, 13326
ESSER, G. 10411
ESSIAMAH, S.K. 13327
ESTEP, M.F. 8492
ETCHEVERS, J.D. 11607
ETEJERE, E.O. 11608
ETHERINGTON, J.R. 13328, 13329
EVANGELOU, V.P. 13171
EVANS, C.E. 6949
EVANS, D.D. 8493, 8494
EVANS, D.W. 13330
EVANS, J.J. 13029
EVANS, K. 10027
EVANS, L.S. 8495, 11609, 13331,
13332
EVANS, M.L. 8875
EVANS, R.A. 14279
EVANS, R.G. 14135, 14186
EVANS, S.E. 9618
EVELING, D.W. 13333
EVENARI, M. 7246, 7793, 7794
EVEN-CHEN, Z. 6990, 11610, 12825
EVENSON, J.P. 7097, 7098
EVERETT, P.H. 9021
EVERSON, E.H. 7118
EVERT, D.R. 11611
EVERT, R.F. 6991, 9899, 11373,
11627
EWEL, K.C. 13434
EWERS, F.W. 13334, 13335
EYSSAUTIER, A. 10494
EZE, J.M.O. 8496, 11612
EZRA, C.E. 13508

F

FABIAN, G. 10028
FADEEVA, L.G. 6992
FAENSEN-THIEBES, A. 11613, 12270,
13336
FAGEIRA, N.K. 8497, 11285

FAGERLUND, E. 10340
FAGERSTEDT, K. 13337
FAHEY, R.C. 8498
FAHEY, T.J. 8840, 8841, 13338,
14288
FAHN, A. 11667
FAILS, B.S. 10029, 10030, 10031
FAIRBOURN, M.L. 10032
FAIZ, S.M.A. 10033
FAIZY, S.E-D.A. 6993
FALCON, M.F. 13339
FALK, K.-E. 9682
FALK, S.O. 8835, 8925
FALKE, H. 6994
FAL'KOVA, T.V. 11553
FALKOWSKI, M. 6995
FALLOON, P.G. 8499
FALTER, J. 8151, 8152
FALUDI-DANIEL, Á. 9588, 10554,
14021
FANGMEIER, D.D. 11630
FANJUL, L. 6996, 10034, 11856,
13340
FAPOHUNDA, H.O. 13341
FARAH, M. 10188, 10244
FARAH, S.M. 8500, 13342
FARDJAH, M. 6997
FARIA, C.M.B., de 11245, 12618
FARIAS, J.M. 11016
FARINEAU, J. 9904, 10035
FARKAS, T. 9588
FARODA, A.S. 10898
FAROOQUE, A.M. 11256
FARQUHAR, G.D. 6998, 9592, 9901,
9921, 10036, 10037, 10038, 11386,
11614, 13005, 13006, 13118, 13343,
13344, 13720, 14552
FARRAH, J.F. 6999, 10635
FARRAR, S.L. 12812
FARRIS, M.A. 13345
FASEHUN, F.E. 7078, 9628
FAUSEY, N.R. 9623, 10911
FAUST, M. 10968, 10969, 10970,
11675, 11676, 12670, 13481
FAVOLA, G. 9537
FAWUSI, M.O.A. 7000, 13346
FAY, M.F. 9124
FAZYLOVA, S. 7001 see also
FAZYLOVA, S.F. 11615
FECHNER, G.H. 8501
FEDCHENKO, P.P. 13733
FEDDES, R.A. 10229, 12843, 13032,
13347
FEDERER, C.A. 7002, 10039, 13155
FEDERMAN, E. 10211
FEDINA, I. 10696
FEDORENKO, I.V. 13261
FEDOROVSKAYA, M.D. 11069
FEDOSEEVA, G.P. 12993
FEDULOV, Yu.P. 8502

FRITZ, E. 7030
FROLOV, A.K. 10071
FROLOV, E.N. 8865, 12222
FROMMHOLD, I. 7031, 10072, 10073, 10074
FRUSCIANTE, L. 11227, 11228
FRY, K.E. 8613, 12326
FRY, W.E. 9015
FRYREAR, D.W. 7663
FU, C.F. 10075
FUCHIGAMI, L.H. 8285, 8539, 8844, 9810, 11021
FUCHS, M. 9042, 11472, 12163
FUCHS, Y. 13278
FUCIK, J.E. 7032
FUEHRING, H.D. 11645
FÜHR, F. 10783, 12450
FÜHRER, E. 7173
FUHRER, J. 7033, 7034, 8540, 11646
FUHRMAN, M.H. 10076
FUHRMANN, U. 7207, 8541, 8542
FUJIEDA, K. 10607
FUJIMOTO, N. 7562
FUJINO, D.W. 11647
FUJINO, M. 10274
FUJIYAMA, H. 12745
FUKUTOKU, Y. 8543, 8544, 10077, 13389
FULBRIGHT, T.E. 13390
FULLER, R.J. 11477
FULLERTON, T.M. 9476
FULTON, J.M. 7943, 8903, 9507, 12688
FUNADA, S. 9112
FUNES, E. 12984
FUNES, F. 13391
FUNT, R.C. 10078
FUQUA, M.C. 7608
FURCILĂ, P. 6627
FÜRST, Z. 8834
FURTULA, V. 11941
FURUKAWA, A. 8545, 8546
FURUYA, S. 7942

G

GÁBORČÍK, N. 11648
GÁBORČÍK, S. 11648
GABR, A.I. 10079
GABRIELS, R. 9855, 11439, 11440
GABRIËLS, R. 6840, 8333
GABRIELSON, R.L. 9980
GACHECHILADZE, N.D. 8547
GACHKOVSKIĬ, V.F. 13392
GADAL, P. 11649, 12790
GADOIEV, N. 12358
GAFF, D.F. 6904, 6905, 7035, 9625, 13393, 14605

GAGNE, J.A. 8035
GAGNON, C. 10769
GAJRI, P.R. 11650
GALATIS, B. 7036, 7037, 10080, 13394
GALE, J. 7038, 7039, 9361, 9751, 11747, 12532, 13174, 14358, 14413
GALE, M.D. 13715
GALES, K. 6816, 6817, 8548, 8549, 11651, 13395
GALILI, D. 12441
GALKIN, V.I. 6806
GALLAGHER, J.L. 13271
GALLAGHER, J.N. 14238
GALLI, E. 14571
GALLI, J. 12703
GALLI, M.G. 8550, 10081
GALSTON, A.W. 10057, 11646, 13364, 13365
GAMAYUN, I.M. 8551
GAMBHIR, P.N. 13396
GAMBLE, P.E. 13397
GAMZIKOVA, O.I. 7041
GANAPATHY, P.S. 14087
GANCHEV, S.P. 11652
GANGADHARA, M. 10253
GANGOPADHYAY, S. 10082
GANNIBAL, B.K. 12072
GANSKOPP, D.C. 8552
GAO, J.-Y. 11653
GAO(KAO), Y.-Z. 11654
GARAB, O. 14021
GARAGORRY, F.L. 10110
GARAY, A. 11277, 11278
GARAY, A.F. 11655
GARBUTT, K. 11656
GARDETTO, P.E. 10319
GARDNER, B.R. 8553, 8554, 8555, 11799, 12320
GARDNER, C.M.K. 13398
GARDNER, E.A. 11701
GARDNER, H.R. 11657
GARDNER, W.K. 13399
GARDNER, W.R. 7432, 10466, 11658, 11659, 12819
GARG, B.K. 8556, 10083, 13400, 13401, 13402, 13403, 13404
GARG, K.K. 8557
GARG, O.P. 7824, 7825, 7826, 13400, 13401, 13402, 14389
GARLAND, J.A. 7042, 10084
GARNER, J.O., Jr. 12916
GARNSEY, S.M. 8107
GARRATT, J.R. 13405
GARREC, J.-P. 10380, 11660, 13778, 14635
GARRETT, H.E. 11549
GARRETT, M.K. 7265, 11902
GARRITY, D.P. 8553, 10085, 10086, 11661, 13406, 13407

GÖRING, H.   9787, 10113, 10114,
10197, 11188, 13441, 13442
GORKOM, H.J., van see Van GORKOM, H.J.
GÖRNY, A.G.   13443
GOROSHKO, V.V.   11697
GORYACHEV, S.N.   8149
GORYSHINA, T.K.   10071
GOSIEWSKI, W.   10115
GOSS, M.J.   13444
GOSSE, G.   9164
GOSSELINK, J.G.   8366
GOTOW, K.   10116, 10870, 12571,
13445
GOTTSCHALK, K.W.   7073, 10576
GOUDRIAAN, J.   7074, 8015, 13446,
13447
GOULART, B.L.   10078
GOULD, W.I.   14054
GOUNOT, M.   7075
GOURDON, F.   10117
GOVINDARAJAN, A.G.   10118
GOVINDJEE   7076, 8109, 8589,
10119, 10120, 10604, 11496
GOWIN, T.   7077
GOYAL, M.R.   11698, 11710
GOYAL, P.   8254
GOZLAN, G.   11349, 11350
GRACE, J.   7078, 7079, 8590,
8591, 9261, 10121, 10122, 10686,
11382, 13258, 13259, 13777, 14148
GRAFE, V.B.   7080
GRAHAM, C.W.   7081
GRAHAM, J.H.   13096, 13802
GRAHAM, L.E.   13933
GRAHAM, M.E.D.   11682, 12755
GRAHAM, P.H.   8045
GRAHAM, R.D.   10645, 11254, 14093
GRAHLE, A.   12233
GRANAT, L.   10152
GRANDIN, M.   8592
GRANGE, A.   7082
GRANGE, R.I.   8135, 11699, 11700
GRANIER, A.   8593, 9716, 12981
GRANT, J.A.   13448
GRANT, W.J.   6713, 9315, 13041
GRANTZ, D.   10604
GRANTZ, D.A.   8626
GRATTAN, S.R.   10432
GRAVES, C.R.   8594
GRAY, D.   10123
GRAY, L.E.   7288
GREACEN, E.L.   11701, 13449
GREB, B.W.   7083, 7084
GRÉC, L.   11971
GRECO, L.   8028
GREE, C.F.   11702
GREEGOR, D.H.   14024
GREEN, A.E.   13450
GREEN, C.W.   10934
GREEN, D.E.   9632, 14653

GREEN, D.G.   11703, 13162
GREEN, G.C.   7457, 7491, 7492,
8999, 9000, 9001, 10500, 11100,
13964
GREEN, J.L.   9130
GREEN, J.M.   8595, 10124
GREEN, T.G.A.   6892, 8298, 8596,
8597, 8598, 9423, 9424, 9425, 9426,
11704, 11705, 12614, 13451, 13452
GREENE, D.M.   7085, 7086
GREENWAY, H.   7087, 9733, 9734,
9964, 10125, 10546, 10547, 10852,
11706, 11707, 12182, 12183, 12546,
13453
GREENWOOD, D.   10212
GREENWOOD, D.J.   8292, 8599, 13193
GREENWOOD, E.A.N.   7088, 7089, 10126,
10127
GREENWOOD, M.S.   8600
GREER, D.H.   12836
GREER, R.L.   14708
GREGERSEN, A.K.   8601
GREGORCZYK, A.   13454
GREGORIOU, C.   13455
GREGORY, E.J.   10810, 13674, 14309
GREGORY, J.M.   8888
GREGORY, P.J.   9444, 12985, 13456,
13457, 13916, 14479
GRENNFELT, P.   13458
GRENOBLE, D.W.   11540
GREWAL, S.S.   13459
GRIBOVSKAYA, I.V.   7643
GRICE, A.C.   11128
GRIEB, B.   14081
GRIER, C.C.   12705, 14556, 14557
GRIERSON, D.   6880
GRIERSON, W.   10128
GRIEVE, A.M.   12805
GRIEVE, C.M.   13460
GRIEVE, P.W.   8602
GRIFFIN, D.M.   8931, 10129, 11143,
11708
GRIFFIN, J.L.   10130
GRIFFIN, R.E.   7090
GRIFFITH, R.L.   12236
GRIFFITHS, H.   14216
GRIFFITHS, J.H.   8603
GRIGNON, C.   9541
GRIGORYUK, I.A.   13461, 14410
GRILLI, I.   6644
GRIME, J.P.   8604, 9929, 10131
GRIMES, D.W.   10132
GRIMME, K.   10010, 13462, 13759
GRINCHENKO, A.L.   8605
GRINEVA, G.M.   8876
GRIP, H.   8630, 9167, 10133
GRISHANOVA, N.P.   7538, 10346
GRIVET, C.   12850
GRODZINSKI, B.   11789, 13463, 13852
GROF, C.P.L.   13644

HARBAOUI, Y. 8418
HARBAUGH, B.K. 8640, 10941, 12352,
12637, 14485
HARBERD, N.P. 13499
HARBINSON, J. 13500
HARDER, H.J. 10167, 10168, 10169,
10170
HARDIE, K. 8641
HARDIE, W.J. 13501
HARDJOAMIDJOJO, S. 10171, 10912,
11735
HARDWICK, K. 11198
HARDY, J.R. 7122
HARGREAVES, G.H. 10172, 10173,
10174
HARI, P. 8642, 11082
HARIVANDI, M.A. 10175, 10176,
10177, 11736
HARKOV, R. 8290
HARLAN, P.W. 10160
HARLEY, P.C. 7955
HARMAN, W.L. 11737
HARMANNY, K. 12841
HARMS, W.R. 8643
HARNEY, P.M. 10434
HARPER, J.E. 13203
HARPER, L.A. 6789
HARRADINE, A.R. 10178
HARRIS, C.E. 10179, 10180, 11738,
13502, 14285
HARRIS, D. 7003
HARRIS, G.A. 8316, 8644
HARRIS, G.C. 11739, 11740
HARRIS, G.P. 9923
HARRIS, H.C. 9420
HARRIS, J.G. 11741
HARRIS, J.M. 7123
HARRIS, K. 10022
HARRIS, M.J. 8645
HARRIS, P.M. 11251
HARRIS, R.F. 10466
HARRIS, R.W. 8131
HARRISON, A.T. 8530, 8840, 8841,
9732
HARRISON, C.R. 6652
HARRISON, D.S. 10869, 12570, 14453
HARRISON, J.G. 7124, 7125, 11742,
13503
HARRISON, P.G. 8979
HARRISON, S.K. 14508
HART, G.E. 7126, 7127
HART, W. 7095
HARTMANN, H. 8397, 10181
HARTSOCK, T. 7607 see also
HARTSOCK, T.L. 8926, 9084, 9085,
10339, 12227, 14018
HARTUNG, W. 9921, 10182, 11408,
11743, 11744, 13026
HARTWIG, E.E. 10193
HARVEY, B.L. 7197

HARVEY, C.N. 12714
HARVEY, D.M. 7128
HARVEY, D.M.R. 13484
HARVEY, R.G. 7203
HARZALLAH-SKHIRI, F. 10183, 13504
HASEGAWA, P.M. 8291, 11391, 11727,
11728, 13505
HASEGAWA, S. 10184, 11236
HASHEMI, F. 10185
HASHIM, O. 14238
HASHIMOTO, Y. 8871, 8872, 12247,
12248, 12249, 12250, 12251, 13506,
14052
HÄSLER, R. 10186, 11745, 13699
HASNAIN, S. 9377
HASPEL-HORVATOVIČ, Ę. 8646 see also
HASPELOVÁ-HORVATOVICOVÁ, A. 7129
HASSAN, W. 10442
HASSON, E. 10187, 10690, 11746,
11747
HASTINGS, D.F. 8621
HASTINGS, S.J. 14035
HASZA, D. 12642
HATATA, M. 10188
HATCHER, A.I. 14603
HATFIELD, J.L. 8711, 8714, 11748,
11749, 11835, 12267, 12803, 13507,
13508, 13509, 13510, 13582, 13583,
14686
HATLITLIGIL, M.B. 13511
HAUCK, R.D. 9307
HAUG, A.R. 13092
HAUGSTAD, M. 9533
HAUN, J.R. 11750
HAUPT, W. 10189
HAUS, M. 13512
HAUSBECK, M.K. 13524
HAUSER, V.L. 13513
HAUSMANN, K. 13514
HAVAS, P. 8707
HAVAUX, M. 11751, 11752
HAVELKA, U.D. 13515, 13516
HAVERKAMP, R. 13517
HAVILL, D.C. 14538
HAVIS, J.R. 7130
HAVRANEK, W.M. 7131, 7936, 8238
HAWARE, M.P. 14613
HAWKER, J.S. 8647, 8648, 9603,
10190
HAWKINS, C.D.B. 8649
HAYAKAWA, T. 14330
HAYAMA, T. 7132, 7255
HAYASHI, M. 13518
HAYASHIDA, Y. 12206
HAYATA, Y. 12949
HAYCOCK, D. 13916
HAYDEN, D.B. 11682
HAYDEN, T. 7664
HAYDOCK, K.P. 13369
HAYES, J.T. 12702, 14545, 14546,
14547, 14548, 14549, 14550

MORGAN, J.M.   7494, 7495, 12878,
13966, 13967, 13968, 13969
MORGAN, M.   6872
MORGAN, P.W.   6909
MORGAN, T.H.   7496
MORI, K.   8971
MORI, S.   9080
MORIKAWA, Y.   8446
MORIN, J.   13969
MORISON, J.I.L.   8745, 12166, 12167,
12168, 13971, 13972
MORITSUGU, M.   11898, 11899
MORIYA, T.   13677
MORIZET, J.   9043, 13973, 13974,
13975
MORK, H.M.   9044
MOROHASHI, Y.   10526, 10527, 13976
MOROT-GAUDRY, J.-F.   7490, 9763
MOROZOV, V.L.   6702
MOROZOVA, E.M.   12221
MORRIS, J.D.   9045
MORRIS, J.R.   10885, 12169, 12170
MORRIS, P.   10528
MORRIS, R.P.   13498
MORRISON, I.N.   6719, 11216
MORRISON, J.E., Jr.   9046, 9047
MORRISON, N.S.   14175
MORROW, C.T.   7623, 13633
MORTECZKA, H.   12899, 12900
MORTENSEN, D.A.   9777, 11335
MORTIMORE, G.   7182
MORTON, F.I.   12171, 13977
MORTON, J.B.   12172
MORVAN, C.   13240
MOSER, T.J.   10568, 12173
MOSTAGHIMI, S.   12174
MOTA, F.S., da see Da MOTA, F.S.
MOTES, J.E.   10996, 10997
MOTHA, R.P.   7497, 9048
MOTOSUGI, H.   11996
MOTT, K.A.   10529, 13978
MOTTERSHEAD, B.E.   7392
MOTTO, M.   14583
MOTYER, M.S.   11682
MOUGOU, A.   12175, 13979
MOURAVIEFF, I.   9049
MOUSSEAU, M.   7077, 9050, 13980,
14129
MOUTONNET, P.   7634, 7658, 9051,
10530, 13077, 13078, 13981
MOWRY, F.L.   13590
MOY, R.M.   7533
MOYA, T.B.   9126, 10628
MOYSE, A.   10269
MOZAFAR, A.   9459
MOZHAEVA, L.V.   9052
MOZHEIKO, G.A.   7498
MROZEK, E., Jr.   9053
MTUI, T.A.   9054

MUCHOW, R.C.   7499, 7500, 7501,
7502, 7503, 9055, 10531, 14721
MUDRIK, V.A.   10532
MUEKSCH, W.M.C.   10533
MUELLER, P.A.   13982
MUELLER, P.W.   14560
MUELLER, T.   11101
MUENDEL, H.-H.   11775
MÜHLE, H.   10534
MUIRHEAD, W.A.   9056, 14127
MUKHERJEE, I.   10535
MUKHERJEE, S.P.   9057, 10536, 12176,
12177
MUKHINA, V.A.   12178
MUKHIYA, Y.K.   12179
MUKHTANOV, I.   7053
MULÁR, J.   9058, 9967
MULDER, J.C.   13184
MULDOON, J.F.   13267, 13799
MULEV, M.   9059
MÜLLER, C.   7910
MÜLLER, H.J.   13983
MÜLLER, H.-M.   10537, 10538, 10539
MÜLLER, J.   10540
MÜLLER, L.   9060
MULLER, R.N.   7504, 10541
MULLER, W.   9061
MULLER, W.H.   9077, 10585
MÜLLER, W.J.   8101, 11168, 11169
MULLINIX, B.G., Jr.   11611
MULLINS, J.A.   9062
MUMFORD, P.M.   7505
MUNDEL, G.   10542, 10543, 12180,
12181, 13756
MUNGSE, H.B.   10544
MUNNECKE, D.E.   10545
MÜNNICH, H.   13442
MUNNS, D.N.   10064
MUNNS, R.   6681, 7087, 7798,
7900, 9063, 9964, 10125, 10546,
10547, 11706, 11707, 12182, 12183,
13984, 13985, 14102, 14217
MUNNS, R.E.   6682, 12184
MUÑOZ, C.E.   13986
MUNRO, D.S.   13987
MUNTAJABUDDIN, K.   13988
MUONEKE, C.O.   12756
MURAKAMI, E.   10563
MURALI, N.S.   13989
MURARI, K.   10644
MURASE, H.   7506
MURATA, M.   7507
MURATA, Y.   10252, 12185
MURAVLEV, A.P.   8520
MURAYAMA, S.   9090
MURET, I.A.   10548
MURISON, R.D.   9544
MURPHY, C.E., Jr.   9064, 12046
MURPHY, E.M.   10549

VERNET, T.   10055
VERTUCCI, C.W.   12784, 14645
VESELKOV, B.M.   14646
VESELOVA, T.V.   8023
VESELOVSKIĬ, V.A.   8023
VETŐ, F.   12785, 12786, 14647
VETTERLEIN, E.   10352, 12787
VEVERKA, K.   14648, 14649
VIALA, G.   13123
VIANE, R.   8024
VICENTE, A.M.   10497
VICENTE-CHANDLER, J.   6608, 12582
VICHERKOVÁ, M.   7307, 12788, 14650
VICHEV, Zh.   12789
VICTOR, D.M.   11090
VICTORIA, R.L.   14300
VIDAL, A.   9586, 9587, 11091,
14651
VIDAL, J.   11649, 12790
VIDAL-MADJAR, D.   14182
VIDAVER, W.E.   8649, 13265, 13266
VIDRAŞCU, P.   7215
VIEIRA, J.V.   9958
VIEIRA, R.de M.   13011
VIEIRA, S.R.   13510
VIEIRA da SILVA, J.   7631, 10681,
11792, 11793, 11985, 12323, 12324,
12791, 12792, 13352, 14472, 14652
VIEIRA da SILVA, J.B.   10199, 12034,
12919
VIENKEN, J.   7784
VIG, A.C.   7845
VIGH, L.   9588
VIJAYA KUMAR, C.S.K.   7929, 8025,
8026
VILELLA, F.   10148, 13489
VILLA, A.   12968
VILLALOBOS-RODRIGUEZ, E.   14653
VILLA NOVA, N.A.   10665, 12309,
12953
VILLAREAL, R.L.   11092
VILLEGAS, A.N.   14090
VINCENT, J.F.V.   12793
VINCZE, L.   13999
VINE, P.N.   11093
VINES, H.M.   7428, 9712, 11249,
12970
VINOGRADOVA, G.B.   13263, 13775,
13776
VINOPAL, R.T.   8027
VINTILĂ, R.   7271
VIRGIN, H.I.   7933
VIRGINIA, R.A.   12224
VIRK, M.S.   13002
VIRMANI, S.M.   7857, 9493
VIRZO de SANTO, A.   8028, 12794,
12795, 12796
VISSER, R., de see De VISSER, R.
VITKOV, M.   6933, 9589
VÍTKOVÁ, A.   9218

VIVOLI, J.   7425, 10457, 11562
VLAKHOVA, M.   9590, 12797, 14654
VLEK, P.L.G.   9591
V"LEV, V.   8173, 9590, 14654
V"LEV, V.T.   9705
VLIET, W.P.A., van see VAN VLIET,
W.P.A.
VODYANIK, A.S.   14655, 14656
VODYANIK, T.M.   14655, 14656
VOETBERG, G.   14657
VOGEL, S.   14658
VOGELLEHNER, D.   14659
VOGT, G.   11094
VOLENEC, J.J.   11095
VOLFOVÁ, A.   13774
VOLKENBURGH, E., van see VAN
VOLKENBURGH, E.
VOLKOVA, N.P.   12798
VOLKOVA, R.I.   8029, 8030
VOLODARSKIĬ, N.I.   8031 see also
VOLODARSKY, N.I.   8032
VOLOKITA, M.   10754, 14241
Von BERNUTH, R.D.   14660
Von CAEMMERER, S.   9592, 10036,
10038, 13118
VONSHAK, A.   9593
VOORHEES, W.B.   9264, 9845
VOROBEĬKOV, G.A.   8033
VOROB'EV, N.V.   12799, 12930
VOSKRESENSKAYA, N.P.   9193, 14661
VOSS, R.E.   12867
VOTRUBA, M.   11971
VOZARI, É.   7387
VOZNESENSKIĬ, V.L.   14365, 14662,
14663
V"RBANOV, M.   14664
VU, T.T.   10554
VUCHELICH, D.   9664
VUČIĆ, N.   9594, 9595
VUCINIĆ, Ž.   10715, 10735
VUGTS, H.F.   12462
VYAS, S.P.   8556, 10083, 13403,
13404

W

WABER, J.   14665
WACH, M.J.   8399
WADDELL, J.W.   14734
WADSWORTH, J.I.   11096
WADSWORTH, R.M.   10305
WAGENET, R.J.   8034, 8318, 10845,
12638, 12800, 13697, 14666
WAGHMODE, A.P.   8776, 9596, 9597,
9598, 12801
WAGLE, D.S.   6850, 8345, 9877,
9878, 11454, 11455
WAGNER, C.E., van see VAN WAGNER, C.E.
WAGNER, D.J.   12722, 14568, 14667

PLANT INDEX
(Cumulative index to volumes 6 – 10)

This index contains plant genera and types interesting as experimental material for physiological, ecological and agricultural studies. The Latin plant names are the main items which present the reference number. English names of the most common plants are cross-indexed.

A

Abelmoschus 7206, 9252, 12391, 12551, 12592

Abies 6660, 6707, 7126, 7187, 7305, 7343, 7378, 7379, 7399, 7555, 7876, 7998, 8226, 8253, 8365, 8742, 8743, 8762, 8800, 8801, 8802, 8839, 8857, 8894, 9132, 9220, 9229, 9358, 9422, 9627, 9778, 9850, 9876, 10087, 10313, 10314, 10363, 10673, 10674, 10837, 11010, 11263, 11301, 11309, 11587, 11897, 12060, 12120, 12317, 12623, 12705, 13090, 13239, 13323, 13334, 13695, 13699, 13701, 13702, 13705, 13736, 13795, 13823, 13856, 13857, 13956, 14392, 14460, 14556, 14557, 14609, 14687

Abronia 7576, 7802, 12557, 13373

Abrus 9318

Abutilon 6823, 8360, 9991, 10076, 10189, 11787, 11920, 12301, 12391, 12873, 13021, 13573

Acacia 6655, 6740, 7035, 7305, 7911, 8579, 8772, 8857, 8861, 8939, 8985, 9183, 9555, 9646, 9743, 9919, 19363, 10729, 10842, 11550, 12234, 12557, 13114, 14014, 14353, 14581, 14681, 14682

Acalypha 9243, 10253, 10633

Acanthus 7521, 11889

Acer 6614, 6823, .6885, 6923, 6974, 7035, 7093, 7184, 7185, 7254, 7305, 7352, 7550, 7555, 7621, 7712, 8037, 8213, 8366,

8433, 8643, 8677, 8709, 8854, 8857, 8860, 8861, 8929, 8933, 9489, 9516, 9843, 9876, 9919, 9991, 10087, 10101, 10162, 10239, 10363, 10364, 10426, 10460, 10541, 10579, 10753, 10880, 11018, 11023, 11027, 11238, 11305, 11387, 11425, 11588, 11758, 11812, 11820, 11920, 11957, 11960, 12066, 12120, 12277, 12433, 12623, 12666, 12841, 12905, 13090, 13327, 13392, 13595, 13747, 13754, 14137, 14560, 14588, 14606, 14719, 14758

Achillea 6641, 7189, 7519, 9094, 13887, 14595, 14596, 14597

Achyranthes 9243

Aconitum 6914, 8187

Actaea 6914, 8187

Actinidia 8504, 8505, 13237, 13295, 13448, 13907

Adansonia 7004, 12391

Adenocarpus 9318

Adonis 6914, 13632, 14166, 14167

Aegiceras 9600, 10036, 13005, 13168

Aegilops 9912, 9913, 9914, 10687, 11559, 13194, 13740

Aeluropus 9596, 9597, 9598, 12539

Aeonium 11004, 13828

Aeschynomene 8153, 9008

Aesculus 7034, 8517, 8915,
9578, 9991, 10541, 11387, 12910,
13119, 13120

Agathis 13788, 14383

Agave 7235, 7305, 7544,
8861, 9085, 9349, 9747, 9901,
10883, 11584, 12228, 12869, 13654,
13840, 14016, 14681, 14718

Ageratum 11943

Agropyron 6641, 6890, 7027,
7083, 7269, 7302, 7305, 7393,
7508, 8065, 8311, 8531, 8552,
8607, 8644, 8796, 8984, 9094,
9204, 9252, 9600, 9646, 9693,
9732, 9817, 9922, 9938, 10032,
10178, 10471, 10516, 10871, 10883,
11283, 11414, 11487, 11614, 11865,
11960, 12111, 12114, 12158, 12159,
12375, 12459, 12488, 12557, 13020,
13158, 13379, 13641, 14026, 14284,
14751

Agrostis 6828, 7870, 8655,
9252, 9600, 9863, 9907, 10177,
11018, 11208, 11515, 12809, 12881,
12882, 14340, 14487, 14643, 14674

Ailanthus 11018

Albizzia 7945, 10189, 11387,
11550, 11998, 13090, 14235, 14581,
14681

Alchemilla 8850, 14003

alder see *Alnus*

alfalfa see *Medicago*

Algae
  Acetabularia 9827, 9996,
  14698
  Anabaena 7343, 8162, 8452,
  8453, 8492, 9234, 9996,
  11104, 11601, 14176, 14230
  Anacystis 7673, 8492, 9593,
  9996
  Aphanothece 8086, 9996,
  14229
  Ascophyllum 7789, 13133
  Bostrychia 8415, 9655, 9952
  Bryopsis 8256, 9652
  Calothrix 14230
  Caulerpa 12253, 13576
  Ceramium 8256
  Chaetomorpha 8256

Chara 7769, 7952, 8155,
8749, 8932, 9330, 9331,
9457, 9538, 10947, 11049,
11123, 12649, 12650, 13060,
14492, 14604, 14697
Chilomonas 13514
Chlamydomonas 7315, 7510,
7671, 8492, 9996, 10038,
11320, 11681
Chlorella 6795, 6818, 7343,
7745, 7888, 8281, 8338,
8492, 9463, 9684, 9821,
9959, 9996, 10125, 10146,
10192, 10547, 10852, 10929,
10967, 11707, 12183, 12546,
12924, 13453, 14712
Codium 6958, 8256, 9780,
9996, 12416
Corallina 8922
Cosmarium 11323
Cyclotella 8281, 9463,
9996
Dunaliella 6708, 6709,
6710, 7510, 7671, 7769,
7888, 7889, 8281, 8492,
8573, 8574, 8815, 9749,
9959, 9996, 10102, 10103,
10864, 10964, 11114, 11308,
11684, 11707, 12033, 13213,
13424, 13425, 13426, 13427,
13512, 14195
Enteromorpha 6787, 9310,
12416, 13808, 14181
Euglena 9784, 10622
Fucus 6715, 7046, 7789,
7790, 7973, 8281, 8683,
8771, 9765, 9993, 9996,
11687, 13133, 13354, 14347
Gelidium 8922, 9996
Gigartina 6958, 8922
Gloeotrichia 13892
Gracilaria 8492, 8830,
13790
Halicystis 8124, 8126,
8256, 9675, 9827
Isochrysis 8884, 9996
Laminaria 8771, 9993,
13354
Lamprothamnium 8257, 8258,
8259, 10331, 11337, 14040,
14041, 14239, 14697
Macrocystis 8922, 9346
Mesotaenium 9821
Microcystis 11601, 12240,
12241
Nannochloris 9821, 11397
Nitella 6696, 7005, 7132,
7255, 7769, 8126, 8256,
8377, 9248, 9562, 9664,
9959, 10714, 11049, 12650,
14211

Algae (cont.)
  Nostoc 7869, 8452, 8492,
  9923, 9996, 11351, 11353,
  11489, 11490, 14175, 14230
  Ochromonas 7260, 7888,
  8281
  Oedogonium 14408
  Oscillatoria 8492
  Padina 13025
  Pelvetia 7789, 7790,
  8158, 9765, 14033, 14347
  Phaeodactylum 8281, 9350,
  9351
  Platymonas 7284, 7286,
  7287, 8281, 9996, 11707,
  11928
  Polysiphonia 12416, 14228
  Porphyra 6958, 7573, 7694,
  7888, 8063, 9256, 9257,
  9258, 9259, 9260, 9633,
  9996, 10060, 10617, 10618,
  10918, 11134, 11632, 12511,
  12840, 13689, 13726, 14455,
  14710
  Porphyridium 8281, 9674,
  9996
  Rhodymenia 8830
  Rivularia 12418
  Sargassum 6958, 13025
  Scenedesmus 6795, 7315,
  9385, 9996, 13064
  Spirogyra 14408
  Stichococcus 6787, 8281
  Synechococcus 7769, 8282,
  8492, 11351, 11352, 13934,
  14176, 14229, 14230
  Trebouxia 7630, 9385,
  10617, 10619, 13564
  Ulothrix 6787
  Ulva 8434, 8492, 8683,
  9765, 9974, 9996, 12416,
  13354
  Valonia 7285, 7769, 7786,
  7888, 8123, 8126, 8256,
  8612, 8749, 9827, 9996,
  11049, 11790, 13106
  Vaucheria 7933

Alhagi 14201

Alisma 8947, 9049

Allium 6791, 6863, 6894,
6944, 7035, 7057, 7067, 7142,
7152, 7299, 7305, 7443, 7591,
7592, 7724, 7727, 7778, 7779,
7780, 7782, 7831, 7950, 7974,
8038, 8115, 8116, 8158, 8182,
8303, 8405, 8442, 9098, 9101,

9135, 9136, 9142, 9184, 9252,
9282, 9312, 9343, 9349, 9356,
9394, 9477, 9626, 9640, 9667,
9668, 9676, 9919, 10022, 10196,
10333, 10334, 10335, 10336, 10337,
10338, 10423, 10574, 10575, 10607,
10631, 10834, 10838, 10866, 10905,
11360, 11473, 12017, 12144, 12239,
12268, 12280, 12318, 12814, 12898,
12936, 13134, 13193, 13199, 13302,
13418, 13423, 13488, 13545, 13546,
13547, 13548, 14515, 14713, 14718

almond see *Amygdalus*

Alnus 6625, 7020, 7049,
7305, 7343, 8677, 8722, 8762,
8850, 8857, 8861, 8933, 10406,
10872, 11238, 11960, 11992, 12321,
12433, 12448, 14360, 14758

Alocasia 9919, 10427, 12828

Aloë 8068, 8397, 12373, 12846,
13408, 13891, 14681

Alopecurus 6704, 7302, 7571,
9252, 9850, 10032, 10966, 12035,
12623, 12719, 12759, 13174, 14517

Alstonia 9159

Alternanthera 9156, 9243, 10658,
13080, 13825, 14636

Althaea 12391, 12551

Altingia 8040

Alysicarpus 6677

Alyssum 7212, 7213, 7214,
13408

Amaranthus 6677, 6717, 6823,
6980, 7070, 7319, 7389, 7559,
7571, 7575, 7788, 7812, 8307,
8561, 8686, 8745, 8857, 9131,
9160, 9243, 9341, 9429, 9617,
9642, 9747, 9829, 9844, 9861,
9862, 9947, 10343, 10370, 10633,
10729, 11002, 11018, 11022, 11407,
11497, 11581, 11614, 11627, 11891,
11920, 12026, 12301, 12368, 12382,
12526, 12537, 12872, 12969, 13021,
13313, 13314, 13644, 13947, 13971,
13972, 14121, 14122, 14169, 14508

Amaryllis 10073

Ambrosia 6823, 7070, 7607,
7939, 8038, 8352, 8939, 8985,
9991, 10529, 11223, 11377, 11920,
12162, 12770, 13021, 13645, 13787,
14338

Amelanchier 9472, 10541, 13663

Ammi 9371

Amomum 8040

Amygdalus 7305, 8175, 8648,
9848, 12825, 13350, 14167

Anabasis 11667, 14681

Anacardium 14510

Anacharis 7698, 7699, 9523,
10948, 12650, 13839, 14339, 14493

Anagallis 13980

Anagyris 9318

Ananas 7305, 7526, 7748,
9090, 9809, 11018, 14681

Anastatica 8245, 8535

Anchusa 7212, 7214

Andromeda 9433, 10416

Andropogon 7226, 7227, 7256,
9006, 9732, 10862, 13158, 13314,
13725, 13846, 14166, 14167

Androsace 14517

Anemia 10235, 14718

Anemone 6914, 8187, 8475,
10600, 13408, 13632

Anethum 9371

Anoda 12301

Anthoceros 8929, 10073, 14718

Anthoxanthum 7534

Anthurium 12314

Anthyllis 8850, 9318, 11223

Antidesma 10253

Antirrhinum 8055, 10528, 11018,
13333, 14154

Apium 6634, 6724, 6863,
6906, 7152, 7273, 8164, 8248,
9630, 9773, 9774, 10040, 11225,
11636, 12691, 13938, 14488, 14559

Apocynum 8092

apple see *Malus*

apricot see *Armeniaca*

Aquilegia 6914, 7344, 7345,
8187, 9094

Arabidopsis 6717, 6863, 7736,
12626

Arabis 9929, 10131

Arachis 7305, 7359, 7409,
7587, 7588, 7713, 7929, 8202,
8241, 8245, 8342, 8370, 8846,
9088, 9239, 9241, 9252, 9317,
9326, 9429, 9443, 9493, 9811,
9814, 10059, 10259, 10641, 10741,
10742, 10883, 10992, 11018, 11236,
11312, 11364, 11450, 11566, 11711,
11769, 11770, 11859, 11883, 11884,
11914, 11915, 12009, 12010, 12195,
12294, 12295, 12381, 12401, 12409,
12573, 12695, 13038, 13053, 13708,
13875, 13938, 14057, 14088, 14089,
14090, 14208, 14223, 14225, 14266,
14312, 14319, 14638, 14639, 14718

Araucaria 11032, 13686, 14575,
14576

arborvitae see *Thuja*

Arbutus 7954, 9514, 9515,
10087, 10387, 10425, 10514, 11003,
11408, 11744, 13254, 13358, 13668,
14600, 14718

Arctium 7465, 9991

Arctostaphylos 7536, 7960, 8365,
8735, 9011, 9099, 9149, 9194,
9195, 10600, 10779, 11144, 11145,
14035, 14263

Arctous 9876

Ardisia 9858

Areca 7060

Arenaria 6809

Arenga 8128

Argemone 8620

Arisaema 6614

Aristida 6677, 8493, 13158,
14099, 14662, 14663

Armeniaca 7129, 7525, 7934,
7956, 7957, 8094, 8117, 8646,
8849, 8951, 11183, 11253, 13350,
13753, 14652

Armeria 9907

Arnica 8850, 9094, 9422,
9660, 10600, 11172, 12892

Arrhenatherum 8435, 8850, 9929,
10131, 12552, 14643

Artemisia 6717, 7010, 7070,
7212, 7213, 7214, 7519, 7524,
7793, 7794, 8065, 8182, 8352,
8636, 8939, 9072, 9094, 9195,
9360, 10069, 10364, 10600, 10653,
10966, 11127, 11166, 11283, 11912,
13641, 13734, 14026, 14517, 14681

Arum 6827, 8475

Asarum 6614, 8877, 13462,
13753, 14078

Asclepias 8092, 9120, 10087,
13585

ash see *Fraxinus*

Asparagus 7212, 7624, 9098,
11018, 12192

aspen see *Populus*

Asperula 11018

Aspidistra 10866, 14681

Aspidosperma 7879

Aster 7982, 8588, 8850,
9635, 9646, 9817, 9907, 10389,
11496, 13146

Astragalus 7212, 7213, 7214,
7612, 9318, 9736, 10032, 10677,
10966, 13373, 13734, 13804, 14166,
14167, 14517

Astrebla 8861, 9883, 11111,
13158

Atriplex 6717, 6731, 7010,
7076, 7089, 7212, 7214, 7222,
7223, 7305, 7343, 7383, 7486,
7544, 7576, 7577, 7671, 7673,
7704, 7710, 7735, 7802, 8007,
8078, 8091, 8168, 8260, 8261,
8589, 8644, 8749, 8772, 8861,
8939, 8982, 8985, 9044, 9122,
9131, 9183, 9184, 9361, 9419,
9600, 9646, 9784, 9817, 9822,
9850, 9907, 9919, 10339, 10520,
10589, 10681, 10792, 10841, 10935,
11101, 11324, 11550, 11555, 11719,
11905, 11951, 12036, 12160, 12162,
12182, 12349, 12434, 12556, 12557,
12623, 12968, 13035, 13269, 13429,
13644, 13709, 14121, 14122, 14358,
14434, 14674, 14681

Atropa 9320, 10069

Avena 6622, 6751, 6920, 6948,
6994, 7023, 7055, 7070, 7081,
7083, 7140, 7203, 7212, 7225,
7343, 7435, 7440, 7796, 7910,
7930, 7968, 8156, 8168, 8182,
8245, 8428, 8472, 8745, 8755,
8833, 8835, 8864, 8951, 8991,
9099, 9120, 9121, 9131, 9299,
9375, 9389, 9461, 9462, 9523,
9571, 9747, 9890, 9892, 9893,
9919, 10010, 10057, 10068, 10091,
10108, 10194, 10555, 10675, 10676,
10712, 10735, 10808, 10827, 10866,
10883, 10901, 10943, 10944, 10950,
11018, 11023, 11076, 11175, 11216,
11317, 11377, 11463, 11559, 11646,
11730, 11732, 11789, 11839, 11891,
11945, 12010, 12113, 12190, 12510,
12578, 12661, 12687, 12770, 12872,
12969, 13092, 13170, 13364, 13365,
13399, 13577, 13876, 13935, 13938,
14586, 14718, 14745, 14764

Avicennia 6658, 9097, 9183,
9729, 9989, 10036, 11495, 11496,
13005, 13006, 13007, 13109, 13167,
13168

avocado see *Persea*

Azalea 8333

B

Baccharis 10520

Bacteria
   Bacillus 9700
   Chromatium 11376, 14174
   Clostridium 9463
   Corynebacterium 9576
   Erwinia 8190, 8584, 8650
   Escherichia 6654, 7769,
   8027, 9463, 10466, 12266,
   13811
   Halobacterium 7355, 7387,
   9284, 9879, 9959, 11396,
   13549
   Lactobacillus 12266, 13070
   Pseudomonas 6971, 8099,
   8190, 8862, 12266, 13649
   Rhizobium 7400, 8297, 8702,
   8904, 8945, 8946, 9155,
   9552, 9665, 9867, 10064,
   10624, 10924, 11226,11803,
   12011, 12374, 12598, 12599,
   14198, 14446, 14715
   Rhodopseudomonas 7547, 10221,
   10256, 12222, 12938
   Rhodospirillum 7861, 7862,
   8865, 10346, 11376, 13430
   Salmonella 6897, 7298,
   9934, 11498
   Streptococcus 9546, 12266
   Xanthomonas 10286, 10939

Balanops 8208

Bambusa 10435, 14681

banana see *Musa*

Banksia 6863, 8579, 9529,
10363

Barbaraea 12389

barberry see *Berberis*

Barbeya 8208

barley see *Hordeum*

Bassia 13429

basswood see *Tilia*

Batis 6647, 11239, 13429

Bauhinia 7692, 9302, 10065

bean see *Phaseolus*

beech see *Fagus*

Begonia 10711, 11550, 13056,
13891, 14681

Berberis 9422, 11700, 14166,
14167

bermudagrass see *Cynodon*

Beta 6640, 6802, 6804, 6829,
6830, 6832, 6863, 6921, 6948,
6969, 7050, 7070, 7207, 7293,
7299, 7305, 7323, 7343, 7370,
7373, 7435, 7473, 7544, 7571,
7645, 7761, 7792, 7801, 7853,
7920, 7966, 7995, 7996, 8005,
8081, 8145, 8147, 8172, 8173,
8182, 8190, 8199, 8244, 8305,
8307, 8313, 8325, 8427, 8488,
8503, 8513, 8516, 8545, 8614,
8673, 8685, 8719, 8742, 8772,
8805, 8861, 8957, 8985, 9034,
9058, 9146, 9200, 9218, 9223,
9252, 9266, 9297, 9382, 9419,
9429, 9430, 9431, 9518, 9551,
9646, 9685, 9704, 9705, 9706,
9755, 9817, 9822, 9926, 9941,
9967, 9981, 10009, 10022, 10049,
10108, 10111, 10112, 10156, 10163,
10185, 10210, 10224, 10243, 10282,
10297, 10344, 10350, 10352, 10366,
10367, 10432, 10485, 10488, 10759,
10832, 10833, 10840, 10847, 11006,
11018, 11233, 11377, 11535, 11537,
11570, 11571, 11572, 11696, 11733,
11768, 11852, 11865, 11871, 11913,
11959, 12007, 12063, 12090, 12107,
12128, 12284, 12341, 12422, 12423,
12436, 12447, 12517, 12609, 12623,
12691, 12745, 12762, 12930, 12931,
12934, 12962, 12964, 12969, 13007,
13193, 13204, 13217, 13222, 13244,
13269, 13272, 13304, 13356, 13368,
13413, 13471, 13488, 13600, 13666,
13757, 13758, 13763, 13947, 13999,
14031, 14138, 14146, 14170, 14180,
14192, 14276, 14364, 14431, 14551,
14552, 14574, 14626, 14648, 14649,
14664, 14665, 14713, 14763

Betula 6673, 6777, 6836, 6837,
7002, 7019, 7305, 7343, 7555,
7578, 7579, 7621, 7791, 7995,
7996, 8012, 8184, 8249, 8250,

Betula (cont.)
8413, 8642, 8677, 8722, 8857,
9034, 9087, 9123, 9220, 9516,
9731, 9876, 10087, 10153, 10236,
10329, 10363, 10364, 10438, 10541,
10591, 10600, 10602, 10615, 10616,
10672, 10691, 10735, 10838, 10840,
10990, 11012, 11018, 11187, 11387,
11992, 12120, 12319, 12433, 12686,
12689, 12905, 13090, 13257, 13323,
13571, 13743, 13746, 13878, 13930,
14022, 14434, 14458, 14588

Bidens   14511

birch   see   *Betula*

Biscutella   12389

blueberry   see   *Vaccinium*

Boerhaavia   9243, 10009, 12537,
14445

Bombax   12391

Borassus   7060

Borya   6905, 7035, 8772,
10207, 10208, 10209, 13393, 14434

Bothriochloa   9925, 11111,
13202, 14099

Bouteloua   7070, 7083, 7269,
8082, 8160, 8293, 8294, 8804,
9321, 9693, 9695, 9884, 9922,
10032, 10194, 10804, 10805, 12158,
12488, 12969, 13158, 13710, 13895,
14099

Brachiaria   7226, 7227, 9021,
10633, 11268, 12269, 13184, 13391,
13846

Brachypodium   7733, 9929, 10131

Brassica   6610, 6643, 6723,
6815, 6855, 6863, 6904, 7035,
7070, 7142, 7273, 7299, 7305,
7313, 7419, 7627, 7702, 7703,
7817, 7822, 7870, 7915, 7916,
7925, 8020, 8180, 8182, 8287,
8307, 8346, 8393, 8394, 8454,
8516, 8535, 8557, 8614, 8655,
8804, 9095, 9098, 9208, 9252,
9282, 9297, 9620, .9764, 9784,
9858, 9889, 9930, 9933, 9980,
9992, 10022, 10108, 10181, 10243,

10276, 10405, 10432, 10478, 10598,
10614, 10633, 10743, 10745, 10819,
10840, 10883, 10889, 10890, 11018,
11086, 11109, 11280, 11284, 11292,
11694, 11778, 11844, 11868, 12013,
12015, 12070, 12122, 12389, 12516,
12563, 12590, 12623, 12687, 12749,
12762, 12770, 12815, 12837, 12843,
12862, 12966, 12969, 13110, 13149,
13193, 13347, 13377, 13412, 13429,
13460, 13488, 13771, 13814, 13848,
13872, 13893, 13894, 13906, 13908,
13909, 13910, 13938, 13971, 13972,
14015, 14207, 14270, 14348, 14444,
14500, 14572, 14630, 14674, 14717

broadleaf   see   *Hyptis*

brome grass   see   *Bromus*

Bromus   6650, 7070, 7302, 7508,
7733, 8182, 8465, 8466, 8607,
8644, 9193, 9252, 19883, 11018,
12192, 12552, 12675, 13551, 13734,
14092, 14259, 14282, 14621

Bruguiera   12193

Brunellia   10992

Brunfelsia   10992

Bryonia   11238

Bryophyllum   7347, 8790, 9673,
10609, 12858

Bryophyta
   Acrocladium   7100
   Anomodon   9210, 12357
   Anthoceros   12357
   Atrichum   8114, 12357
   Aulacomnium   12357, 12948
   Barbula   7738, 10522, 12357
   Bazzania   12357
   Brachythecium   10672
   Bryum   6635, 8245, 11580,
   12357
   Ceratodon   9210, 12357, 14617
   Climatium   12357
   Conocephalum   9210, 12357
   Cratoneuron   8245, 8431, 9765,
   10955, 14045
   Dicranum   9210, 10672, 12357,
   12948, 14617
   Diphyscium   8114
   Drepanocladus   12357
   Fissidens   8619

Bryophyta (cont.)
  Funaria 7531, 9823, 12480,
    12481, 12482
  Grimmia 6635, 7738, 9210,
    13791
  Hedwigia 12357
  Hylocomiun 9412, 12357,
    12604, 12605, 12859, 14617
  Hypnum 12948
  Leucobryum 9210, 12357,
    13890
  Lophocolea 8114, 12357
  Lophozia 12948
  Lunularia 9210, 12357
  Marchantia 9210, 11705,
    11789, 12357
  Mnium 7544, 9210, 10073,
    12357, 14718
  Neckera 8245, 12357
  Nowellia 8114
  Pellia 9210, 12357
  Plagiochila 12357
  Plagiomnium 12357
  Plagiothecium 12357, 12948
  Pleurozium 9412, 10672,
    12357, 12604, 12605, 12859,
    12948, 14617
  Pogonatum 14617
  Pohlia 10672, 12948
  Polytrichum 6873, 8245,
    8954, 9411, 9412, 9490,
    10670, 10672, 12133, 12357,
    12604, 12605, 12734, 12948,
    13791, 13879, 14617
  Porella 12357
  Preissia 12357
  Pseudoscleropodium 12357
  Ptilium 12859
  Reboulia 9210, 12357
  Rhacomitrium 12357, 14617
  Rhytidiadelphus 12357
  Scapania 12357
  Sphagnum 7079, 8825, 9298,
    9411, 9412, 9433, 10355,
    12357, 12604, 12605, 12722,
    12948, 14427, 14568, 14633,
    14667
  Thamnobryum 12357
  Thuidium 12357, 13890
  Tortula 6635, 7907, 8245,
    8431, 8772, 9210, 9353,
    9354, 9765, 10606, 10884,
    10955, 11327, 12357, 12860,
    14044, 14045, 14046, 14439
  Trichocolea 8114, 12357
  Ulota 12357
  Weissia 6635
  Zygodon 13137

Buchloë 11616, 12158, 12809,
  13158, 13348

Buddleia 10236

Bumelia 7917, 8464, 10992

Bunias 12389

Bupleurum 9371, 13408

Bursera 10529, 14034, 14235

C

cacao see *Theobroma*

Caesalpinia 7692, 7911, 11550

Cajanus 6688, 9242, 9378, 9607,
  9968, 10268, 11539, 11800, 12223,
  12384, 12401, 12593, 13370, 13657,
  13658, 14641, 14721

Cakile 7212, 7214, 10792, 14718

Calamagrostis 6997, 7302, 7649,
  7650, 10600, 10602, 13408, 13749

Callianthemum 6914

Callistephus 8055

Callitriche 8804, 9005

Callitris 9183, 12979

Calluna 7079, 8850, 8851, 8929,
  10416, 10497, 10671, 10672, 11500,
  11514, 13387, 14678

Caltha 6914, 7343, 8187, 9094,
  13408

Camelina 11238

Camellia 7557, 7561, 9066,
  9081, 9329, 9442, 9511, 10556,
  10993, 11018, 11201, 12623, 14295,
  14296

Camissonia 7486, 7802, 13131,
  13373, 14122

Campanula 14718, 14757

Canavalia 7333

Canna 7343, 10608

Cannabis 6960, 7868, 11018

Capparis 7386, 11550, 12004

Capsella 10721, 10777

Capsicum 6988, 7163, 7305,
7328, 7803, 8126, 8182, 8217,
8413, 8568, 8652, 8812, 9037,
9252, 9377, 9603, 9618, 9850,
10018, 10391, 10398, 10431, 10435,
10552, 10729, 10939, 11018, 11315,
11826, 12008, 12174, 12341, 12382,
12448, 12474, 12476, 12623, 12650,
13488, 13524, 13906, 13938, 14289,
14493, 14718

Caragana 9076, 9219, 11283

Carex 6697, 6838, 6863, 7068,
7212, 7214, 7302, 7304, 7305,
7343, 7371, 7649, 7650, 8065,
8182, 8588, 8850, 8851, 8929,
9032, 9094, 9135, 9413, 9531,
10356, 10600, 10966, 11139, 11283,
12072, 12158, 12192, 12602, 12719,
12720, 13132, 13337, 13587, 13641,
13959, 14247, 14517

Carica 10252, 10412, 10413,
11298, 14681

Carissa 7397, 7683, 7684,
9244, 9254

Carlina 11018

Carpinus 6974, 7020, 7184, 7185,
7621, 7998, 8184, 8474, 8709,
10239, 10460, 11018, 11812, 12433,
12623, 14168, 14758

carrot see *Daucus*

Carthamus 7645, 7849, 8154,
8409, 9252, 9313, 9403, 10432,
10900, 11718, 12409, 12837, 13349,
13399, 14208, 14388, 14638, 14639

Carum 9371, 13734, 14691

Carya 6614, 6754, 8689,
9407, 10364, 10541, 10650, 11156,
12149, 12277, 12433, 12874

Caryota 8128

Casearia 7911, 12126

Cassia 7692, 8402, 9243, 9244,
9255, 10189, 10659, 11550, 12301,
13154, 13219, 14681

Cassiope 10600, 10601, 14247,
14517

Castanea 6923, 7305, 8689,
8857, 9336, 10425, 11387, 13090,
13175, 14306, 14758

Castanopsis 8040

Casuarina 11465, 12448, 13306,
14496, 14681

Catalpa 14717

Catharanthus 12100, 14445

Ceanothus 7536, 7556, 7775,
8365, 8735, 9011, 9099, 9194,
10437, 10779, 11144, 11145, 11700,
12061, 12143, 14263

cedar see *Tamarix*

Cedrela 7192

Cedrus 9717, 10053, 10087,
11263

Celosia 6677, 7000, 8055,
12537, 13346

Celtis 6614, 8762, 10067,
12659

Cenchrus 6650, 6677, 7383,
7995, 7996, 8073, 8522, 11111,
12854, 12856, 12857, 13158, 13846,
14099, 14392, 14394

Centaurea 7214, 7887, 12192

Centaurium 14336

Cerastium 6704

Cerasus 6613, 8857, 9212, 9701,
10094, 10610, 10611, 11513, 13753,
13789, 14219

Ceratonia 7891, 8676, 9849,
9973, 10514, 11550, 12234

Ceratophyllum 9005, 10283

Cercis 10363, 10541, 13119,
13120

Cereus 7453, 13149, 13840

Ceriops 8776, 9596

Cestrum 10552, 10992

Chamaecyparis 7530, 8127, 8857,
9627, 10087, 10363, 11029, 11030,
11538, 12243, 12726, 12727, 12728,
14687

Chamaedorea 12335

Chamaenerion 10672, 11238

Chamaerops 10087

Chamerion 13329

Chelidonium 14167

Chenopodium 7070, 7272, 7273,
7946, 8126, 8182, 8301, 8302,
8307, 8557, 8561, 8816, 9050,
9160, 9844, 9865, 10998, 11002,
11407, 12650, 12969, 13107, 13276,
13277, 13429, 13947, 14121, 14122,
14493, 14674

cherry see *Cerasus*

Chloris 6650, 6677, 9252,
10327, 11111, 12623, 12928, 13184,
14412

Chlorophytum 8116, 9667, 9668,
10216, 10486, 11179, 12898, 14718

Chrysanthemum 6632, 7212, 7213,
7214, 7291, 7519, 7574, 8640,
8804, 9073, 9130, 9387, 9919,
10250, 10435, 10941, 11018, 11593,
12352, 12376, 12637, 12815, 12817,
13524, 13607, 13649, 13845, 13851,
13891, 14485, 14514, 14718

Chrysopogon 13632, 14166, 14167

Cicer 6688, 7410, 7512, 7523,
8409, 8521, 8557, 9313, 9318,
9392, 9393, 9398, 9607, 9642,
10268, 10626, 10865, 10892, 11261,
11453, 11901, 12588, 12590, 12969,
13370, 13459, 13536, 14087, 14207,
14362, 14613

Cichorium 7212, 7214, 11811,
13117

Cinnamomum 9081, 14681

Circaea 10087

Cirsium 13794

Cissus 8028, 12721, 12795,
12796, 14043

Cistus 10497, 10976, 12234

Citrullus 7070, 7343, 10893,
12537, 12822, 12969

Citrus 6717, 6744, 6755, 6835,
6966, 7006, 7025, 7028, 7032,
7093, 7305, 7306, 7365, 7415,
7467, 7481, 7482, 7491, 7492,
7505, 7561, 7645, 7801, 7918,
8108, 8358, 8412, 8510, 8622,
8857, 8860, 8861, 8951, 8955,
8999, 9033, 9252, 9349, 9355,
9429, 9446, 9494, 9495, 9496,
9497, 9498, 9499, 9636, 9646,
9658, 9663, 9775, 9931, 9932,
10022, 10330, 10363, 10398, 10654,
10774, 10789, 10838, 10841, 10886,
10971, 10972, 10973, 11018, 11099,
11100, 11232, 11309, 11315, 11333,
11463, 11472, 11725, 11774, 11799,
11807, 11862, 11954, 12018, 12019,
12020, 12021, 12022, 12080, 12099,
12163, 12213, 12290, 12362, 12452,
12513, 12587, 12673, 12674, 12804,
12805, 12844, 12884, 13058, 13096,
13138, 13172, 13173, 13218, 13269,
13350, 13351, 13409, 13488, 13557,
13747, 13802, 13964, 14048, 14191,
14326, 14328, 14329, 14501, 14522,
14523, 14580, 14621, 14746

Clarkia 9099

Claytonia 14517

Cleistogenes 11283

Clematis 6914, 8187, 11553

Cleome 6677

Clethra 10992

Clivia 8995, 9345, 10228,
10504, 10633, 10838, 12006

clover see *Trifolium*

Clusia 7879, 10992

Coccoloba 8464

Cochlearia 9907

Cochlospermum 14235

Cocos 7035, 7060, 7343,
10469, 11187, 11605

Codiaeum 13056

Coelogyne 12314

Coffea 7566, 8951, 9688,
10159, 10363, 10373, 10757, 10775,
10846, 11294, 11866, 12086, 12623,
13431, 14242, 14555, 14681

Coix 7300, 7942, 8846,
14681

Colchicum 12393

Coleus 8055, 9387, 9908,
10846, 11602, 11603, 11604, 12752,
13333, 14155

Collomia 12145

Colocasia 12235, 12659, 12665,
12813, 13144, 14405, 14681

Colutea 9318

Comarum 12602

Commelina 6912, 6942, 6943,
6946, 7210, 7346, 7402, 7403,
7466, 7666, 7812, 7974, 8038,
8060, 8071, 8158, 8441, 8442,
8533, 8654, 8745, 8940, 8941,
8942, 8943, 8947, 8948, 8950,
8951, 8952, 8953, 9139, 9230,
9231, 9232, 9294, 9295, 9312,
9359, 9429, 9543, 9618, 9626,
9640, 9873, 9904, 9919, 9982,
10042, 10097, 10274, 10275, 10296,
10481, 10631, 10669, 10828, 10926,
10927, 11129, 11130, 11340, 11546,
11629, 11660, 11873, 12054, 12055,
12056, 12057, 12058, 12065, 12078,
12079, 12086, 12168, 12208, 12268,
12310, 12311, 12312, 12313, 12368,
12410, 12851, 12872, 12898, 13059,

13165, 13166, 13520, 13553, 13849,
13850, 13878, 14130, 14357, 14464,
14561, 14562, 14563, 14635, 14718

Comptonia 10872

Consolida 6914

Convallaria 8475, 10251

Convolvulus 6614, 6677, 7212,
7213, 7214, 10254, 10397, 12537,
13174

Coptis 6914

Corchorus 6677, 6688, 7405,
7605, 7685, 7686, 7804, 9104,
9138, 9947, 11018, 12296, 12391,
12537, 12551, 13346, 14366

Cordia 6756, 7660, 7911,
8872, 10363, 13090, 14235

Coreopsis 7645

Coriandrum 9371, 14369

cornel see *Cornus*

Cornus 6780, 6831, 7130,
7156, 7183, 7305, 7599, 7712,
8037, 8289, 8492, 8677, 8709,
8857, 9273, 9352, 10363, 10364,
10541, 10880, 11021, 11700, 11771,
11813, 12277, 12433, 12623, 13911,
14434, 14717

Coronilla 8857, 9318, 11865,
14166, 14167

Corydalis 8475

Corylus 6998, 8929, 9360,
10363, 10841, 11700, 13500, 14600,
14602

Cotinus 11700, 12946

Cotoneaster 8131

cotton see *Gossypium*

cottonwood see *Populus*

Cotula 7519

cowpea see *Vigna*

Daucus   7142, 7152, 7514, 7724,
7727, 7801, 8303, 8490, 8588,
9098, 9103, 9252, 9282, 9371,
9026, 9958, 9992, 10022, 10108,
10123, 10951, 10974, 11109, 11598,
11774, 12144, 12643, 13130, 13193,
13204, 13205, 13848, 13872, 14398,
14704

Delphinium   6914, 8187

Dendrobium   10075, 14433, 14434,
14435

Deschampsia   7668, 8850, 8929,
9094, 10253, 10677, 10966, 11943,
14517, 14643

Descurainia   8688

Desmodium   7333, 7870, 8146,
8669, 8929, 9243, 11160

Deutzia   11700

Dianthus   6626, 7301, 9800,
10633, 10743, 10745, 11920, 12059,
13524, 13607, 13809, 13891, 14514

Diapensia   10600, 10602

Dicentra   6614

Dichantium   12146, 13158, 14020,
14099

Dieffenbachia   7150, 9196, 10633,
10693, 11311, 12335, 13056, 13891

Digera   6677, 12537

Digitalis   9929, 10131, 11161,
12883

Digitaria   6650, 6677, 6745,
7356, 7942, 8041, 9021, 11497,
13184, 13924, 14543

Dionaea   9650, 10189, 11939

Dioscorea   11518

Diospyros   8455, 9244, 9678,
10364

Diplotaxis   12389

Dipsacus   9991

Dipterix   14034

Dipterocarpus   9159

Distichlis   6915, 8749, 8810,
8811, 8939, 9960, 11221, 13270,
14246

dogwood see *Cornus*

Dolichos   8557

Dorstenia   7448

Dorycnium   9318

Douglas fir see *Pseudotsuga*

Draba   9929, 10131

Drimys   7486

Drosera   10189

Dryas   7995, 7996, 10600, 10601,
10602, 14247, 14517

Dryobalanops   9159

Dupontia   6842, 8929, 9099,
14350

E

Ecbalium   7212, 7214, 8791

Echeveria   10435, 13851

Echinochloa   6650, 6948, 7256,
7571, 7870, 7942, 7945, 7978,
8058, 10629, 11041, 11209, 11908,
12146, 13182, 13183, 13391, 13591,
14105

Echinops   7212, 7214

Echinopsis   8827

Ehretia   8040, 9244

Eichhornia   7729, 12411, 13604,
13605, 13951

Elaeagnus   7186, 9703

Elaeis   7060, 8128, 10363, 11480,
12623, 12792, 12919

Gladiolus 9144, 9283, 10104, 10251, 10780, 10781, 12388, 12393, 12440, 12441, 12449

Glaucium 7454, 7455

Glaux 8592, 9907

Glechoma 6646, 8330, 11588, 14166, 14167

Gleditsia 7945, 8689, 10101, 10364, 11387

Glinus 13052

Gloriosa 12393

Glyceria 7343, 9850, 12623, 12759, 13208

Glycine 6631, 6644, 6696, 6720, 6722, 6738, 6769, 6770, 6778, 6799, 6818, 6823, 6834, 6863, 6874, 6945, 6947, 7003, 7013, 7070, 7083, 7136, 7137, 7138, 7154, 7181, 7190, 7191, 9229, 7230, 7241, 7266, 7273, 7288, 7294, 7305, 7311, 7343, 7383, 7385, 7400, 7418, 7419, 7423, 7424, 7441, 7463, 7480, 7504, 7515, 7544, 7575, 7576, 7577, 7587, 7610, 7616, 7644, 7645, 7646, 7696, 7713, 7806, 7807, 7808, 7841, 7843, 7847, 7853, 7855, 7865, 7896, 7918, 7927, 7928, 7947, 7948, 7949, 7955, 7961, 7995, 7996, 8051, 8052, 8064, 8083, 8167, 8197, 8198, 8203, 8240, 8267, 8274, 8283, 8297, 8306, 8307, 8312, 8360, 8378, 8395, 8404, 8406, 8543, 8544, 8578, 8594, 8653, 8702, 8704, 8705, 8711, 8742, 8772, 8800, 8853, 8894, 8896, 8910, 8944, 8945, 8959, 8966, 8967, 8968, 8985, 8998, 8999, 9001, 9040, 9093, 9102, 9154, 9158, 9198, 9199, 9252, 9262, 9325, 9363, 9364, 9417, 9429, 9437, 9447, 9483, 9523, 9552, 9553, 9554, 9558, 9586, 9587, 9594, 9595, 9632, 9637, 9643, 9646, 9651, 9690, 9708, 9742, 9794, 9830, 9831, 9838, 9842, 9850, 9853, 9860, 9919, 9977, 9988, 10009, 10022, 10024, 10025, 10077, 10108, 10117, 10162, 10165, 10193, 10212, 10243, 10269, 10187, 10288,

10291, 10306, 10307, 10328, 10347, 10392, 10393, 10394, 10419, 10456, 10464, 10465, 10485, 10488, 10500, 10501, 10597, 10599, 10605, 10648, 10649, 10652, 10659, 10662, 10665, 10729, 10734, 10750, 10751, 10783, 10787, 10823, 10843, 10846, 10862, 10876, 10883, 10910, 10916, 10924, 10928, 10994, 10999, 11002, 11014, 11018, 11023, 11042, 11050, 11088, 11091, 11098, 11118, 11137, 11175, 11190, 11191, 11204, 11236, 11252, 11273, 11276, 11277, 11278, 11365, 11370, 11377, 11379, 11380, 11422, 11423, 11452, 11458, 11460, 11463, 11479, 11484, 11499, 11501, 11566, 11579, 11609, 11614, 11644, 11655, 11658, 11694, 11698, 11724, 11731, 11732, 11753, 11755, 11769, 11775, 11791, 11796, 11797, 11803, 11861, 11865, 11878, 11920, 11929, 11935, 11937, 11946, 11947, 11948, 11953, 11954, 12010, 12062, 12076, 12083, 12095, 12119, 12123, 12153, 12179, 12229, 12300, 12301, 12309, 12325, 12341, 12368, 12401, 12409, 12450, 12451, 12460, 12483, 12535, 12540, 12542, 12543, 12564, 12589, 12598, 12599, 12616, 12617, 12621, 12625, 12668, 12691, 12693, 12694, 12697, 12714, 12753, 12779, 12784, 12792, 12837, 12863, 12864, 12870, 12871, 12873, 12909, 12911, 12913, 12935, 12969, 13010, 13017, 13037, 13084, 13108, 13123, 13139, 13157, 13171, 13196, 13244, 13285, 13307, 13313, 13326, 13332, 13356, 13361, 13370, 13389, 13414, 13477, 13478, 13483, 13509, 13515, 13519, 13544, 13590, 13602, 13631, 13652, 13653, 13660, 13662, 13684, 13712, 13724, 13750, 13796, 13876, 13885, 13886, 13888, 13893, 13894, 13931, 13932, 13949, 14027, 14088, 14089, 14090, 14118, 14134, 14169, 14205, 14208, 14210, 14268, 14269, 14308, 14320, 14345, 14359, 14367, 14424, 14446, 14447, 14467, 14516, 14518, 14541, 14542, 14612, 14638, 14645, 14651, 14653, 14676, 14677, 14715, 14721, 14738, 14742, 14751, 14765

Glycyrrhiza 14201

Gnetum 14034

Gomphrena 13463, 14636

Gordonia 8857

Gossypium 6611, 6696, 6765,
6786, 6863, 6875, 6876, 6907,
6909, 6945, 7024, 7070, 7134,
7135, 7138, 7229, 7248, 7305,
7374, 7475, 7383, 7411, 7442,
7459, 7544, 7576, 7622, 7631,
7637, 7645, 7663, 7751, 7766,
7801, 7850, 7881, 7918, 7947,
7971, 7997, 8129, 8132, 8133,
8134, 8138, 8145, 8182, 8240,
8319, 8326, 8368, 8482, 8613,
8693, 8745, 8985, 9034, 9042,
9108, 9116, 9169, 9224, 9225,
9226, 9252, 9265, 9323, 9324,
9373, 9429, 9443, 9446, 9556,
9575, 9642, 9681, 9685, 9735,
9747, 9759, 9807, 9850, 9903,
9919, 9936, 9937, 9957, 10009,
10022, 10042, 10066, 10079, 10132,
10136, 10137, 10162, 10194, 10195,
10222, 10223, 10243, 10244, 10245,
10363, 10432, 10464, 10468, 10476,
10478, 10584, 10589, 10630, 10656,
10707, 10712, 10718, 10719, 10733,
10809, 10810, 10841, 10846, 10854,
10862, 10883, 11018, 11023, 11044,
11196, 11203, 11212, 11329, 11332,
11433, 11644, 11669, 11670, 11694,
11769, 11790, 11792, 11793, 11809,
11864, 11865, 11878, 11920, 11931,
11963, 11964, 11978, 11979, 12034,
12038, 12122, 12188, 12198, 12245,
12267, 12300, 12316, 12323, 12324,
12325, 12326, 12327, 12349, 12358,
12359, 12368, 12378, 12391, 12394,
12395, 12396, 12398, 12399, 12401,
12409, 12495, 12536, 12551, 12623,
12627, 12628, 12679, 12693, 12695,
12753, 12762, 12789, 12862, 12872,
12898, 12925, 12969, 13011, 13157,
13313, 13350, 13352, 13356, 13411,
13469, 13470, 13488, 13509, 13519,
13581, 13582, 13583, 13587, 13588,
13589, 13606, 13666, 13703, 13724,
13770, 13793, 13812, 13971, 13972,
14006, 14013, 14075, 14142, 14170,
14196, 14197, 14210, 14218, 14275,
14309, 14383, 14434, 14471, 14472,
14508, 14509, 14600, 14638, 14639,
14652, 14686, 14699

Goupia 7660

grape vine see *Vitis*

groundnut see *Arachis*

Guizotia 14421

H

Haematoxylon 7911

Halimione 9907, 10298

Halimium 10497

Halocnemum 11194

Halophila 6694, 11048

Haloxylon 7001, 10653, 11416,
12537, 12539, 14662, 14681

Hamamelis 11700

Hammada 6717, 7383, 7576,
7577, 7793, 7794, 7995, 7996,
8772, 8951, 9099, 9360, 9761,
9919, 11166, 14681

Haplopappus 8038, 8550, 10081,
13429

Haworthia 8288, 8397, 10435,
14681

hawthorn see *Crataegus*

hazel see *Corylus*

Hedera 6691, 6717, 6966, 7305,
7343, 8475, 8709, 8895, 9349,
9355, 9850, 9919, 10087, 10225,
10228, 10238, 11298, 12623, 13056,
13500, 14168, 14169, 14757

Hedysarum 7186, 9318, 11416,
12624

Helianthemum 9929, 13408

Helianthus 6627, 6642, 6644,
6688, 6696, 6733, 6739, 6818,
6823, 6832, 6886, 6887, 6896,
6918, 6945, 6960, 7070, 7074,
7094, 7144, 7169, 7215, 7229,
7233, 7234, 7240, 7262, 7305,
7343, 7349, 7381, 7383, 7438,
7576, 7645, 7646, 7659, 7689,
7690, 7729, 7764, 7801, 7834,
7853, 7881, 7918, 7951, 7995,
7996, 7997, 8015, 8020, 8031,
8038, 8101, 8122, 8141, 8182,
8194, 8222, 8306, 8307, 8325,
8352, 8354, 8391, 8392, 8424,
8508, 8545, 8546, 8561, 8577,

Helianthus (cont.)
8614, 8635, 8657, 8694, 8717,
8772, 8800, 8804, 8857, 8861,
8871, 8898, 8902, 8951, 8985,
8992, 9043, 9052, 9089, 9111,
9214, 9445, 9453, 9528, 9551,
9556, 9557, 9558, 9563, 9596,
9617, 9642, 9670, 9761, 9784,
9801, 9818, 9830, 9837, 9910,
9919, 9985, 10009, 10033, 10087,
10146, 10194, 10233, 10243, 10432,
10498, 10530, 10544, 10548, 10622,
10634, 10641, 10694, 10706, 10717,
10729, 10737, 10738, 10817, 10841,
10846, 10866, 10883, 10949, 10959,
10980, 10981, 11018, 11044, 11053,
11063, 11064, 11069, 11113, 11168,
11169, 11192, 11212, 11213, 11328,
11344, 11379, 11436, 11463, 11583,
11609, 11682, 11690, 11692, 11694,
11703, 11717, 11737, 11755, 11759,
11778, 11891, 11920, 11953, 11954,
11961, 12039, 12062, 12092, 12102,
12112, 12248, 12249, 12250, 12251,
12297, 12337, 12359, 12361, 12368,
12377, 12401, 12402, 12403, 12409,
12443, 12444, 12445, 12506, 12566,
12609, 12619, 12620, 12623, 12625,
12639, 12707, 12732, 12753, 12754,
12761, 12770, 12807, 12811, 12822,
12837, 12872, 12915, 12934, 13010,
13064, 13093, 13107, 13111, 13191,
13211, 13246, 13263, 13313, 13350,
13399, 13429, 13465, 13471, 13479,
13487, 13488, 13506, 13571, 13608,
13651, 13750, 13751, 13752, 13760,
13832, 13878, 13893, 13894, 13900,
13901, 13920, 13939, 13940, 13958,
13971, 13972, 13973, 13974, 13975,
13978, 14053, 14060, 14062, 14063,
14064, 14065, 14108, 14127, 14146,
14208, 14217, 14264, 14304, 14390,
14425, 14434, 14466, 14565, 14600,
14601, 14602, 14638, 14639, 14668,
14683, 14684, 14718

Heliconia 7520

Heliotropium 6677, 6717, 7486,
7737, 10791, 12537, 13429

Helleborus 6914

Hemerocallis 10251

hemlock see *Tsuga*

hemp see *Cannabis*

Hepatica 6914, 8877

Heracleum 6701, 6702, 11682

Hesperis 11238

Heteromeles 8076, 8077, 9761,
9784, 10363, 11144, 12007, 12160,
13119, 13120, 13358

Heteropogon 7303, 7995, 7996,
8073, 8493, 8522, 12854, 12857

Hevea 7158, 7343, 7648, 7753,
7754, 8090, 8582, 9850, 10253,
10363, 11159, 11442, 12623, 13750

Hewardia 12393

Hibiscus 7383, 7404, 7499,
7500, 7501, 7502, 7503, 9055,
9243, 9947, 10435, 11012, 12391,
12551, 12756, 12822, 13493, 13859

Hieracium 8850

Hilaria 7545, 9082, 12228,
12437

Hippeastrum 10251

Hippocrepis 13827

Hippophaë 8547

Hirschfeldia 12389

Holcus 6989, 7534, 7743, 7870,
8046, 8534, 11018, 12820, 14643

holly see *Ilex*

Hopea 9159

Hordeum 6623, 6624, 6643,
6650, 6656, 6663, 6674, 6700,
6749, 6750, 6751, 6803, 6806,
6850, 6863, 6868, 6903, 6920,
6921, 6941, 6945, 6955, 6981,
6994, 7015, 7023, 7062, 7070,
7081, 7083, 7094, 7115, 7116,
7117, 7118, 7119, 7139, 7140,
7197, 7256, 7295, 7302, 7305,
7317, 7335, 7357, 7381, 7394,
7395, 7433, 7434, 7435, 7483,
7523, 7642, 7645, 7651, 7652,
7653, 7655, 7705, 7720, 7759,
7812, 7817, 7842, 7844, 7850,
7853, 7860, 7901, 7907, 7910,
7914, 7918, 7933, 8007, 8015,

Hordeum (cont.)
8033, 8034, 8038, 8091, 8105,
8111, 8116, 8118, 8119, 8140,
8168, 8169, 8177, 8182, 8185,
8190, 8247, 8318, 8319, 8345,
8350, 8354, 8409, 8416, 8417,
8421, 8428, 8436, 8437, 8459,
8491, 8511, 8526, 8536, 8537,
8565, 8571, 8639, 8646, 8678,
8679, 8681, 8737, 8742, 8774,
8803, 8842, 8867, 8901, 8970,
8985, 8996, 9037, 9071, 9089,
9119, 9161, 9182, 9184, 9189,
9193, 9206, 9207, 9223, 9236,
9252, 9266, 9313, 9337, 9375,
9390, 9396, 9401, 9403, 9414,
9446, 9532, 9558, 9572, 9582,
9591, 9639, 9646, 9668, 9669,
9720, 9769, 9820, 9822, 9835,
9836, 9869, 9872, 9877, 9878,
9929, 9964, 9997, 9998, 10019,
10022, 10035, 10042, 10057, 10072,
10091, 10099, 10100, 10131, 10145,
10218, 10231, 10243, 10260, 10267,
10268, 10272, 10317, 10340, 10371,
10378, 10432, 10439, 10440, 10451,
10525, 10534, 10546, 10626, 10635,
10695, 10701, 10722, 10723, 10730,
10759, 10765, 10766, 10810, 10874,
10883, 10887, 10903, 10917, 10919,
11018, 11020, 11070, 11076, 11094,
11105, 11109, 11113, 11165, 11195,
11254, 11267, 11269, 11270, 11317,
11377, 11412, 11431, 11445, 11454,
11455, 11459, 11469, 11478, 11524,
11537, 11558, 11559, 11572, 11644,
11651, 11690, 11694, 11721, 11730,
11751, 11774, 11794, 11805, 11865,
11867, 11871, 11899, 11920, 11945,
11953, 11966, 11974, 11975, 11990,
11995, 12002, 12003, 12038, 12045,
12048, 12068, 12082, 12110, 12115,
12179, 12199, 12287, 12305, 12332,
12338, 12354, 12368, 12377, 12380,
12436, 12485, 12493, 12499, 12503,
12519, 12575, 12576, 12623, 12648,
12687, 12695, 12758, 12766, 12778,
12807, 12830, 12872, 12901, 12904,
12969, 12972, 13013, 13026, 13027,
13031, 13047, 13072, 13163, 13170,
13204, 13223, 13245, 13269, 13272,
13285, 13301, 13315, 13316, 13337,
13376, 13382, 13395, 13399, 13413,
13438, 13443, 13456, 13457, 13465,
13475, 13485, 13486, 13488, 13499,
13571, 13638, 13639, 13659, 13665,
13674, 13720, 13749, 13805, 13839,
13843, 13851, 13860, 13861, 13893,
13894, 13938, 13971, 13972, 13984,

13985, 14007, 14030, 14102, 14136,
14169, 14183, 14184, 14185, 14194,
14199, 14200, 14207, 14252, 14309,
14321, 14362, 14378, 14382, 14451,
14456, 14482, 14489, 14493, 14565,
14572, 14577, 14657, 14661, 14668,
14671, 14672, 14673, 14731, 14756,
14757

hornbeam see *Carpinus*

horsechestnut see *Aesculus*

Hoya 7885, 12404

Humiria 7662

Humulus 12812

Hyacinthus 12771, 12949

Hybanthus 6659

Hydnophytum 12314

Hydrangea 9542

Hydrocotyle 8926, 9129

Hyoscyamus 8182, 11018

Hyparrhenia 7226, 7227, 14679

Hypericum 9929, 10131, 13749

Hypochoeris 7534, 9580, 14204,
14315

Hyptis 7544, 8857, 8929,
10589, 14014

I

Ilex 7093, 7712, 8857,
9273, 10087, 10363, 10638, 10639,
10880, 10992, 11018, 11700, 11887,
11953, 12987, 13500, 14434, 14681

Impatiens 7838, 8330, 9662,
9712, 11602, 13333, 13408, 14681

Indigofera 6677, 12537

Inula 7212, 9103, 14079

Ipomoea 6677, 6895, 7608,
7635, 8052, 9252, 9713, 9768,
9833, 9991, 10155, 10641, 11018,

Ipomoea (cont.)
11092, 11677, 11678, 11734, 12087,
12537, 12916, 13769

Iresine 6827

Iris 8325, 10251, 11018,
13208, 14167

Isopyrum 8475

Ixia 9144

Ixora 9244, 9254

J

Jatropha 8420, 8872, 10253,
11550

Juglans 6614, 6699, 7305,
7343, 8677, 8857, 8962, 9578,
10363, 10426, 11523, 11540, 11685,
11686, 12277, 12333, 12433, 12730,
14306

Juncus 6972, 6989, 7343,
8588, 9120, 9646, 9817, 9907,
10105, 14336

Juniperus 7305, 7876, 8677,
9578, 9627, 10087, 11686, 11988,
12243, 12277, 14353

Jute see *Corchorus*

K

Kadsura 8208

Kalanchoë 7305, 7343, 7428,
7885, 7906, 7974, 8079, 8126,
8739, 8757, 8758, 8792, 8814,
8837, 8838, 8861, 8987, 9286,
9357, 9458, 9626, 9901, 10435,
10479, 10609, 10633, 11018, 11410,
11543, 11783, 11825, 11874, 12371,
12372, 12373, 12650, 13056, 13105,
13149, 13302, 13851, 13884, 13891,
14032, 14121, 14457, 14493, 14718

Kalmia 10087, 10880, 11568

Kleinia 8688

Kobresia 6705, 7302, 7371,
14247.

Kochia 7212, 7213, 7214,
11283, 12748

Koeleria 8065, 9732, 11283,
13641

Krameria 8939, 13787

L

Lablab 13370, 14721

Laburnum 9318

Lactuca 7142, 7152, 7162,
7273, 7305, 7725, 7727, 7752,
7757, 7915, 7930, 7941, 8182,
8245, 8322, 8384, 8470, 8610,
8721, 8734, 8806, 8852, 9098,
9252, 9287, 9900, 10022, 10166,
10198, 10243, 10399, 10587, 10800,
11018, 11109, 11426, 11427, 11503,
11567, 11777, 11790, 11818, 11833,
12010, 12496, 12544, 13034, 13176,
13193, 13198, 13199, 13417, 13938,
14311, 14371, 14379, 14380, 14404,
14521, 14718

Lagenaria 11715, 12822

Lagestroemia 10065

Lallemantia 10562

Lamium 8475, 10087, 10562,
10777, 14078

larch see *Larix*

Larix 7019, 7065, 7077,
7146, 7343, 7378, 8238, 8324,
8742, 8762, 8857, 9216, 10363,
10841, 11018, 11187, 11463, 12632,
12988, 13090, 13207, 14351, 14591,
14687, 14758

Larrea 6717, 6731, 7383,
7486, 7577, 7607, 7612, 7802,
8260, 8352, 8861, 8939, 8985,
9044, 9269, 9605, 9761, 9850,
10339, 10363, 10364, 10520, 11555,
11719, 11960, 12306, 12623, 12968,
13787, 14014, 14121, 14122, 14338,
14350, 14383

Lathyrus 8475, 9318, 9683,
9771, 11238, 13117, 13734

Liriodendron 6753, 7305, 7693, 7712, 8037, 8740, 8857, 9991, 10065, 10363, 10364, 10524, 10541, 12109, 12623

Lobelia 13024, 13333, 14681

loblolly pine see *Pinus*

locust see *Robinia*

Loiseleuria 8850, 8851, 8929, 10600, 10602, 12529

Lolium 6650, 6656, 6766, 6779, 6828, 6951, 6960, 6995, 7011, 7231, 7232, 7240, 7265, 7305, 7356, 7480, 7534, 7544, 7563, 7568, 7571, 7593, 7654, 7743, 7799, 7870, 7920, 7921, 7985, 8018, 8041, 8046, 8182, 8354, 8435, 8548, 8655, 8669, 8742, 8746, 8929, 8937, 9252, 9333, 9570, 9591, 9719, 9779, 9784, 9850, 10177, 10178, 10283, 10291, 10294, 10295, 10430, 10592, 10593, 10594, 10595, 10732, 10840, 10883, 10920, 11016, 11018, 11121, 11152, 11306, 11738, 11865, 11902, 12007, 12108, 12155, 12254, 12271, 12302, 12447, 12507, 12581, 12623, 12708, 12717, 12753, 12793, 12809, 12820, 12928, 12983, 13187, 13188, 13253, 13502, 13570, 13803, 13895, 13972, 14259, 14260, 14285, 14341, 14434, 14487, 14621, 14706

Lonicera 6809

Loranthus 8709, 11691, 14354

Lotus 7439, 7440, 7870, 8214, 9077, 9078, 9318, 9929, 10131, 10585, 11658, 13012, 13174, 13804

Luffa 10189, 11715, 14513

Lumnitzera 8776, 9596

Lunaria 13408

lupine see *Lupinus*

Lupinus 6643, 6696, 7583, 7603, 7604, 7633, 7758, 7870, 7913, 8038, 8525, 8682, 8904, 9099, 9137, 9154, 9318, 9463, 9469, 9470, 9708, 10009, 10220, 10602, 10677, 11087, 11590, 11633,

11634, 12162, 12368, 12770, 12969, 13373, 13413, 13527, 13783, 13984, 14030, 14102, 14626, 14718

Luzula 6704, 8850, 14517

Lychnis 7035, 11018, 14618

Lycium 7607, 8939, 13787

Lycopersicon 6687, 6774, 6775, 6776, 6789, 6863, 6880, 6881, 6884, 6949, 6960, 6988, 6993, 7000, 7051, 7096, 7152, 7153, 7201, 7262, 7275, 7305, 7343, 7429, 7540, 7559, 7569, 7638, 7645, 7679, 7801, 7881, 7893, 7918, 7943, 7958, 7959, 8002, 8056, 8100, 8163, 8182, 8291, 8329, 8351, 8393, 8400, 8545, 8551, 8568, 8638, 8779, 8793, 8804, 8861, 8960, 8985, 9005, 9014, 9098, 9178, 9252, 9300, 9305, 9322, 9355, 9448, 9449, 9460, 9473, 9474, 9504, 9506, 9507, 9523, 9570, 9602, 9698, 9806, 9908, 9962, 10017, 10096, 10108, 10115, 10146, 10243, 10312, 10391, 10398, 10399, 10420, 10432, 10435, 10489, 10518, 10552, 10622, 10656, 10684, 10838, 10880, 10996, 10997, 11018, 11023, 11073, 11074, 11078, 11109, 11113, 11187, 11227, 11244, 11245, 11250, 11372, 11383, 11384, 11385, 11386, 11391, 11416, 11418, 11441, 11477, 11494, 11689, 11710, 11727, 11728, 11777, 11790, 11920, 11953, 12028, 12059, 12106, 12150, 12174, 12175, 12234, 12251, 12261, 12282, 12283, 12349, 12364, 12448, 12453, 12456, 12465, 12467, 12474, 12479, 12557, 12618, 12693, 12694, 12744, 12745, 12753, 12914, 12940, 12947, 13218, 13238, 13269, 13272, 13333, 13435, 13468, 13505, 13509, 13524, 13540, 13584, 13617, 13644, 13666, 13707, 13772, 13781, 13883, 13886, 13933, 13938, 13943, 13979, 13991, 14027, 14049, 14080, 14091, 14113, 14129, 14163, 14304, 14352, 14383, 14397, 14434, 14528, 14529, 14602, 14621

Lysimachia 8929, 11238, 12602

M

Maba 7397, 7683, 7684, 9244,
9254

Macaranga 10253

Macroptilium 8146, 8515, 8522,
8560, 8929, 10054, 12043, 12854,
13314, 13370, 13837, 13838, 13971,
13972, 14394

Magnolia 7130, 10065, 10541

maize see *Zea*

Mallotus 10253

Malus 6680, 6910, 6996, 7159,
7171, 7305, 7325, 7329, 7331,
7343, 7422, 7427, 7467, 7485,
7618, 7746, 7995, 7996, 8110,
8199, 8341, 8388, 8450, 8479,
8480, 8506, 8507, 8611, 8634,
8689, 8743, 8745, 8772, 8804,
8813, 8857, 8894, 9003, 9037,
9061, 9209, 9252, 9349, 9429,
9485, 9512, 9625, 9754, 9786,
9810, 9850, 10023, 10034, 10063,
10108, 10144, 10161, 10228, 10284,
10285, 10363, 10382, 10396, 10448,
10590, 10704, 10744, 10968, 10969,
10970, 11018, 11187, 11240, 11325,
11492, 11508, 11540, 11675, 11676,
11723, 11856, 11858, 11950, 11968,
11972, 11976, 11996, 12069, 12157,
12189, 12201, 12366, 12378, 12433,
12436, 12554, 12585, 12670, 12671,
12731, 12745, 12957, 12975, 12976,
13049, 13050, 13340, 13415, 13428,
13437, 13633, 13647, 13648, 13782,
13784, 13848, 13946, 14008, 14135,
14186, 14187, 14330, 14335, 14350,
14414, 14520, 14587

Malva 10189, 11180, 11181,
12898

Mamillaria 8595, 9085, 10124,
13296

mangrove see *Rhizophora*

Mangifera 13455

Manihot 6863, 6871, 8367,
8481, 8715, 8716, 9850, 9902,
10190, 10247, 10248, 10253, 10255,
10642, 11533, 11678, 12278, 12279,

12504, 12505, 12746, 13311, 13312,
13313, 13314, 13609, 13612, 13677,
14086, 14681

maple see *Acer*

Maranta 9196, 10608, 11292,
12898, 14681

Marrumbium 7213, 7214, 10562

Matricaria 7519, 8182, 10069,
12035

Matthiola 12389

Medicago 6627, 6648, 6728,
6748, 7070, 7090, 7109, 7212,
7214, 7343, 7356, 7456, 7472,
7480, 7563, 7593, 7594, 7640,
7645, 7657, 7665, 7667, 7672,
7700, 7715, 7729, 7730, 7761,
7765, 7801, 7855, 7865, 7917,
7922, 8085, 8111, 8142, 8182,
8204, 8205, 8214, 8275, 8276,
8277, 8299, 8304, 8320, 8348,
8354, 8580, 8607, 8711, 8712,
8714, 8731, 8761, 8983, 9023,
9048, 9186, 9204, 9252, 9282,
9303, 9318, 9324, 9339, 9376,
9407, 9525, 9576, 9601, 9646,
9678, 9710, 9764, 9795, 9796,
9802, 9845, 9850, 9908, 10022,
10032, 10108, 10138, 10149, 10183,
10185, 10242, 10243, 10279, 10294,
10295, 10345, 10432, 10444, 10451,
10468, 10485, 10499, 10622, 10716,
10769, 10801, 10810, 10812, 10883,
11008, 11018, 11023, 11076, 11136,
11141, 11195, 11290, 11316, 11358,
11399, 11400, 11428, 11429, 11430,
11537, 11552, 11561, 11635, 11658,
11730, 11731, 11732, 11756, 11777,
11830, 11835, 11845, 11878, 11892,
11893, 11895, 11896, 11926, 12062,
12070, 12131, 12151, 12192, 12322,
12420, 12472, 12561, 12615, 12623,
12691, 12693, 12735, 12762, 12767,
12839, 12872, 12913, 12928, 12969,
13012, 13020, 13022, 13043, 13221,
13244, 13269, 13330, 13356, 13504,
13509, 13629, 13691, 13697, 13750,
13804, 13896, 13957, 13971, 13972,
14019, 14092, 14138, 14140, 14149,
14152, 14153, 14203, 14274, 14284,
14303, 14309, 14364, 14413, 14428,
14448, 14490, 14572, 14594, 14622,
14708, 14728

Melaleuca   8882, 9183, 9369,
12448, 13746, 14530

Melica   14168

Melilotus   9318, 9608, 11223,
12969

Melothria   6677

Mentha   6859, 11525, 11526,
11527, 14718

Menyanthes   14718

Mercurialis   8476, 11588, 13408,
13462, 13759, 14168

Merremia   12537

Mesembryanthemum   7577, 7671,
7982, 8080, 8126, 8674, 8861,
10435, 11052, 11531, 11873, 11951,
12182, 12204, 12861, 14032, 14493,
14681

Metasequoia   12206

Metrosideros   8075

millet   see   *Panicum*

Mimosa   7933, 9673, 10189,
10641

Mimulus   9099, 10051, 11127

Minuartia   9094, 10966, 14517

Miscanthus   6624

Molinia   12203, 13387, 14148

Mollugo   7270, 14636

Momordica   7967

Monarda   9006

Monstera   8509, 14681

Moricandia   12966

Morinda   13769

Morus   7951, 9850, 12623,
12659

mosses   see   *Bryophyta*

mulberry   see   *Morus*

mung bean   see   *Vigna*

Musa   6608, 7305, 7624, 7645,
7994, 8883, 9544, 10095, 10226,
10227, 10363, 10533, 10608, 11018,
11671, 11779, 11780, 11781, 14048,
14681

mustard   see   *Sinapis*

Myosotis   9929, 10131, 14517

Myosurus   8187

Myrica   8688, 8857, 10101,
10872, 10992, 12533

Myrmecodia   12314

Myrothamnus   6905, 7035, 8772,
11379, 13393

Myrsine   10992, 13451

Myrtus   10425, 11408

N

Naravelia   8187

Nasturtium   7436

Nectandra   10992

Nelumbo   9120

Nepentes   13839

Nepeta   10562

Nerium   6717, 6731, 6732, 7035,
7076, 7389, 7673, 7769, 8260,
8448, 8589, 9061, 9429, 9761,
9782, 9783, 10087, 10296, 10387,
10425, 10698, 11310, 11555, 11744,
11873, 11844, 12349, 12872, 13062,
13093, 14121, 14600, 14602

Nicotiana   6717, 6733, 6746,
6758, 6784, 6960, 6998, 7161,
7188, 7229, 7258, 7289, 7300,
7305, 7337, 7338, 7343, 7388,
7645, 7856, 8158, 8182, 8343,
8363, 8413, 8427, 8675, 8701,
8720, 8794, 8795, 8846, 8985,
9065, 9312, 9443, 9463, 9569,
9571, 9667, 9784, 9828, 9850,

Nicotiana (cont.)
9919, 10009, 10042, 10055, 10072,
10074, 10087, 10162, 10213, 10318,
10364, 10374, 10435, 10528, 10552,
10622, 10633, 10680, 10841, 10867,
10883, 10988, 11018, 11023, 11046,
11106, 11153, 11154, 11798, 11837,
11942, 12190, 12263, 12268, 12270,
12274, 12368, 12401, 12528, 12568,
12623, 12762, 12807, 12898, 13010,
13034, 13064, 13285, 13321, 13333,
13336, 13419, 13698, 13750, 13878,
14131, 14307, 14371, 14668, 14713,
14718

Nidularium 7522

Nigella 6914, 8187

Nitraria 7671, 11550, 14681

Nolana 7487, 7737

Nothofagus 7614, 7936, 11700,
11943, 12120, 13451, 13452, 14392

Notonia 6848

Nuphar 10241, 10242

Nymphaea 9905, 13839, 14764

Nyssa 8366, 8643, 10101,
10364, 10541, 13951

O

oak see *Quercus*

oat see *Avena*

Ocotea 7879

Oenothera 7612, 10633, 12162

oil palm see *Elaeis*

Olea 6616, 7035, 7156, 7305,
7343, 8676, 8857, 9252, 9727,
10087, 10363, 10388, 10514, 10582,
11595, 11985, 12715, 12736, 12988,
13350, 14306, 14681

olive see *Olea*

onion see *Allium*

Onobrychis 7212, 7213, 7214, 7302,
9376, 10032, 12072, 13174, 13804

Ononis 9318

Opuntia 4373, 7546, 8477,
8595, 8861, 8872, 8939, 9085,
9629, 9771, 9850, 10348, 10435,
10883, 11127, 11199, 11719, 12031,
12225, 12226, 12227, 12373, 13009,
13103, 13296, 14018, 14068

Orchis 11018

Origanum 11166

Oryza 6624, 6683, 6686, 6688,
6689, 6812, 6899, 6900, 6901,
6945, 6964, 7035, 7070, 7170,
7176, 7179, 7194, 7195, 7196,
7256, 7300, 7343, 7359, 7582,
7605, 7606, 7611, 7634, 7669,
7670, 7672, 7674, 7680, 7701,
7749, 7770, 7845, 7905, 7919,
7942, 7951, 7962, 7978, 7979,
7988, 8069, 8103, 8104, 8157,
8182, 8183, 8206, 8207, 8398,
8399, 8407, 8497, 8518, 8545,
8585, 8586, 8668, 8724, 8725,
8726, 8729, 8736, 8756, 8768,
8778, 8786, 8797, 8843, 8846,
8870, 8993, 9005, 9025, 9036,
9091, 9117, 9125, 9126, 9127,
9217, 9221, 9235, 9252, 9391,
9397, 9509, 9653, 9711, 9753,
9784, 9822, 9824, 9885, 9987,
9994, 10184, 10194, 10200, 10214,
10217, 10237, 10257, 10292, 10293,
10299, 10326, 10408, 10458, 10511,
10530, 10533, 10569, 10570, 10627,
10628, 10629, 10756, 10767, 10818,
10819, 10848, 10849, 10850, 10869,
10883, 10888, 10933, 10986, 10987,
11018, 11031, 11041, 11043, 11052,
11096, 11210, 11211, 11236, 11237,
11285, 11296, 11359, 11369, 11443,
11462, 11463, 11497, 11500, 11509,
11510, 11761, 11762, 11769, 11772,
11773, 11827, 11865, 11891, 11936,
11977, 12073, 12074, 12075, 12105,
12129, 12190, 12202, 12265, 12368,
12428, 12486, 12487, 12500, 12501,
12570, 12580, 12582, 12600, 12623,
12662, 12683, 12703, 12717, 12750,
12751, 12798, 12799, 12872, 12913,
12926, 12939, 13061, 13077, 13078,
13079, 13142, 13153, 13212, 13314,
13326, 13362, 13363, 13371, 13483,
13518, 13521, 13533, 13543, 13554,
13555, 13586, 13593, 13594, 13613,
13616, 13622, 13623, 13640, 13675,
13677, 13727, 13728, 13807, 13819,
13868, 13870, 13871, 13874, 13971,

Persica (cont.)
9508, 9649, 9661, 9945, 9955,
10078, 10094, 10158, 10443, 10512,
10984, 10985, 11173, 11253, 11444,
11611, 11750, 11788, 11881, 12148,
12246, 12365, 12366, 12468, 12469,
12875, 13236, 13353, 13753, 13797,
13986, 14189, 14219

Petroselinum 7059, 7152

Petunia 6978, 6979, 8055,
10622, 11018, 11023, 11248, 12623,
13333

Peucedanum 14167

Phacelia 7802, 9606, 13373

Phalaris 6948, 7068, 7343,
7392, 7821, 7870, 8531, 9252,
9582, 9850, 10295, 10863, 10883,
11076, 11960, 12166, 12203, 12623,
13020, 13187, 13971, 13972, 14058,
14059, 14623

Pharbitis 7769, 8545, 11481,
11482, 11774

Phaseolus 6618, 6619, 6627,
6667, 6672, 6677, 6685, 6688,
6727, 6733, 6764, 6785, 6786,
6825, 6832, 6833, 6863, 6971,
7033, 7038, 7070, 7082, 7094,
7097, 7098, 7113, 7166, 7174,
7252, 7305, 7321, 7333, 7336,
7343, 7429, 7432, 7438, 7447,
7484, 7523, 7567, 7570, 7571,
7576, 7577, 7587, 7598, 7602,
7607, 7628, 7637, 7645, 7673,
7760, 7809, 7810, 7811, 7824,
7825, 7833, 7850, 7853, 7854,
7863, 7882, 7883, 7884, 7897,
7937, 7987, 8016, 8038, 8045,
8057, 8161, 8182, 8199, 8240,
8325, 8351, 8408, 8411, 8413,
8488, 8495, 8496, 8514, 8516,
8647, 8656, 8670, 8734, 8742,
8770, 8793, 8821, 8847, 8868,
8985, 8997, 9005, 9007, 9056,
9079, 9106, 9147, 9151, 9176,
9177, 9185, 9249, 9250, 9252,
9267, 9282, 9318, 9337, 9361,
9373, 9419, 9429, 9450, 9467,
9468, 9570, 9571, 9577, 9592,
9646, 9708, 9715, 9722, 9747,
9771, 9772, 9784, 9797, 9803,
9807, 9812, 9850, 9858, 9868,
9908, 9919, 9948, 9949, 9975,
9990, 10016, 10036, 10038, 10049,

10064, 10084, 10091, 10146, 10162,
10189, 10205, 10206, 10234, 10241,
10243, 10428, 10433, 10435, 10453,
10478, 10490, 10510, 10589, 10622,
10637, 10641, 10647, 10653, 10697,
10729, 10814, 10841, 10880, 10883,
10961, 11018, 11019, 11023, 11024,
11045, 11080, 11171, 11184, 11189,
11214, 11281, 11298, 11361, 11362,
11363, 11371, 11413, 11424, 11517,
11566, 11574, 11575, 11592, 11594,
11609, 11612, 11613, 11614, 11620,
11631, 11644, 11673, 11714, 11722,
11767, 11774, 11821, 11822, 11823,
11828, 11831, 11882, 11890, 11898,
11951, 11953, 12007, 12010, 12025,
12059, 12062, 12104, 12106, 12112,
12116, 12137, 12152, 12212, 12267,
12270, 12329, 12340, 12341, 12345,
12349, 12368, 12374, 12442, 12447,
12514, 12540, 12544, 12609, 12622,
12623, 12677, 12693, 12694, 12710,
12717, 12753, 12762, 12770, 12773,
12800, 12803, 12822, 12828, 12835,
12853, 12862, 12872, 12898, 12909,
12961, 12969, 13036, 13040, 13055,
13064, 13081, 13118, 13255, 13269,
13272, 13288, 13299, 13313, 13314,
13317, 13318, 13319, 13331, 13333,
13336, 13344, 13442, 13456, 13483,
13488, 13578, 13579, 13598, 13666,
13690, 13713, 13749, 13751, 13752,
13769, 13774, 13818, 13838, 13851,
13855, 13864, 13878, 13886, 13903,
13938, 13941, 13945, 13958, 14098,
14112, 14159, 14169, 14170, 14171,
14172, 14210, 14300, 14304, 14331,
14358, 14378, 14426, 14429, 14430,
14434, 14462, 14468, 14627, 14688,
14695

Phillyrea 8676, 9059, 10514,
11408

Philodendron 7624, 8509, 8650,
11292, 11628

Phleum 6828, 7562, 7870, 8951,
9094, 9135, 9723, 9784, 10294,
10295, 10982, 11587, 12155, 12192,
12271, 12793, 13020, 14092, 14166,
14167, 14718

Phlox 11656, 14356

Phoenix 9252, 14681

Phragmites 7677, 7729, 7747,
8934, 9115, 9273, 9817, 10178,
11018, 13208, 13748, 14339

Pisum (cont.)
8413, 8414, 8423, 8425, 8426,
8499, 8545, 8772, 8848, 8906,
8951, 8963, 8964, 8969, 9109,
9110, 9203, 9318, 9553, 9554,
9573, 9626, 9635, 9692, 9708,
9744, 9784, 9804, 9852, 9976,
9981, 9983, 10035, 10049, 10134,
10135, 10187, 10189, 10191, 10195,
10232, 10243, 10273, 10360, 10461,
10468, 10526, 10527, 10622, 10641,
10690, 10754, 10790, 10883, 10929,
10930, 11018, 11109, 11132, 11174,
11286, 11376, 11406, 11484, 11485,
11551, 11566, 11574, 11732, 11746,
11747, 11769, 11834, 11849, 11938,
12010, 12062, 12112, 12154, 12273,
12280, 12359, 12385, 12386, 12387,
12456, 12494, 12544, 12591, 12650,
12669, 12695, 12700, 12717, 12743,
12753, 12837, 12898, 12942, 12969,
13034, 13192, 13222, 13250, 13333,
13361, 13368, 13399, 13400, 13456,
13571, 13730, 13732, 13816, 13862,
13893, 13894, 13971, 13972, 14021,
14049, 14083, 14084, 14100, 14115,
14371, 14429, 14442, 14493, 14499,
14588, 14600, 14655, 14656, 14685,
14726, 14727

Pittosporum 9183, 9646, 10106,
10992, 11943

Plantago 6863, 6888, 6889,
6985, 6986, 7216, 7323, 7534,
8264, 8379, 8588, 8886, 8887,
9120, 9517, 9606, 9646, 9817,
9907, 9911, 9919, 9929, 9991,
10043, 10131, 10199, 10975, 11187,
11621, 11940, 12737, 13232, 13233,
13234, 13235, 13373, 13734, 14003,
14169, 14332, 14333

Platanus 6614, 6823, 7254,
8451, 8545, 8857, 9991, 10101,
10438, 10989, 11920, 12910, 13743,
13746, 14261

Plectranthus 7544, 9084, 10435,
10589, 11018, 14564

Poa 6650, 6697, 7068, 7070,
7302, 7544, 8067, 8182, 8607,
8850, 9094, 9938, 10073, 10177,
10295, 10434, 10613, 10622, 10777,
10883, 10966, 11150, 11306, 11616,
11865, 11943, 12072, 12192, 12281,
12641, 12719, 12809, 13020, 13348,
13734, 14517, 14578, 14643, 14702

Podocarpus 8354, 10992, 12120,
13090, 13451, 13937

Podophyllum 12178

Polygonatum 11588, 14166, 14167

Polygonum 6614, 6806, 7070,
8804, 8850, 9094, 9844, 9991,
10600, 10777, 12873, 12969, 13021,
13174, 13373, 13626, 14245, 14517

Polypogon 12146

poplar see *Populus*

Populus 6614, 6781, 6790,
6823, 6839, 6841, 6928, 6930,
7020, 7073, 7186, 7268, 7305,
7343, 7447, 7460, 7544, 7555,
7585, 7586, 7621, 7645, 7829,
7858, 7859, 8142, 8501, 8517,
8545, 8677, 8740, 8801, 8802,
8804, 8809, 8823, 8857, 8859,
8861, 8981, 8985, 9037, 9072,
9139, 9140, 9141, 9306, 9429,
9550, 9703, 9739, 9753, 9776,
9850, 9856, 9857, 9859, 9919,
9939, 9991, 10087, 10101, 10313,
10314, 10315, 10363, 10364, 10529,
10576, 10577, 10753, 10784, 10880,
10907, 11012, 11023, 11027, 11052,
11053, 11177, 11242, 11243, 11387,
11609, 11920, 11992, 12120, 12147,
12201, 12321, 12433, 12586, 12623,
12663, 12841, 12905, 13067, 13090,
13098, 13141, 13294, 13416, 13595,
13694, 13731, 13747, 14009, 14010,
14231, 14232, 14233, 14234, 14236,
14306, 14434, 14476, 14588, 14758

Portulaca 7035, 7070, 7237,
7296, 7297, 7571, 8307, 8790,
9829, 10349, 10370, 10633, 11407,
12537, 12872, 12873, 13891, 13947,
14543, 14544, 14636

Portulacaria 13472, 13473

Posidonia 9559, 11048, 14603

Potamogeton 7343, 10283, 10375,
10713, 10838, 11369, 14339

potato see *Solanum*

Potentilla 9094, 10966, 11652,
13734, 14517

Poterium 14538

Primula 6838, 8850, 11545, 12834, 12835

privet see *Ligustrum*

Prosopis 6740, 7047, 7120, 7430, 7488, 7621, 8644, 8934, 8985, 9793, 10041, 10164, 10363, 10586, 11617, 11618, 11619, 12224, 12537, 14014, 14681

Protea 12142

Prunella 10562

Prunus 6614, 6753, 7021, 7171, 7305, 7343, 7383, 7620, 7660, 7793, 7794, 7999, 8009, 8038, 8182, 8199, 8285, 8310, 8539, 8844, 8857, 8929, 9211, 9212, 9252, 9273, 9360, 9443, 9521, 9578, 9761, 9910, 10065, 10094, 10108, 10250, 10363, 10468, 10746, 10811, 10841, 11027, 11298, 11387, 11530, 11610, 11719, 11887, 11954, 12008, 12100, 12363, 12376, 12433, 12957, 12988, 13128, 13358, 13481, 13488, 13607, 13746, 14035, 14048, 14166, 14167, 14350, 14351, 14600

Pseudotsuga 6661, 6736, 7093, 7305, 7350, 7352, 7378, 7535, 7621, 7739, 8262, 8263, 8359, 8443, 8446, 8742, 8743, 8762, 8763, 8857, 8899, 8913, 8929, 9132, 9139, 9229, 9438, 9439, 9440, 9505, 9627, 9716, 10001, 10280, 10363, 10364, 10390, 10401, 10438, 10482, 10483, 10484, 10549, 10673, 10674, 10688, 10840, 10879, 10880, 11018, 11110, 11263, 11446, 11587, 11848, 11950, 12120, 12286, 12317, 12623, 12894, 12895, 12905, 12913, 12981, 13700, 13702, 13705, 13736, 13823, 13826, 13929, 14255, 14432, 14460, 14478, 14687, 14705

Psoralea 13370

Psychotrichia 10992

Pteridophyta
  Adianthum 7380, 11155, 11647, 14017, 14681
  Alsophila 11155
  Aneimia 7380
  Angiopteris 9616

Aspidium 7380
Asplenium 6844, 7380, 7641, 8024, 8254, 14017, 14681
Azolla 9005, 12541, 13116
Blechnum 7380, 14017
Botrychium 11155
Ceterach 8254, 12133, 12531, 14355
Cibotium 14017
Dennstaedtia 14017
Dicksonia 14017
Dipteris 7274
Drynaria 7380
Dryopteris 7380, 8787, 13408, 13891
Elaphoglossum 7380
Equisetum 6802, 12192, 12602, 14718
Humada 8040
Isoëtes 11903
Lycopodium 6802, 10073, 14681
Marsilea 11155, 14718
Matteuccia 10699, 10700, 12350, 12351, 14179
Nephrolepis 9196, 10435, 10633, 12335, 13851, 14017
Notholaena 8939, 11155
Onoclea 6879, 14017
Ophioglossum 9349, 9616, 14718
Phyllitis 7380, 8024, 8254
Platycerium 7380, 13891
Polypodium 7035, 7380, 8787, 8929, 9060, 9616, 9640, 9919, 12086, 12133, 14017, 14681, 14718
Polystichum 7380, 14017
Pteridium 7711, 9616, 11609, 14017
Pteris 7380, 9616, 14017, 14681
Pyrrosia 14433, 14434, 14435
Selaginella 6802, 6905, 6919, 6970, 8040, 8065, 8911, 10011, 11318, 11586, 14681, 14718

Puccinellia 6704, 7903, 7921, 8050, 8091, 9600, 9646, 9817, 9907, 10176, 10177, 11306, 11736, 12760, 13132, 14578

Pulicaria 6677, 10397

Pulmonaria 6975, 8475, 8877, 11558, 13305, 13753

Pulsatila 6914, 14166

Q

Quercus 6614, 6754, 6836,
6913, 6931, 6974, 7002, 7019,
7020, 7035, 7184, 7185, 7187,
7305, 7340, 7343, 7348, 7351,
7474, 7475, 7536, 7555, 7621,
7695, 7924, 7998, 8184, 8330,
8445, 8470, 8474, 8569, 8570,
8629, 8676, 8677, 8689, 8709,
8743, 8857, 8860, 8933, 9034,
9038, 9059, 9081, 9123, 9220,
9514, 9515, 9519, 9578, 9671,
9717, 9738, 9785, 9843, 9854,
9919, 9973, 10071, 10087, 10101,
10239, 10240, 10363, 10364, 10387,
10425, 10426, 10438, 10460, 10514,
10541, 10560, 10561, 10650, 10753,
10779, 10977, 10991, 11018, 11023,
11027, 11144, 11145, 11186, 11229,
11242, 11243, 11264, 11339, 11387,
11408, 11548, 11549, 11583, 11609,
11685, 11686, 11691, 11758, 11771,
11812, 12047, 12078, 12120, 12277,
12315, 12353, 12433, 12438, 12579,
12629, 12666, 12680, 12742, 12828,
12841, 12967, 12988, 13004, 13063,
13090, 13257, 13258, 13323, 13381,
13416, 13562, 13571, 13592, 13595,
13702, 13743, 13746, 13747, 13897,
13936, 13953, 14035, 14166, 14167,
14168, 14202, 14209, 14263, 14354,
14539, 14588, 14758

Quillaja 10363, 12140

R

Ramondia 8787

Randia 7684, 7911, 9244

Ranunculus 6614, 6704, 6914,
8187, 8850, 11298, 11943, 12162,
13117, 13734, 14517, 14718

Raphanus 6733, 7561, 8038,
8182, 8361, 8846, 8915, 9252,
9477, 10550, 11038, 11238, 11410,
11491, 12010, 12272, 12701, 12709,
12710, 12716, 12896, 13285, 13361,
13468, 13972, 14155, 14352

Ravenala 10608

Remijia 7879

Reseda 12192

Rhamnus 7035, 11127, 11387,
12849, 13358

Rheum 8182

Rhizophora 9097, 9183, 13124,
13167, 13168

Rhodiola 14517

Rhododendron 6840, 7058, 7130,
7895, 8266, 8730, 8784, 8857,
8895, 9855, 9858, 10087, 10101,
11439, 11441, 11568, 11700, 11943,
11953, 11954, 12489, 13056, 13500,
13669, 13670, 14273

Rhus 7821, 9011, 9099, 9195,
9850, 11127, 12433, 14035, 14681

Ribes 6614, 6806, 9094, 10108,
11144, 11145, 11514, 11857

rice see *Oryza*

Ricinus 6818, 7035, 7144,
7469, 7871, 7872, 7873, 7874,
8545, 8657, 8820, 9005, 9037,
9418, 10022, 10146, 10489, 11187,
11213, 11759, 12010, 12409, 12944,
13673, 14216, 14223, 14430, 14434,
14638, 14639

Robinia 7186, 8038, 8517,
8861, 9005, 9489, 10361, 10362,
10363, 10364, 10438, 10480, 10541,
11018, 11238, 11416, 11527, 14166,
14167, 14758

Rosa 6924, 8021, 8429, 8752,
9422, 10551, 10883, 10957, 11089,
11475, 13222, 13249, 13333, 13524,
13599, 14025

rose see *Rosa*

Rosmarinus 10497

Rubus 7079, 8040, 8182, 8642,
9252, 9433, 10600, 10672, 10885,
11514, 11540, 12053, 13265, 13266

Rudbeckia 9006

Rumex 7273, 7534, 8182, 8861,
8951, 9991, 10090, 10565, 11185,
12203, 13285, 13345, 13429, 14315,
14517, 14718

Senecio 6701, 6702, 7737, 8952, ·9094, 9422, 9819, 9929, 10131, 10435, 10633, 10866, 10966, 12086, 12492, 12845, 12846, 13197, 13851, 14114, 14241, 14517, 14681, 14718

Sequoia 9627, 12243, 14687

Sequoiadendron 9627, 12243

Serratula 11283

service tree see *Sorbus*

Sesamum 6717, 7305, 8881, 8951, 9242, 9761, 10196, 10432, 10841, 10883, 11960, 12409, 12537, 13488, 14208, 14638

Seseli 7212, 7214, 7465, 9371, 11223

Sesleria 8850

Setaria 6948, 7070, 7083, 7480, 7575, 8669, 9124, 9252, 9493, 9844, 11018, 11060, 11111, 11341, 11840, 11920, 13184

Shorea 9159

Sibbaldia 9094

Sicyos 12076

Sida 8469, 8872, 9243, 9947, 11578, 12301, 12391, 12537

Silene 7212, 7304, 8850, 9929, 10131, 10677

Silphium 8857

Silybum 7290

Simmondsia 6696, 7881, 7997, 8861, 8939, 9919, 10363, 11435, 11960, 13488, 14014

Sinapis 6621, 6960, 8066, 8182, 8409, 8737, 9638, 9850, 10644, 10846, 11018, 11483, 12389, 12409, 12776, 12831, 14208, 14293, 14430, 14630, 14632, 14638

Sisymbrium 8561

Sitanion 7010

Sium 9120

Smilax 10387, 10425

snap bean see *Vicia*

Solanum 6624, 6698, 6713, 6763, 6806, 6824, 6849, 6866, 6929, 7070, 7083, 7177, 7207, 7208, 7305, 7343, 7368, 7412, 7435, 7506, 7514, 7532, 7539, 7584, 7590, 7645, 7660, 7708, 7752, 7792, 7801, 7853, 7865, 7901, 7943, 7958, 7959, 8029, 8030, 8144, 8182, 8190, 8354, 8373, 8374, 8460, 8461, 8742, 8872, 8907, 8927, 8974, 9027, 9028, 9098, 9168, 9201, 9252, 9276, 9297, 9375, 9449, 9478, 9482, 9507, 9510, 9620, 9659, 9680, 9685, 9686, 9748, 9751, 9850, 9978, 9979, 10009, 10022, 10027, 10108, 10118, 10146, 10243, 10325, 10332, 10397, 10432, 10552, 10596, 10641, 10708, 10729, 10759, 10773, 10838, 10883, 10910, 10992, 10996, 11018, 11238, 11251, 11319, 11374, 11392, 11426, 11540, 11573, 11696, 11715, 11932, 11953, 11954, 11994, 12016, 12088, 12138, 12236, 12242, 12264, 12341, 12422, 12432, 12436, 12479, 12537, 12549, 12550, 12572, 12691, 12723, 12762, 12807, 12843, 12867, 12872, 12909, 12928, 12969, 12993, 12999, 13002, 13033, 13089, 13128, 13169, 13181, 13244, 13413, 13456, 13495, 13561, 13735, 13906, 13921, 14068, 14134, 14243, 14313, 14378, 14398, 14458, 14475, 14528, 14535, 14546, 14577, 14584, 14620

Solidago 6717, 9120, 9846, 10633, 10841, 12346, 14177

Sonchus 8557, 8561

Sonneratia 11889

Sophora 9611, 14306

Sorbus 6923, 7156, 7555, 8857, 9273, 9731, 11771, 13258

Sorghum 6609, 6612, 6620, 6633, 6650, 6666, 6696, 6717, 6773, 6786, 6832, 6863, 6912, 6945, 6960, 7000, 7070, 7083, 7116, 7162, 7193, 7219, 7233,

Sorghum (cont.)
7236, 7242, 7300, 7305, 7309,
7343, 7376, 7383, 7458, 7477,
7480, 7501, 7502, 7576, 7577,
7645, 7681, 7729, 7800, 7850,
7865, 7870, 7918, 7923, 7931,
7932, 7995, 7996, 7997, 8041,
8072, 8168, 8182, 8242, 8286,
8315, 8413, 8553, 8561, 8605,
8681, 8727, 8772, 8846, 8894,
8950, 8985, 9018, 9019, 9070,
9096, 9238, 9252, 9282, 9290,
9384, 9408, 9429, 9446, 9493,
9556, 9567, 9568, 9617, 9656,
9677, 9764, 9822, 9838, 9850,
9870, 9919, 9932, 10022, 10042,
10085, 10086, 10087, 10162, 10194,
10268, 10301, 10302, 10303, 10323,
10432, 10435, 10464, 10503, 10531,
10572, 10636, 10646, 10729, 10748,
10807, 10841, 10853, 10862, 10902,
10952, 11018, 11039, 11044, 11115,
11117, 11231, 11236, 11330, 11342,
11348, 11377, 11509, 11520, 11565,
11569, 11576, 11589, 11649, 11658,
11659, 11661, 11711, 11730, 11731,
11732, 11748, 11769, 11777, 11801,
11865, 11877, 11878, 11880, 11891,
11898, 11916, 11917, 11920, 11954,
11962, 11963, 11964, 11997, 12038,
12068, 12092, 12207, 12255, 12267,
12285, 12291, 12368, 12377, 12379,
12400, 12401, 12409, 12414, 12436,
12447, 12547, 12623, 12654, 12691,
12753, 12762, 12764, 12790, 12876,
12877, 12878, 12885, 12931, 13010,
13082, 13083, 13252, 13264, 13314,
13350, 13369, 13378, 13406, 13407,
13456, 13460, 13493, 13507, 13509,
13560, 13570, 13593, 13655, 13767,
13768, 13851, 13866, 13880, 13881,
13893, 13894, 13914, 13931, 13938,
13971, 13972, 13995, 13996, 14060,
14062, 14063, 14065, 14110, 14206,
14223, 14275, 14375, 14376, 14421,
14434, 14495, 14512, 14532, 14601,
14619, 14622, 14638, 14668, 14669,
14670, 14693

soybean see *Glycine*

Sparganium 9581

Spartina 6717, 6915, 6972,
7367, 8422, 8576, 8749, 9038,
9053, 9600, 9646, 9817, 9960,
10105, 10729, 11067, 11437, 12036,
12623, 12838, 12872

Spartium 9318

Spergula 12737

Spergularia 9103, 9646, 9907,
14042

spinach see *Spinacia*

Spinacia 6715, 6818, 6940,
7142, 7366, 7467, 7557, 7561,
8091, 8109, 8182, 8380, 8419,
8754, 8780, 8781, 8782, 8783,
8836, 8951, 8958, 8986, 9005,
9252, 9332, 9359, 9642, 9646,
9757, 9758, 9915, 10035, 10049,
10108, 10146, 10231, 10296, 10578,
10696, 10752, 10979, 11011, 11047,
11109, 11162, 11256, 11298, 11320,
11321, 11322, 11354, 11376, 11379,
11381, 11496, 11681, 11729, 11739,
11740, 11744, 11873, 11874, 11875,
12110, 12242, 12446, 12508, 12531,
12653, 12681, 12685, 12824, 12872,
13034, 13073, 13552, 13637, 13668,
13803, 13906, 14267, 14317, 14355,
14371, 14521

Spiraea 11018

Spirodela 13869, 14372

Sporobotus 6904, 6905, 7035,
7040, 12539, 13158

spruce see *Picea*

Stachys 8219, 8857, 10562

Stachytarpheta 7645

Stanleya 7612

Stapelia 10435, 13851

Staphylea 14168

Stellaria 10777, 14168

Stipa 6650, 7935, 8065, 8552,
9732, 10032, 11283, 12192, 12946,
13158, 13390, 13408, 14166, 14167

strawberry see *Fragaria*

Stylosanthes 6696, 6863, 7016,
8978, 8979, 12246, 12247, 12994,
13370, 13924

Vaccinium (cont.)
12954, 12955, 13041, 13226, 14155,
14247, 14681

Valeriana 9919, 10966, 14517

Valerianella 8929, 9919, 11743,
11744

Vallisneria 9309, 10799

Vanda 12314

Veratrum 8850, 12393

Verbascum 7212, 9120, 9991,
13408

Verbena 9120, 12096

Veronica 6677, 10777

Viburnum 7130, 7156, 7716,
7717, 8040, 10087, 10992, 11771,
14167, 14560

Vicia 6719, 6733, 6827, 6927,
6946, 7124, 7125, 7212, 7213,
7214, 7247, 7249, 7250, 7300,
7307, 7318, 7335, 7343, 7383,
7389, 7558, 7559, 7560, 7636,
7645, 7721, 7778, 7779, 7780,
7781, 7782, 7783, 7784, 7812,
7831, 8038, 8041, 8097, 8115,
8116, 8125, 8158, 8161, 8182,
8186, 8190, 8413, 8500, 8660,
8770, 8772, 8800, 8846, 8916,
8944, 8951, 8952, 8990, 9005,
9074, 9099, 9101, 9128, 9154,
9223, 9237, 9252, 9285, 9296,
9297, 9311, 9312, 9318, 9340,
9347, 9348, 9429, 9513, 9524,
9570, 9571, 9608, 9612, 9613,
9626, 9667, 9668, 9673, 9708,
9784, 9799, 9807, 9919, 10052,
10098, 10116, 10146, 10157, 10189,
10235, 10275, 10305, 10376, 10377,
10380, 10470, 10481, 10604, 10631,
10703, 10790, 10808, 10834, 10835,
10866, 10870, 10883, 10906, 10913,
10914, 10915, 11018, 11108, 11116,
11122, 11174, 11228, 11238, 11338,
11367, 11398, 11507, 11519, 11660,
11664, 11674, 11742, 11754, 11829,
11842, 11971, 12007, 12065, 12086,
12124, 12154, 12251, 12268, 12269,
12308, 12359, 12368, 12521, 12524,
12560, 12571, 12630, 12660, 12753,
12770, 12778, 12780, 12788, 12816,

12851, 12880, 12898, 12969, 13064,
13117, 13193, 13445, 13494, 13503,
13734, 13878, 13891, 13971, 13972,
14076, 14077, 14134, 14238, 14302,
14310, 14342, 14357, 14473, 14474,
14536, 14572, 14696, 14713, 14718,
14727

Vigna 6618, 6818, 7036, 7037,
7070, 7099, 7104, 7105, 7305,
7566, 7609, 7673, 7787, 7832,
7990, 7991, 7992, 7993, 8141,
8221, 8626, 8879, 8923, 8951,
8972, 9008, 9057, 9154, 9184,
9205, 9252, 9349, 9360, 9665,
9740, 9741, 9822, 9850, 9970,
10139, 10243, 10258, 10268, 10300,
10392, 10393, 10394, 10468, 10536,
10836, 10841, 10875, 10875, 10883,
11018, 11118, 11125, 11212, 11214,
11236, 11271, 11449, 11679, 11695,
11714, 11732, 11766, 11823, 12010,
12176, 12177, 12191, 12401, 12442,
12447, 12456, 12491, 12528, 12530,
12540, 12547, 12577, 12580, 12623,
12902, 12907, 12908, 12912, 12969,
13111, 13143, 13341, 13370, 13488,
13610, 13711, 13712, 13796, 13938,
13939, 13971, 13972, 13976, 13998,
14039, 14088, 14089, 14090, 14143,
14144, 14198, 14305, 14344, 14350,
14376, 14572, 14600, 14602, 14638,
14652, 14718, 14721, 14737

Vinca 11018

Viola 11588, 13214, 13215,
14166

Virola 14034

Viscum 14354, 14681

Vitis 6670, 6671, 6959, 7029,
7305, 7383, 7446, 7452, 7576,
7597, 7797, 7801, 7827, 7996,
7997, 8165, 8182, 8199, 8200,
8201, 8325, 8449, 8457, 8458,
8861, 8985, 8994, 9069, 9252,
9409, 9429, 9604, 9696, 9697,
9725, 9871, 9999, 10000, 10363,
10382, 10383, 10541, 10883, 10978,
11018, 11023, 11186, 11187, 11275,
11387, 11421, 11463, 11547, 11554,
11556, 11557, 11585, 11639, 11640,
11641, 11934, 11960, 12169, 12170,
12186, 12262, 12470, 12471, 12553,
12608, 12623, 12657, 12706, 12745,
12828, 12951, 12958, 13008, 13131,
13179, 13261, 13262, 13269, 13274,

SUBJECT INDEX

(Cumulative index to volumes 6 - 10)

This index contains a selection of primary items chosen according to their interest for water relation researchers and to their relative importance and occurrence.

A

Abaxial and adaxial leaf epidermes 6611, 6618, 6650, 6662, 6684, 6737, 6739, 6793, 6794, 6809, 6843, 6844, 6859, 6914, 6919, 6947, 7031, 7151, 7173, 7189, 7227, 7256, 7263, 7276, 7305, 7333, 7405, 7419, 7461, 7520, 7521, 7522, 7557, 7582, 7585, 7587, 7602, 7645, 7660, 7683, 7690, 7692, 7736, 7807, 7813, 7843, 7929, 7974, 8066, 8067, 8092, 8128, 8167, 8177, 8185, 8197, 8221, 8233, 8240. 8285, 8309, 8461, 8467, 8475, 8557, 8579, 8645, 8686, 8699, 8744, 8775, 8791, 8818, 8855, 8985, 8997, 8999, 9016, 9074, 9106, 9109, 9139, 9141, 9237, 9243, 9304, 9318, 9372, 9373, 9374, 9382, 9422, 9443, 9543, 9585, 9597, 9600, 9718, 9719, 9768, 9809, 9859, 9897, 9954, 9983, 10013, 10076, 10100, 10111, 10112, 10180, 10200, 10202, 10234, 10253, 10374, 10397, 10476, 10501, 10510, 10514, 10562, 10630, 10651, 10669, 10686, 10756, 10878, 10889, 10907, 11000, 11068, 11198, 11291, 11292, 11386, 11388, 11413, 11494, 11519, 11573, 11622, 11627, 11722, 11739, 11790, 11798, 11811, 11824, 11842, 11855, 11932, 12004, 12034, 12079, 12086, 12191, 12305, 12310, 12311, 12312, 12313, 12410, 12421, 12427, 12439, 12454, 12537, 12551, 12555, 12593, 12595, 12622, 12623, 12651, 12683, 12694, 12696, 12699, 12701, 12704, 12714, 12765, 12878, 12922, 12988, 13067, 13141, 13175, 13176, 13181, 13194, 13229, 13312, 13408, 13535, 13589, 13615, 13616, 13635, 13747, 13749, 13774, 13783, 13820, 13830, 13885, 13902, 13945, 13949, 13978, 13982, 14104, 14126, 14130, 14143, 14170, 14171, 14172, 14177, 14207, 14233, 14234, 14236, 14253, 14262, 14277, 14292, 14301, 14319, 14389, 14422, 14468, 14476, 14508, 14601, 14685, 14718, 14719, 14738

Abscisic acid see Antitranspirants; Growth substances, hormones, inhibitors etc.

Absorption of water see Water absorption ...

Aerodynamic methods, bioclimatological methods (sampling, measurement of wind, rain, dew etc.) 9089, 12219, 13247, 14416

Age of leaf see Age of plant ...; Leaf insertion level ...

Age of plant and conductance for water vapour and carbon dioxide transfer 6612, 6833, 7241, 7421, 7485, 7881, 7883, 8238, 8250, 8460, 8537, 8627, 8729, 8768, 8843, 9329, 9408, 9443, 9475, 9494, 9499, 9604, 9741, 9987, 10013, 10149, 10225, 10465, 10510, 10697, 11100, 11155, 11199, 11265, 11296, 11765, 11886, 11896, 11962, 11963, 12165, 12191, 12329, 12345, 12430, 12622, 12623, 12742, 12863, 12908, 12923, 13171, 13175, 13515, 13535, 13616, 13862, 13965, 13996, 14071, 14170, 14171, 14466, 14468, 14479

Age of plant and stomata 9329, 10510, 11000, 11717, 14536, 14641

Age of plant and transpiration 6619, 6621, 6642, 6643, 6700, 6737, 7049, 7603, 7657, 7665, 7689, 7744, 7979, 7992, 8252, 8390, 8407, 9326, 9375, 9979, 10303, 10332, 10419, 10420, 10462, 10463, 10530, 10645, 10821, 10899, 11085, 11155, 11251, 11252, 11280, 11296, 11364, 11478, 11510, 11606, 11661, 11722, 12144, 12165, 12191, 12530, 12570, 12577, 12623, 12662, 12699, 12864, 12908, 12920, 12965, 12974, 12985, 12993, 13406, 13457, 13608, 13995, 14030, 14124, 14125, 14638, 14641, 14683

Availability of soil water (continued) 12561, 12581, 12588, 12590, 12615, 12638,
          12639, 12645, 12652, 12654, 12663, 12665, 12652, 12654, 12663, 12665, 12696,
          12713, 12729, 12749, 12761, 12762, 12768, 12813, 12843, 12867, 12876, 12886,
          12913, 12974, 12985, 13014, 13018, 13030, 13032, 13037, 13038, 13043, 13051,
          13086, 13125, 13129, 13154, 13177, 13189, 13199, 13212, 13263, 13268, 13300,
          13310, 13325, 13326, 13342, 13375, 13376, 13387, 13414, 13444, 13449, 13457,
          13482, 13550, 13557, 13566, 13581, 13592, 13596, 13614, 13615, 13619, 13620,
          13621, 13625, 13643, 13651, 13662, 13675, 13691, 13697, 13712, 13724, 13768,
          13775, 13776, 13787, 13793, 13841, 13856, 13860, 13863, 13896, 13913, 13936,
          13943, 13948, 13954, 13970, 13987, 13995, 13998, 14002, 14064, 14074, 14120,
          14143, 14155, 14182, 14188, 14202, 14238, 14256, 14278, 14279, 14289, 14302,
          14315, 14316, 14333, 14337, 14356, 14366, 14384, 14387, 14388, 14394, 14443,
          14488, 14500, 14558, 14619, 14660, 14673, 14730

B

Beta gauge see Water saturation deficit, methods

Biological clock see Diurnal changes ...

Biliproteins see Deuterium oxide, tritium oxide ...; Flooding ...; Irrigation
          and biliproteins

Books on plant water relation see General aspects on plant water relations

Bound water 6671, 6759, 7006, 7054, 7186, 7215, 7320, 7331, 7332, 7336, 7446, 7465,
          7525, 7528, 7538, 7677, 7712, 8073, 8086, 8090, 8096, 8098, 8225, 8235, 8689,
          8805, 8855, 8976, 9011, 9067, 9139, 9248, 9250, 9309, 9363, 9379, 9471, 9659,
          9664, 9827, 9875, 10016, 10018, 10079, 10094, 10129, 10346, 10403, 10424,
          10443, 10446, 10689, 10735, 10789, 10873, 10961, 11159, 11217, 11232, 11274,
          11365, 11716, 11953, 11972, 12378, 12644, 12696, 12856, 12905, 12990, 13120,
          13261, 13272, 13315, 13461, 13780, 13848, 14140, 14157, 14201, 14409, 14410,
          14411, 14432, 14452, 14484, 14569, 14645, 14717

Bound water, methods 7511, 8225, 8398, 10789

Boundary layer of air see Conductance for water vapour and carbon dioxide transfer,
          leaf boundary layer

C

C$_3$, C$_4$, CAM pathways see Comparison of plants with different types of carbon
          metabolism

Canopy architecture see Drought ...; Humidity of air ...; Irrigation ...; Pre-
          cipitation, dew ...; Salinity ...; Soil moisture ...; Water status in
          plant and canopy architecture

Canopy model see Model of canopy

Canopy structure, method 11500, 11709, 11836, 11855, 12107, 12217, 12607, 13413,
          13507, 13733, 14024

Canopy water vapour profiles see Humidity of air, gradients in canopy

Carbohydrates see Saccharides ...

Carbon dioxide and conductance for water vapour and carbon dioxide transfer 6889,
          7151, 7153, 7209, 7490, 7545, 7596, 7808, 7883, 7996, 8361, 8677, 8744, 8745,
          8799, 8913, 8929, 8951, 8952, 8969, 9016, 9134, 9225, 9226, 9372, 9373, 9404,
          9761, 9901, 10087, 10251, 10413, 10424, 10587, 10659, 10819, 10883, 11144,
                                                                              (continued)

Conductance for water vapour and carbon dioxide transfer, oscillations 7170, 7209, 7225, 7809, 9661, 9919, 10076, 12383, 13721, 14231, 14234, 14602, 14615

Conductance for water vapour and carbon dioxide transfer, seasonal course 6625, 6647, 6726, 6859, 6893, 6911, 6974, 7241, 7621, 7695, 7711, 7739, 7743, 7840, 7932, 8093, 8127, 8128, 8197, 8250, 8289, 8385, 8460, 8512, 8615, 8667, 8676, 8727, 8921, 9041, 9152, 9195, 9209, 9306, 9321, 9360, 9494, 9498, 9514, 9606, 9607, 9763, 9889, 9902, 9924, 9954, 9960, 10152, 10169, 10314, 10315, 10419, 10425, 10448, 10549, 10576, 10586, 10733, 10779, 10805, 10841, 10937, 11067, 11144, 11152, 11332, 11414, 11429, 11514, 11582, 11686, 11758, 11811, 11855, 11858, 11994, 12100, 12142, 12224, 12277, 12437, 12623, 12699, 12821, 12954, 12955, 12957, 12984, 13049, 13063, 13119, 13175, 13178, 13236, 13297, 13305, 13406, 13473, 13592, 13626, 13994, 14010, 14035, 14116, 14135, 14143, 14189, 14262, 14286, 14308, 14359, 14394, 14460, 14479, 14556

Conductance for water vapour and carbon dioxide transfer, stomata 6611, 6612, 6620, 6625, 6641, 6649, 6660, 6664, 6680, 6685, 6698, 6706, 6712, 6726, 6733, 6737, 6739, 6766, 6774, 6779, 6780, 6818, 6832, 6833, 6839, 6840, 6841, 6859, 6871, 6876, 6877, 6889, 6907, 6908, 6911, 6912, 6931, 6943, 6946, 6947, 6974, 6975, 6976, 6977, 6978, 6987, 6990, 6998, 7002, 7003, 7006, 7013, 7027, 7033, 7047, 7058, 7073, 7092, 7104, 7106, 7113, 7117, 7130, 7135, 7151, 7153, 7156, 7163, 7170, 7227, 7232, 7241, 7258, 7259, 7266, 7268, 7269, 7270, 7276, 7278, 7281, 7282, 7297, 7304, 7305, 7319, 7321, 7335, 7343, 7352, 7356, 7362, 7365, 7370, 7375, 7384, 7397, 7419, 7420, 7421, 7460, 7461, 7468, 7485, 7490, 7491, 7492, 7501, 7502, 7504, 7540, 7544, 7545, 7572, 7578, 7582, 7587, 7595, 7609, 7612, 7615, 7616, 7621, 7638, 7645, 7646, 7654, 7682, 7683, 7684, 7689, 7693, 7695, 7696, 7704, 7711, 7723, 7739, 7740, 7741, 7742, 7753, 7754, 7797, 7806, 7807, 7808, 7811, 7827, 7840, 7858, 7864, 7876, 7877, 7883, 7918, 7932, 7936, 7937, 7939, 7954, 7955, 7957, 7961, 7976, 7996, 8010, 8061, 8075, 8077, 8078, 8079, 8083, 8093, 8104, 8127, 8128, 8130, 8131, 8133, 8134, 8137, 8140, 8141, 8159, 8167, 8189, 8190, 8192, 8197, 8198, 8204, 8216, 8221, 8224, 8233, 8239, 8240, 8244, 8249, 8250, 8251, 8252, 8260, 8262, 8263, 8265, 8284, 8289, 8290, 8296, 8306, 8309, 8315, 8325, 8326, 8332, 8333, 8334, 8359, 8360, 8361, 8374, 8383, 8385, 8387, 8406, 8411, 8413, 8416, 8417, 8433, 8445, 8449, 8457, 8460, 8461, 8474, 8494, 8495, 8531, 8536, 8537, 8538, 8540, 8546, 8549, 8590, 8591, 8603, 8606, 8615, 8627, 8641, 8644, 8645, 8662, 8664, 8665, 8667, 8670, 8677, 8710, 8727, 8729, 8740, 8742, 8743, 8744, 8745, 8750, 8751, 8768, 8769, 8770, 8776, 8799, 8800, 8801, 8809, 8810, 8811, 8813, 8814, 8818, 8823, 8826, 8831, 8833, 8838, 8839, 8857, 8860, 8861, 8891, 8894, 8895, 8912, 8913, 8914, 8921, 8923, 8925, 8926, 8927, 8929, 8944, 8949, 8951, 8952, 8953, 8960, 8985, 8997, 8999, 9007, 9016, 9032, 9033, 9034, 9041, 9042, 9050, 9054, 9066, 9071, 9074, 9086, 9096, 9109, 9110, 9129, 9134, 9141, 9157, 9158, 9160, 9164, 9166, 9178, 9209, 9225, 9237, 9250, 9253, 9261, 9267, 9273, 9278, 9280, 9288, 9290, 9297, 9306, 9321, 9327, 9337, 9360, 9372, 9376, 9395, 9397, 9401, 9402, 9404, 9407, 9408, 9422, 9428, 9435, 9443, 9444, 9445, 9460, 9475, 9494, 9496, 9497, 9498, 9499, 9506, 9510, 9517, 9518, 9522, 9523, 9534, 9550, 9569, 9570, 9571, 9577, 9582, 9599, 9604, 9606, 9622, 9623, 9627, 9628, 9639, 9659, 9660, 9661, 9681, 9695, 9698, 9718, 9719, 9728, 9729, 9741, 9747, 9761, 9762, 9763, 9784, 9805, 9811, 9822, 9829, 9830, 9848, 9849, 9850, 9855, 9856, 9857, 9859, 9868, 9870, 9871, 9901, 9902, 9919, 9924, 9925, 9940, 9947, 9950, 9951, 9953, 9976, 9985, 9987, 9991, 10000, 10013, 10034, 10037, 10051, 10061, 10076, 10087, 10090, 10111, 10112, 10117, 10118, 10120, 10121, 10149, 10152, 10153, 10156, 10158, 10162, 10165, 10169, 10170, 10200, 10202, 10212, 10219, 10225, 10230, 10234, 10245, 10269, 10280, 10314, 10315, 10332, 10339, 10343, 10363, 10368, 10373, 10374, 10376, 10391, 10392, 10401, 10402, 10404, 10413, 10418, 10419, 10424, 10425, 10426, 10428, 10437, 10447, 10448, 10450, 10465, 10476, 10482, 10483, 10484, 10510, 10529, 10549, 19575, 10577, 10580, 10586, 10587, 10591, 10600, 10609, 10616, 10629, 10630, 10635, 10636, 10651, 10659, 10661, 10682, 10687, 10704, 10708, 10712, 10718, 10730, 10733, 10756, 10757, 10764, 10767, 10775, 10778, 10779, 10791, 10804, 10805, 10812, 10822, 10841, 10842, 10856, 10871, 10878, 10883, 10932, 10963, 10968, 10969, 10970, 10984, 10985, 10989, 10990, 10991, 11000, 11003, 11006, 11019, 11023, 11024, 11026, 11028, 11031, 11035, 11065, 11072, 11084, 11095, 11099, 11109, 11118, 11138, 11139, 11144, 11145, 11152,
(continued)

Drainage see Farming practices ...

Drought and canopy architecture 8317, 10603, 11252, 11855, 12607, 12699, 12751,
    14002, 14709

Drought and carbon dioxide influx 6786, 7094, 7231, 7292, 7296, 7297, 7312, 7343,
    7351, 7357, 7386, 7421, 7486, 7723, 7755, 7794, 7802, 7918, 7996, 8371, 8467,
    8506, 8627, 8861, 8901, 8922, 8981, 9044, 9134, 9506, 9583, 9605, 9696, 9761,
    9882, 10120, 10124, 10204, 10250, 10349, 10363, 10404, 10476, 10560, 10653,
    10841, 10883, 11014, 11118, 11199, 11278, 11317, 11332, 11436, 11442, 11499,
    11576, 11782, 11809, 11962, 11963, 11964, 11987, 11990, 12279, 12286, 12287,
    12392, 12405, 12532, 12574, 12696, 12699, 12935, 13298, 13607, 14004, 14005,
    14016, 14018, 14109, 14144, 14394, 14667, 14265, 14684

Drought and carbon fixation pathways 7292, 7296, 8467, 8973, 9406, 9721, 10349,
    11138, 11204, 11318, 11436, 12372, 12405, 12406, 12861, 14349, 14435

Drought and carotenoids 6747, 14595, 14597

Drought and chlorophyll 6747, 6762, 7703, 7802, 8857, 8973, 9057, 9134, 9188, 9385,
    10618, 10619, 10848, 11204, 11752, 12532, 12664, 12704, 12941, 13393, 13403,
    13489, 14595, 14597, 14667, 14723

Drought and chloroplasts 9134, 9188, 9323, 10809, 10816, 10883, 11047, 11455, 12023,
    12495, 12678, 13807

Drought and conductance for water vapour and carbon dioxide transfer 6611, 7002,
    7130, 7231, 7232, 7268, 7351, 7377, 7449, 7582, 7621, 7645, 7723, 7996, 8406,
    8627, 8645, 8667, 8744, 8809, 8840, 9134, 9141, 9404, 9506, 9741, 9882, 9901,
    10061, 10120, 10349, 10363, 10404, 10653, 10712, 10756, 10841, 10878, 10883,
    11199, 11400, 11429, 11442, 11516, 11809, 11855, 11923, 11925, 11955, 11962,
    11963, 11964, 11990, 12019, 12120, 12279, 12287, 12321, 12405, 12574, 12595,
    12606, 12655, 12663, 12696, 13067, 13186, 13395, 13431, 13794, 14004, 14018,
    14069, 14080, 14262, 14265, 14308, 14389, 14394, 14419, 14435, 14466

Drought and electron transport chain 8032, 8973, 9385, 10618, 10619, 10809, 10995,
    11752, 12023, 12532

Drought and growth, productivity 6625, 6636, 6651, 6659, 6682, 6696, 6783, 6792,
    6822, 6834, 6853, 6869, 6973, 6992, 7014, 7015, 7016, 7041, 7051, 7084, 7116,
    7117, 7135, 7149, 7231, 7235, 7238, 7268, 7279, 7292, 7312, 7326, 7327, 7334,
    7356, 7357, 7408, 7449, 7452, 7483, 7503, 7509, 7621, 7663, 7668, 7691, 7696,
    7702, 7709, 7730, 7799, 7801, 7847, 7891, 7913, 7918, 7919, 7938, 7970, 7971,
    7990, 7991, 7993, 7996, 8059, 8134, 8202, 8243, 8278, 8294, 8317, 8347, 8382,
    8403, 8404, 8409, 8433, 8435, 8494, 8506, 8511, 8548, 8552, 8554, 8565, 8604,
    8608, 8613, 8624, 8626, 8627, 8653, 8671, 8684, 8712, 8727, 8751, 8761, 8768,
    8774, 8809, 8901, 8927, 8939, 8962, 8994, 9019, 9029, 9071, 9098, 9118, 9123,
    9132, 9133, 9166, 9234, 9271, 9301, 9325, 9392, 9404, 9453, 9480, 9483, 9493,
    9502, 9506, 9557, 9581, 9583, 9587, 9608, 9737, 9738, 9745, 9765, 9795, 9796,
    9845, 9867, 9869, 9881, 9882, 9891, 9917, 9938, 9950, 10000, 10001, 10017,
    10024, 10025, 10026, 10044, 10053, 10086, 10132, 10168, 10224, 10236, 10263,
    10295, 10326, 10327, 10363, 10377, 10404, 10424, 10455, 10464, 10476, 10525,
    10553, 10574, 10585, 10592, 10593, 10594, 10595, 10623, 10630, 10646, 10653,
    10675, 10676, 10677, 10692, 10712, 10726, 10734, 10747, 10756, 10820, 10827,
    10848, 10862, 10883, 10902, 10916, 10928, 10940, 10942, 10977, 11039, 11050,
    11092, 11118, 11135, 11205, 11209, 11210, 11215, 11267, 11280, 11301, 11317,
    11328, 11329, 11332, 11359, 11390, 11400, 11416, 11428, 11490, 11533, 11544,
    11569, 11579, 11599, 11617, 11659, 11664, 11679, 11703, 11731, 11735, 11754,
    11766, 11785, 11796, 11801, 11803, 11833, 11836, 11855, 11865, 11904, 11923,
    11927, 11932, 11936, 11950, 11955, 11956, 11963, 12016, 12017, 12022, 12053,
    12068, 12073, 12074, 12076, 12088, 12105, 12107, 12285, 12326, 12363, 12379,
    12400, 12472, 12521, 12537, 12574, 12583, 12590, 12595, 12606, 12625, 12634,
    12654, 12663, 12678, 12680, 12688, 12696, 12698, 12699, 12702, 12704, 12709,
                                                                    (continued)

Drought and growth, productivity (continued) 12741, 12751, 12791, 12803, 12854,
        12941, 12947, 13002, 13152, 13162, 13197, 13225, 13246, 13268, 13290, 13310,
        13371, 13403, 13404, 13489, 13501, 13539, 13750, 13787, 13794, 13815, 13830,
        13887, 13916, 13925, 13932, 13974, 14002, 14059, 14069, 14073, 14080, 14160,
        14227, 14235, 14261, 14282, 14284, 14299, 14308, 14310, 14340, 14375, 14403,
        14419, 14466, 14479, 14570, 14630, 14631, 14637, 14640, 14660, 14667, 14703,
        14706, 14709, 14723, 14725, 14761

Drought and leaf anatomy 6682, 6696, 6853, 6901, 7231, 7232, 7268, 7344, 7360, 7452,
        7737, 7996, 8129, 8280, 8471, 8644, 8809, 8844, 8901, 9002, 9081, 9506, 9736,
        9882, 10008, 10243, 10327, 10476, 10592, 10593, 10594, 10595, 10653, 10883,
        10890, 11088, 11400, 11499, 11865, 11963, 12285, 12295, 12363, 12405, 12574,
        12678, 12696, 12709, 12941, 12947, 13000, 13220, 13290, 13345, 13403, 13404,
        13410, 13422, 13431, 13489, 13925, 14067, 14080, 14282, 14466, 14570, 14725

Drought and other processes than above and below 11467, 11529, 11793, 11837, 12217,
        12363, 12458, 12525, 12680, 12776, 12831, 13246, 13404, 13930, 14109, 14160,
        14284, 14637, 14657, 14746

Drought and photorespiration 8861, 11014, 12574, 14684

Drought and respiration 6786, 7117, 7231, 7357, 9173, 10883, 11014, 11671, 11855,
        12287, 12574, 13316, 14667, 14684, 14723

Drought and stomata 6611, 6696, 6792, 7232, 7360, 7377, 7449, 7645, 7918, 7996,
        8413, 8744, 9141, 9882, 10653, 10712, 11108, 11516, 11955, 12655, 12704,
        12745, 12941, 13079, 14109, 14419, 14536

Drought and transpiration 6619, 7014, 7105, 7231, 7268, 7296, 7334, 7351, 7360,
        7377, 7535, 7657, 7723, 7755, 7865, 7918, 7996, 8171, 8277, 8506, 8605, 8981,
        9044, 9479, 9483, 9506, 9621, 9696, 9736, 10009, 10106, 10124, 10404, 10574,
        10630, 10878, 10883, 11014, 11278, 11287, 11301, 11419, 11584, 11658, 11662,
        11731, 12019, 12022, 12120, 12279, 12286, 12363, 12379, 12405, 12515, 12574,
        12606, 12696, 12699, 12828, 13067, 13079, 13395, 13779, 14004, 14005, 14080,
        14393, 14630, 14703, 14765

Drought and water absorption by plant 7142, 7232, 8218, 9479, 10184, 10299, 11328,
        11658, 12575, 12696, 13916

Drought and water status in plant 6611, 6648, 6681, 6926, 7002, 7186, 7232, 7351,
        7358, 7360, 7503, 7582, 7621, 7755, 7895, 7905, 7996, 8073, 8153, 8239, 8241,
        8406, 8840, 8962, 9141, 9392, 9450, 9882, 10028, 10106, 10124, 10363, 10476,
        10565, 10574, 10628, 10653, 10692, 10756, 10843, 10853, 11007, 11118, 11202,
        11400, 11429, 11442, 11760, 11855, 11923, 12019, 12022, 12023, 12073, 12092,
        12105, 12223, 12295, 12334, 12378, 12547, 12561, 12575, 12589, 12595, 12620,
        12696, 12698, 12699, 12806, 12882, 12919, 13037, 13186, 13197, 13315, 13395,
        13594, 13612, 13794, 13925, 13939, 14004, 14005, 14080, 14109, 14262, 14375,
        14394, 14419, 14435, 14466, 14630, 14684

Drought and water transport in cells 7531, 7769, 9134, 9188, 9740, 13315, 14491,
        14723

Drought and water transport in plant 9932, 10883, 12841,

Drought and wilting 6730, 6935, 6952, 7228, 7992, 7996, 8406, 8751, 9141, 9882,
        10653, 10883, 11007, 13489

Drought resistance 6651, 6660, 6682, 6725, 6730, 6743, 6761, 6768, 6792, 6808, 6821,
        6836, 6899, 6900, 6901, 6904, 6905, 6910, 6958, 6993, 7017, 7035, 7041, 7052,
        7083, 7102, 7105, 7106, 7116, 7117, 7118, 7149, 7183, 7186, 7212, 7213, 7214,
        7215, 7219, 7221, 7235, 7236, 7245, 7279, 7312, 7338, 7358, 7364, 7390, 7399,
        7439, 7495, 7535, 7542, 7594, 7621, 7631, 7661, 7663, 7701, 7702, 7704, 7710,
        7712, 7755, 7770, 7789, 7790, 7797, 7822, 7904, 7905, 7907, 7918, 7923, 7945,
                                                                              (continued)

Drought resistance (continued)   7965, 7988, 7990, 7991, 7992, 7993, 7996, 8002, 8003,
        8038, 8062, 8068, 8119, 8137, 8149, 8182, 8194, 8221, 8235, 8245, 8255, 8271,
        8306, 8346, 8347, 8350, 8367, 8382, 8390, 8398, 8408, 8431, 8435, 8477, 8494,
        8502, 8511, 8535, 8556, 8567, 8583, 8604, 8605, 8626, 8627, 8661, 8671, 8677,
        8693, 8727, 8747, 8748, 8761, 8772, 8800, 8807, 8809, 8831, 8855, 8857, 8877,
        8906, 8927, 8928, 8931, 8939, 8994, 9005, 9006, 9071, 9094, 9095, 9116, 9118,
        9125, 9127, 9134, 9137, 9139, 9141, 9151, 9181, 9217, 9219, 9251, 9271, 9272,
        9278, 9293, 9303, 9317, 9325, 9329, 9341, 9342, 9354, 9393, 9403, 9441, 9492,
        9520, 9558, 9580, 9586, 9587, 9611, 9647, 9665, 9711, 9746, 9761, 9765, 9774,
        9779, 9803, 9882, 9887, 9888, 9894, 9925, 9950, 9972, 10028, 10081, 10091,
        10092, 10093, 10094, 10106, 10128, 10162, 10213, 10249, 10279, 10282, 10351,
        10363, 10371, 10390, 10392, 10393, 10394, 10404, 19424, 10498, 10503, 10522,
        10535, 10553, 10556, 10569, 10574, 10595, 10612, 10627, 10630, 10645, 10653,
        10673, 10674, 10705, 10712, 10726, 10736, 10748, 10756, 10757, 10767, 10816,
        10820, 10837, 10871, 10883, 10892, 10940, 10953, 10981, 10996, 11009, 11010,
        11033, 11035, 11039, 11045, 11091, 11118, 11185, 11210, 11222, 11226, 11241,
        11267, 11284, 11318, 11345, 11347, 11348, 11349, 11350, 11359, 11364, 11379,
        11409, 11416, 11434, 11454, 11514, 11516, 11533, 11550, 11553, 11563, 11569,
        11659, 11661, 11662, 11666, 11668, 11722, 11760, 11771, 11774, 11792, 11843,
        11855, 11864, 11865, 11866, 11922, 11923, 11925, 11927, 11955, 11960, 11974,
        11976, 12016, 12017, 12019, 12034, 12043, 12105, 12123, 12141, 12175, 12190,
        12203, 12211, 12244, 12255, 12258, 12277, 12286, 12291, 12370, 12371, 12373,
        12392, 12454, 12492, 12522, 12557, 12563, 12583, 12595, 12606, 12681, 12627,
        12628, 12634, 12664, 12665, 12673, 12696, 12699, 12711, 12753, 12794, 12796,
        12808, 12809, 12882, 12904, 12919, 12920, 12941, 12968, 12999, 13000, 13011,
        13069, 13084, 13120, 13162, 13267, 13290, 13300, 13379, 13390, 13393, 13411,
        13422, 13467, 13505, 13532, 13560, 13588, 13589, 13606, 13615, 13642, 13655,
        13710, 13718, 13719, 13736, 13737, 13811, 13830, 13846, 13866, 13892, 13940,
        13966, 13973, 13974, 13975, 14014, 14059, 14075, 14089, 14098, 14142, 14193,
        14199, 14200, 14275, 14284, 14297, 14298, 14299, 14308, 14347, 14355, 14370,
        14471, 14472, 14480, 14484, 14512, 14536, 14582, 14583, 14589, 14613, 14629,
        14630, 14631, 14640, 14651, 14652, 14667, 14737, 14756

Drought resistance, methods 7583, 9071, 9127, 12190, 12919, 13736, 13737, 14193

Dry matter production see Growth, productivity ...

E

Ear removal see Defoliation, decapitation, ear, root removal ...

Ecotypes, geographical types and conductance for water vapour and carbon dioxide
        transfer 6839, 6889, 7305, 7383, 7996, 8238, 8289, 8306, 8385, 8809, 8850,
        9041, 9129, 9274, 9280, 9606, 9857, 9991, 10186, 10388, 10576, 10878, 10937,
        11855, 11943, 12161, 13214, 13215, 13230, 13937, 14034

Ecotypes, geographical types and stomata 6889, 7383, 7524, 7996, 8850, 9374, 10992,
        12686, 12927, 13239, 13408, 13570, 14253, 14483

Ecotypes, geographical types and transpiration 6839, 7088, 7302, 7870, 7979, 8001,
        8278, 8688, 8850, 9129, 9423, 9517, 9606, 10186, 10388, 10404, 10878, 11819,
        12161, 13488, 13937, 14303

Ecotypes, geographical types and water status in plant 7015, 7214, 8289, 8306, 8688,
        8714, 8824, 8850, 9041, 9280, 9423, 9731, 11208, 11226, 11553, 12620, 13215,
        13230, 13488, 13570, 13798, 13830

Ecotypes, geographical types and water transport in plant 9100

Ecotypes, geographical types and wilting 6660, 6696, 7015, 7212, 7302, 7535, 7996,
        8255, 8748, 9280, 10390

Electron transport chain see Deuterium oxide, tritium oxide ...; Drought ...;
     Humidity of air ...; Osmotically active substances ...; Salinity ...; Soil
     moisture ...; Water status in plant and electron transport chain

Enzyme inhibitors and conductance for water vapour and carbon dioxide transfer 9372,
     9985

Enzyme inhibitors and stomata 7666, 9232, 9294, 10098, 11130, 14357, 14372

Enzyme inhibitors and transpiration 14675

Enzyme inhibitors and water absorption by plant 6798, 8308, 9052

Enzyme inhibitors and water status in plant 9799, 10331

Enzyme inhibitors and water transport in cells 11054

Enzyme inhibitors and water transport in plant 9182

Enzyme inhibitors and wilting 9182

Enzymes and stomata 10157, 10835, 13851

Enzymes and water absorption by plant 8153, 8297, 13976

Enzymes and water status in plant 7260, 8153, 9134, 9256, 9403, 9710, 10883, 11430,
     11673, 12042, 12699, 12711, 13035, 13223, 13364, 13365, 13612, 13673

Enzymes and wilting 6884, 7823, 7982, 8153, 8245, 9134, 9378, 9403, 9710, 10599,
     10955, 12711, 13397, 14538

Epidermal conductance (resistance) see Conductance for carbon dioxide transfer,
     epidermis; Conductance for water vapour transfer, epidermis

Epidermis see Abaxial and adaxial epidermes; Anatomical structure of epidermis;
     Conductance for carbon dioxide transfer, epidermis; Conductance for water
     vapour transfer, epidermis; Stomata and epidermis, heterogeneity of single
     leaf blade

EPR, NMR (methods and results) 9682, 9687, 10346, 10461, 10578, 11071, 11496, 12221

Evaporation 6687, 6700, 6811, 6867, 6920, 6956, 6963, 7021, 7029, 7042, 7080, 7089,
     7111, 7162, 7205, 7386, 7392, 7399, 7423, 7424, 7426, 7433, 7434, 7435, 7456,
     7470, 7493, 7525, 7613, 7614, 7684, 7743, 7747, 7763, 7764, 7938, 7943, 7969,
     7979, 7992, 8196, 8198, 8217, 8300, 8393, 8394, 8410, 8416, 8456, 8485, 8494,
     8558, 8562, 8569, 8617, 8685, 8688, 8710, 8723, 8743, 8770, 8788, 8883, 8979,
     9035, 9152, 9191, 9195, 9208, 9212, 9252, 9270, 9360, 9375, 9408, 9420, 9447,
     9483, 9487, 9555, 9582, 9607, 9625, 9694, 9834, 9842, 9905, 9909, 9930, 9945,
     10058, 10110, 10126, 10127, 10133, 10184, 10224, 10285, 10310, 10351, 10404,
     10463, 10728, 10824, 10943, 10956, 11137, 11165, 11193, 11224, 11252, 11282,
     11297, 11325, 11342, 11370, 11393, 11428, 11478, 11541, 11552, 11616, 11657,
     11662, 11683, 11699, 11701, 11702, 11714, 11731, 11845, 11855, 11901, 11914,
     11918, 11994, 12039, 12119, 12130, 12149, 12151, 12163, 12201, 12254, 12255,
     12296, 12304, 12353, 12377, 12413, 12414, 12436, 12460, 12468, 12498, 12569,
     12577, 12608, 12656, 12663, 12691, 12699, 12715, 12762, 12802, 12856, 12913,
     12928, 12946, 12955, 12977, 12978, 12981, 13032, 13114, 13126, 13134, 13143,
     13196, 13264, 13267, 13274, 13355, 13356, 13384, 13398, 13405, 13459, 13475,
     13476, 13497, 13563, 13566, 13614, 13615, 13674, 13776, 13788, 13841, 13861,
     13867, 13915, 13916, 13919, 13927, 13928, 13950, 13977, 13995, 14088, 14093,
     14120, 14124, 14143, 14149, 14150, 14182, 14224, 14235, 14240, 14266, 14281,
     14287, 14315, 14334, 14346, 14373, 14387, 14396, 14404, 14405, 14416, 14417,
     14444, 14462, 14467, 14497, 14498, 14572, 14638, 14678, 14722, 14733, 14745,
     14748

Evapotranspiration 6608, 6621, 6627, 6637, 6642, 6643, 6645, 6661, 6684, 6697, 6700,
        6711, 6713, 6748, 6811, 6829, 6837, 6936, 6056, 6962, 6963, 6969, 6997, 7011,
        7042, 7111, 7114, 7115, 7126, 7138, 7162, 7193, 7198, 7205, 7217, 7224, 7230,
        7283, 7311, 7316, 7318, 7342, 7392, 7410, 7416, 7424, 7455, 7457, 7470, 7471,
        7473, 7474, 7483, 7497, 7515, 7536, 7539, 7543, 7565, 7593, 7605, 7649, 7650,
        7696, 7720, 7759, 7763, 7771, 7772, 7799, 7800, 7801, 7804, 7805, 7832, 7865,
        7880, 7884, 7918, 7932, 7943, 7948, 7978, 7979, 7992, 8012, 8118, 8173, 8181,
        8195, 8207, 8218, 8236, 8263, 8275, 8276, 8277, 8303, 8356, 8390, 8403, 8404,
        8410, 8456, 8458, 8472, 8485, 8487, 8488, 8493, 8494, 8511, 8532, 8538, 8553,
        8562, 8563, 8569, 8630, 8637, 8638, 8659, 8709, 8710, 8731, 8732, 8740, 8743,
        8850, 8851, 8856, 8857, 8897, 8908, 8927, 8966, 8994, 8999, 9001, 9012, 9036,
        9041, 9055, 9152, 9164, 9166, 9207, 9270, 9290, 9324, 9360, 9364, 9375, 9438,
        9439, 9440, 9453, 9476, 9479, 9481, 9507, 9508, 9510, 9537, 9555, 9583, 9594,
        9621, 9634, 9671, 9678, 9694, 9716, 9722, 9818, 9824, 9834, 9854, 9870, 9883,
        9893, 9905, 9909, 9919, 9935, 9936, 9937, 9942, 9943, 9979, 9984, 10002,
        10003, 10004, 10026, 10032, 10039, 10050, 10058, 10059, 10086, 10110, 10133,
        10138, 10160, 10172, 10174, 10184, 10185, 10229, 10240, 10262, 10263, 10299,
        10303, 10310, 10325, 10337, 10355, 10408, 10411, 10445, 10447, 10456, 10458,
        10462, 10463, 10477, 10500, 10506, 10508, 10533, 10542, 10543, 10544, 10636,
        10647, 10665, 10702, 10726, 10728, 10750, 10751, 10759, 10810, 10821, 10824,
        10869, 10887, 10899, 10913, 10914, 10915, 10916, 10921, 10942, 10943, 10952,
        10960, 11005, 11017, 11056, 11072, 11073, 11074, 11075, 11078, 11102, 11127,
        11141, 11165, 11193, 11195, 11224, 11235, 11236, 11237, 11251, 11252, 11264,
        11265, 11273, 11277, 11282, 11287, 11288, 11292, 11293, 11296, 11316, 11325,
        11342, 11358, 11365, 11370, 11390, 11393, 11394, 11428, 11478, 11511, 11512,
        11528, 11532, 11541, 11544, 11552, 11572, 11593, 11606, 11616, 11628, 11662,
        11683, 11701, 11712, 11722, 11730, 11731, 11749, 11769, 11775, 11776, 11797,
        11814, 11815, 11818, 11830, 11835, 11855, 11860, 11877, 11895, 11901, 11927,
        11932, 11959, 11965, 11905, 11909, 12038, 12047, 12068, 12105, 12108, 12119,
        12120, 12130, 12137, 12144, 12151, 12165, 12171, 12180, 12181, 12187, 12201,
        12219, 12244, 12267, 12315, 12364, 12377, 12409, 12411, 12412, 12414, 12420,
        12423, 12436, 12452, 12503, 12515, 12537, 12557, 12561, 12570, 12577, 12581,
        12597, 12608, 12640, 12654, 12656, 12663, 12665, 12668, 12679, 12688, 12691,
        12696, 12699, 12702, 12759, 12768, 12802, 12813, 12823, 12826, 12843, 12867,
        12876, 12884, 12913, 12928, 12931, 12933, 12934, 12945, 12950, 12963, 12978,
        12980, 12981, 12985, 13016, 13032, 13043, 13058, 13104, 13126, 13127, 13155,
        13212, 13216, 13221, 13275, 13280, 13310, 13324, 13326, 13341, 13347, 13348,
        13356, 13371, 13374, 13375, 13376, 13395, 13399, 13414, 13437, 13444, 13449,
        13456, 13457, 13475, 13476, 13490, 13496, 13497, 13509, 13511, 13521, 13537,
        13550, 13557, 13563, 13566, 13581, 13604, 13605, 13614, 13631, 13636, 13650,
        13674, 13675, 13691, 13692, 13695, 13702, 13739, 13748, 13756, 13768, 13779,
        13786, 13787, 13812, 13830, 13861, 13922, 13928, 13929, 13938, 13952, 13954,
        13957, 13963, 13975, 13977, 13981, 13995, 14003, 14013, 14024, 14030, 14051,
        14072, 14103, 14119, 14124, 14143, 14146, 14147, 14169, 14202, 14203, 14209,
        14220, 14222, 14224, 14238, 14240, 14254, 14271, 14272, 14274, 14281, 14295,
        14296, 14303, 14309, 14377, 14378, 14386, 14387, 14388, 14396, 14402, 14403,
        14404, 14405, 14415, 14416, 14417, 14449, 14453, 14478, 14488, 14495, 14497,
        14498, 14504, 14507, 14512, 14518, 14545, 14546, 14547, 14548, 14549, 14550,
        14572, 14610, 14619, 14638, 14639, 14678, 14689, 14702, 14733

Evapotranspiration, methods, evaporimeters and lysimeters 6811, 6816, 6817, 6835,
        6951, 6953, 7138, 7224, 7828, 7943, 7978, 8166, 8313, 8462, 8563, 8731, 8994,
        9072, 9152, 9167, 9200, 9201, 9263, 9324, 9507, 9808, 10070, 10138, 10159,
        10206, 10291, 10824, 10956, 11370, 11372, 11404, 11542, 11616, 11683, 11767,
        12171, 12201, 12304, 12364, 12420, 12842, 12953, 12977, 13241, 13268, 13275,
        13405, 13450, 13509, 13722, 13748, 13926, 13964, 14107, 14254, 14258, 14273,
        14387, 14404, 14416, 14550

Evapotranspiration, methods, other 7114, 7115, 7224, 7605, 7649, 7650, 7747, 7763,
        8043, 8390, 8559, 8743, 9075, 9166, 9167, 9270, 9439, 9893, 9909, 10174,
        10408, 10810, 10844, 11017, 11224, 11273, 11279, 11542, 11642, 11712, 11749,
        11814, 11835, 12046, 12120, 12219, 12420, 12515, 12577, 12656, 12826, 12963,
        12981, 13016, 13042, 13126, 13127, 13241, 13295, 13405, 13450, 13475, 13476,
                                                                      (continued)

Evapotranspiration, methods, other (continued) 13497, 13509, 13563, 13739, 13928, 13929, 13977, 14030, 14051, 14103, 14220, 14224, 14271, 14272, 14303, 14387, 14402, 14416, 14485, 14689, 14705, 14716

Exhaust gases see Pollutants, ozone ...

Exposure chambers see Transpiration chambers

Extension see Growth, productivity ...

Exudation see Root pressure, exudation

F

Farming practices and conductance for water vapour and carbon dioxide transfer 6625, 6664, 6680, 7092, 7278, 8727, 9042, 11343, 11678, 12984, 13395, 14064

Farming practices and stomata 6793, 6891, 7092, 7562, 7563

Farming practices and transpiration 6664, 6891, 7092, 7179, 7195, 7416, 7934, 8278, 9512, 9567, 9896, 10740, 10750, 11343, 11522, 12377, 12379, 12581, 12696, 12762, 13104, 13160, 13395, 14388

Farming practices and water absorption by plant 6767. 6860, 6950, 7566, 7581, 8045, 8120, 8858, 8972, 9389, 10910, 10911, 11124, 12581, 13936, 13970, 14388

Farming practices and water status in plant 6625, 6767, 7092, 7278, 7674, 8304, 8727 10565, 10750, 10762, 11343, 11678, 11804, 13138, 13395, 14376

Farming practices and wilting 6606, 7518, 10740

Fatty acids see Lipids, fatty acids ...

Fine structure see Stomata fine structure; Transpiration ...; Water absorption by plant and fine structure

Flooding and biliproteins 9743

Flooding and carbon dioxide influx 7983, 8022, 8981, 9397, 9805, 9882, 10369, 11384, 11385, 11782, 11908, 12894, 13328, 13479, 13627, 13744, 13747, 14061, 14137, 14237, 14684

Flooding and carbon fixation pathways 11903

Flooding and carotenoids 8845

Flooding and chlorophyll 7669, 7984, 8845, 9547, 10396, 10918, 11996, 13013, 13227, 13328, 13337, 13627, 13744, 13772, 14022, 14237

Flooding and chloroplasts 12193, 12811

Flooding and conductance for water vapour and carbon dioxide transfer 6774, 7033, 7377, 8265, 8284, 9369, 9397, 9623, 11384, 11385, 11677, 11834, 11853, 11855, 12193, 12674, 12957, 13037, 13063, 13226, 13395, 13479, 13627, 13743, 13744, 13747, 14022, 14531, 14684

Flooding and growth, productivity 6683, 6703, 6796, 6815, 6816, 6817, 6972, 6989, 7066, 7135, 7201, 7240, 7356, 7367, 7668, 7670, 7680, 7796, 7833, 7983, 8019, 8033, 8056, 8057, 8103, 8214, 8229, 8265, 8284, 8366, 8381, 8593, 8616, 8643, 8724, 8725, 8726, 8842, 8845, 8882, 8885, 8916, 8993, 9008, 9173, 9235, 9367, 9368, 9369, 9525, 9547, 9637, 9882, 9907, 9918, 10002, 10150, 10171, 10222, 10223, 10257, 10364, 10369, 10396, 10494, 10580, 10601, 10813, 10945, 10953,
(continued)

Flooding and growth, productivity (continued) 10986, 10989, 10990, 10991, 11041,
    11411, 11515, 11735, 11780, 11853, 11996, 12087, 12129, 12453, 12582, 12602,
    12674, 12689, 12690, 12732, 12739, 12799, 12894, 12939, 12956, 12957, 13013,
    13031, 13156, 13183, 13226, 13227, 13231, 13329, 13548, 13565, 13572, 13622,
    13627, 13684, 13743, 13744, 13745, 13746, 13772, 13815, 13830, 13874, 13923,
    13951, 13953, 14022, 14060, 14061, 14062, 14063, 14065, 14137, 14193, 14218,
    14237, 14321, 14336, 14501, 14591, 14700

Flooding and leaf anatomy 6703, 7034, 7983, 8033, 8229, 9698, 10139, 10990, 11435,
    11834, 11996, 12689, 13063, 13328, 13744, 13746, 14022, 14062, 14063, 14137

Flooding and other processes than above and below 9927, 11435, 11831, 11903, 12212,
    12216, 12610, 12689, 12690, 12732, 13012, 13156, 13227, 13231, 13337, 13499,
    13548, 13701, 13743, 13744, 13772, 13951, 13953, 14063, 14164, 14237, 14336,
    14501, 14530, 14531

Flooding and photorespiration 14684

Flooding and respiration 9882, 10369, 10404, 11384, 11425, 12156, 13208, 13316,
    13328, 13772, 14684

Flooding and stomata 7066, 7377, 7881, 9806, 11384, 13565, 13744, 14237

Flooding and transpiration 7066, 7377, 7755, 7983, 8056, 8381, 8876, 8981, 9397,
    9547, 9623, 9806, 10869, 11834, 12894, 13395, 13744, 13747, 14137

Flooding and water absorption by plant 7066, 9547, 9806, 9882, 10369, 11285, 12739,
    13772, 14061

Flooding and water status in plant 6774, 7755, 7983, 8153, 8265, 8381, 8593, 8701,
    9422, 9623, 9698, 9765, 9806, 10991, 11677, 11834, 11855, 12193, 12212, 12213,
    12674, 12894, 13037, 13063, 13226, 13315, 13395, 13451, 13747, 14061, 14531,
    14684

Flooding and water transport in cells 13315

Flooding and water transport in plant 12674

Flooding and wilting 8284, 8701, 9806, 10404, 11031, 11834, 13063

Fluorine see Pollutants, ozone ...

Foliage see Canopy architecture ...

Frost resistance 6634, 6730, 6736, 6778, 6797, 6810, 6842, 6849, 6923, 7023, 7048,
    7101, 7158, 7267, 7401, 7406, 7446, 7590, 7594, 7597, 7599, 7623, 7717, 7718,
    7859, 7904, 7922, 8090, 8094, 8106, 8174, 8200, 8201, 8226, 8255, 8289, 8310,
    8412, 8529, 8602, 8622, 8634, 8730, 8784, 8806, 8895, 8903, 8928, 9095, 9287,
    9291, 9336, 9454, 9455, 9471, 9472, 9553, 9554, 9560, 9561, 9588, 9614, 9649,
    9658, 9725, 9746, 9894, 9895, 10087, 10128, 10141, 10153, 10293, 10365, 10403,
    10407, 10424, 10446, 10451, 10468, 10473, 10553, 10638, 10654, 10673, 10674,
    10689, 10759, 10769, 10789, 10837, 10937, 10971, 11035, 11140, 11142, 11232,
    11253, 11275, 11302, 11303, 11467, 11468, 11574, 11575, 11672, 11723, 11760,
    11826, 11857, 11987, 12027, 12037, 12132, 12197, 12225, 12256, 12290, 12317,
    12366, 12428, 12508, 12583, 12612, 12644, 12669, 12673, 12703, 12844, 12971,
    13024, 13170, 13242, 13353, 13416, 13461, 13633, 13663, 13670, 13753, 13766,
    13830, 14025, 14056, 14092, 14157, 14191, 14193, 14199, 14219, 14273, 14322,
    14368, 14369, 14491, 14506, 14529, 14579, 14594, 14606, 14717, 14723, 14746,
    14749, 14750.

Frost resistance, methods 7101, 12669, 14193

Fungal diseases see Pathogens ...

G

Gas exchange see Carbon dioxide influx ...; Transpiration ...

Gases organic and conductance for water vapour and carbon dioxide transfer 12059, 12099

Gases organic and stomata 12059, 12099

Gases organic and water absorption by plant 8437

Gasometric methods see Transpiration rate, methods, gasometric systems ...

General aspects on plant water relations, books, reviews 6638, 6726, 6729, 6838, 7026, 7071, 7111, 7164, 7217, 7292, 7343, 7360, 7361, 7424, 7697, 7722, 7801, 7864, 7918, 7996, 8185, 8230, 8243, 8246, 8327, 8733, 8735, 8744, 8759, 8857, 9010, 9026, 9122, 9134, 9166, 9202, 9289, 9362, 9433, 9566, 9615, 9815, 9834, 9851, 9882, 9906, 9919, 9928, 9946, 10056, 10087, 10119, 10120, 10238, 10363, 10364, 10385, 10403, 10404, 10414, 10415, 10424, 10447, 10449, 10474, 10491, 10532, 10553, 10583, 10883, 11062, 11066, 11112, 11206, 11388, 11668, 11754, 11855, 11869, 11953, 11982, 11983, 12030, 12032, 12081, 12086, 12125, 12233, 12262, 12342, 12343, 12344, 12392, 12471, 12509, 12537, 12538, 12696, 12699, 12810, 12813, 12905, 13048, 13185, 13206, 13220, 13272, 13374, 13550, 13569, 13601, 13618, 13773, 13741, 13742, 13744, 13817, 13830, 13882, 13983, 13992, 13993, 14067, 14124, 14173, 14215, 14277, 14365, 14377, 14387, 14409, 14502, 14554, 14566, 14572, 14586, 14593, 14681, 14718

Genetics and conductance for water vapour and carbon dioxide transfer 7027, 8309, 8460, 8461, 8831, 9054, 9141

Genetics and drought resistance 6725, 6743, 7035, 7117, 7221, 7236, 7390, 7770, 7907 7918, 7996, 8626, 8761, 8805, 8831, 9125, 9134, 9219, 9291, 9492, 9803, 10627 10709, 10757, 10767, 10883, 11033, 11327, 11496, 12367, 12370, 14613

Genetics and stomata 6720, 6800, 6806, 6855, 6959, 7027, 7173, 7189, 7206, 7338, 7404, 7422, 7489, 7585, 7586, 7648, 7692, 7736, 7931, 8191, 8309, 8420, 8463, 8797, 9139, 9219, 9503, 9768, 9864, 9866, 9897, 10847, 11176, 11368, 11663, 12096, 12363, 12393, 13194, 13635, 13715, 13902, 14536

Genetics and transpiration 6800, 7027, 7342, 7681, 8805, 8831, 9092, 9186, 10687, 10817, 11757, 12322, 12696, 12909, 13067, 13367, 13560, 13777, 13973, 14685

Genetics and water absorption by plant 7041, 7197, 7850

Genetics and water status in plant 6630, 7062, 7118, 7197, 7248, 7494, 7495, 7678, 7996, 8530, 8805, 9054, 9141, 10817, 10853, 10857, 11095, 11115, 11349, 11442, 11917, 12293, 12620, 13203, 13367, 13740, 13973, 14157, 14221

Growth analysis, methods 9150, 11709, 12107, 13162

Growth, productivity see Drought ...; Flooding ...; Humidity of air ...; Irri-gation ...; Osmotically active substances ...; Precipitation, dew ...; Root, underground part ...; Salinity ...; Soil Moisture ...; Water status in plant and growth, productivity

Growth substances, hormones, inhibitors etc. and conductance for water vapour and carbon dioxide transfer 6611, 6780, 6871, 6946, 7055, 7151, 7258, 7384, 7397 7431, 7609, 7808, 7859, 7864, 8093, 8378, 8413, 8457, 8628, 8664, 8665, 8667, 8677, 8719, 8744, 8950, 8951, 8953, 9005, 9134, 9226, 9681, 9784, 9805, 9951, 10042, 10200, 10212, 10251, 10280, 10404, 10424, 10448, 19474, 10549, 10708, 10718, 10822, 10856, 11206, 11371, 11383, 11385, 11386, 11408, 11481, 11482, 11516, 11575, 11765, 11790, 11809, 11923, 11925, 12368, 12510, 12696, 12954, 13222, 13236, 13288, 13308, 13465, 13515, 13530, 13535, 13715, 13783, 13833, 13885, 13935, 14049, 14194, 14196, 14683

Growth substances, hormones, inhibitors etc. and stomata 6611, 6615, 6656, 6871,
        6912, 6946, 6988, 7055, 7210, 7218, 7300, 7346, 7397, 7429, 7466, 7557, 7574,
        7659, 7684, 7812, 7864, 7968, 7996, 8038, 8060, 8092, 8413, 8664, 8665, 8744,
        8757, 8758, 8846, 8871, 8940, 8942, 8950, 8951, 8953, 8991, 9005, 9134, 9162,
        9225, 9237, 9262, 9359, 9513, 9534, 9612, 9613, 9641, 9673, 9681, 9810, 9921,
        9951, 9986, 10042, 10074, 10098, 10116, 10212, 10261, 10273, 10275, 10283,
        10424, 10474, 10486, 10669, 10708, 10718, 10782, 10822, 10835, 10926, 10927,
        11057, 11077, 11108, 11109, 11116, 11122, 11130, 11163, 11206, 11340, 11369,
        11371, 11386, 11398, 11516, 11594, 11674, 11743, 11790, 11829, 11923, 11939,
        11967, 12040, 12055, 12078, 12079, 12124, 12125, 12136, 12310, 12311, 12312,
        12410, 12524, 12571, 12696, 12770, 12816, 12817, 12833, 12898, 13047, 13065,
        13066, 13165, 13166, 13442, 13530, 13715, 13764, 13830, 13877, 13898, 14124,
        14130, 14464, 14562, 14718

Growth substances, hormones, inhibitors etc. and transpiration 6808, 6907, 6988,
        7151, 7240, 7295, 7300, 7431, 7557, 7561, 8171, 8319, 8605, 8667, 8756, 8805,
        9005, 9092, 9134, 9162, 9262, 9359, 9429, 9615, 9638, 9755, 9784, 9970, 10042,
        10072, 10448, 10474, 11077, 11206, 11317, 11383, 11608, 11645, 11688, 11967,
        12040, 12086, 12368, 12379, 12450, 12608, 12699, 12904, 12954, 12970, 13047,
        13577, 13744, 13747, 13764, 13941, 14683

Growth substances, hormones, inhibitors etc. and water absorption by plant 6872,
        7767, 7768, 7923, 8008, 8437, 8550, 8577, 9076, 9182, 9262, 9357, 9540, 9951,
        10136, 11516, 11692, 11890, 12770, 13768

Growth substances, hormones, inhibitors etc. and water status in plant 6611, 6656,
        6808, 6909, 6912, 6946, 7257, 7384, 7450, 7552, 7637, 7859, 7864, 7996, 8038,
        8093, 8096, 8182, 8283, 8391, 8413, 8424, 8457, 8496, 8508, 8666, 8667, 8668,
        8744, 8827, 8871, 8951, 9005, 9134, 9176, 9177, 9221, 9227, 9322, 9357, 9461,
        9462, 9615, 9681, 9951, 10016, 11079, 10114, 10141, 10249, 10302, 10303,
        10311, 10449, 10491, 10549, 10906, 10950, 10974, 11202, 11284, 11383, 11451,
        11482, 11485, 11516, 11673, 11763, 11765, 11900, 11923, 11925, 12072, 12136,
        12334, 12368, 12369, 12370, 12407, 12608, 12661, 12770, 12912, 12954, 13047,
        13093, 13138, 13192, 13285, 13308, 13465, 13623, 13768, 13816, 13885, 13968,
        13969, 14196, 14308, 14477, 14752

Growth substances, hormones, inhibitors etc. and water transport in cells 7417, 7445,
        7714, 7769, 8437, 8577, 8794, 9299, 9951, 10081, 10283, 10423, 11206, 11369,
        11485, 11774, 12478, 12770, 13192, 14626

Growth substances, hormones, inhibitors etc. and water transport in plant 7417, 8514,
        9182, 9541, 11383, 11485, 11516, 12029, 12648, 12770, 12956, 12959, 13834,
        14586, 14626

Growth substances, hormones, inhibitors etc. and wilting 6611, 6615, 6634, 6656,
        6665, 6820, 6822, 6909, 6912, 7017, 7035, 7175, 7201, 7384, 7467, 7535, 7552,
        7574, 7625, 7661, 7858, 7859, 7864, 7923, 7996, 8038, 8089, 8113, 8665, 8666,
        8668, 8693, 8805, 8953, 9005, 9134, 9221, 9262, 9528, 9602, 9645, 9681, 9807,
        10162, 10200, 10249, 10280, 10283, 10424, 10472, 10583, 10662, 10708, 10710,
        10883, 11116, 11122, 11178, 11203, 11206, 11516, 11612, 11744, 11760, 11761,
        11762, 11763, 11765, 11774, 11786, 11790, 11973, 11990, 12078, 12136, 12367,
        12368, 12369, 12370, 12770, 12897, 13000, 13001, 13026, 13393, 13531, 13532,
        13533, 13534, 13615, 13768, 13835, 13877, 13885, 13990, 14685, 14715

Guard cells see Stomata ...; Stomatal ...

Guttation 6870, 6890, 8436, 8437, 8505, 8857, 11855, 12608, 13638, 13985

H

H$_2$ isotopes see Deuterium oxide, tritium oxide ...

Heat see Transpiration rate, theoretical background

Herbicides see Pesticides, herbicides ...

Heterogeneity of single leaf blade see Conductance for water vapour and carbon
        dioxide transfer ...; Stomata ...; Transpiration ...; Water status in
        plant ...; Water transport in cells, heterogeneity of single leaf blade

Hormones see Growth substances, hormones, inhibitors etc. ...

Humidity of air and canopy architecture 8732, 8971, 11276, 11342, 12621, 12699,
        13600, 13696, 14085, 14579

Humidity of air and carbon dioxide influx 6620, 6624, 6726, 6925, 6998, 7065, 7104,
        7113, 7254, 7579, 7809, 7841, 8252, 8619, 8742, 8793, 8831, 8961, 8965, 9044,
        9129, 9397, 9514, 9551, 9829, 9831, 9843, 9856, 10030, 10120, 10186, 10205,
        10332, 10339, 10376, 10413, 10479, 10482, 10568, 10582, 10729, 10818, 10886,
        10890, 11298, 11299, 11407, 11581, 11610, 11623, 11772, 11783, 11827, 11984,
        12037, 12161, 12162, 12173, 12191, 12321, 12336, 12528, 12730, 12845, 12846,
        12850, 13005, 13107, 13108, 13202, 13298, 13311, 13314, 13554, 13562, 13641,
        13677, 13720, 13763, 13788, 13813, 13836, 13937, 14017, 14244, 14335, 14350,
        14383, 14539, 14564, 14600, 14615, 14662, 14663, 14736

Humidity of air and carbon fixation pathways 11150, 14412

Humidity of air and carotenoids 11799

Humidity of air and chlorophyll 6744, 7835, 8117, 10208, 10209, 10715, 10918, 11799,
        12766, 13094, 13618

Humidity of air and chloroplasts 12456, 12766

Humidity of air and conductance for water vapour and carbon dioxide transfer 6620,
        6726, 6733, 6889, 6934, 6974, 6993, 7092, 7104, 7209, 7365, 7383, 7431, 7501,
        7502, 7579, 7695, 7739, 7809, 7937, 7996, 8127, 8198, 8238, 8252, 8262, 8307,
        8359, 8417, 8662, 8677, 8744, 8745, 8763, 8799, 8801, 8855, 8860, 8913, 8929,
        8944, 8951, 9064, 9129, 9134, 9139, 9141, 9164, 9297, 9360, 9370, 9397, 9514,
        9570, 9571, 9627, 9628, 9703, 9728, 9741, 9784, 9829, 9856, 9925, 10000,
        10034, 10051, 10076, 10087, 10120, 10202, 10313, 10314, 10315, 10332, 10339,
        10376, 10401, 10413, 10424, 10425, 10479, 10482, 10484, 10549, 10590, 10591,
        10729, 10779, 10818, 10841, 10866, 10883, 10972, 11003, 11118, 11144, 11242,
        11259, 11296, 11407, 11458, 11623, 11624, 11633, 11790, 11848, 11855, 11856,
        11934, 11943, 11960, 11976, 12037, 12060, 12086, 12161, 12162, 12166, 12191,
        12321, 12438, 12488, 12528, 12572, 12663, 12696, 12717, 12836, 12850, 13005,
        13063, 13093, 13108, 13121, 13173, 13178, 13202, 13230, 13311, 13314, 13466,
        13488, 13490, 13554, 13562, 13582, 13616, 13616, 13650, 13677, 13694, 13721,
        13765, 13788, 13830, 13878, 13937, 13961, 13977, 13978, 13994, 14017, 14118,
        14135, 14143, 14189, 14262, 14288, 14335, 14383, 14393, 14423, 14467, 14556,
        14557, 14600, 14615, 14705

Humidity of air and electron transport chain 7835, 12376, 12456, 127666

Humidity of air and growth, productivity 6621, 6777, 7030, 7348, 7438, 7451, 7480,
        7635, 7795, 7970, 8114, 8293, 8384, 8752, 8788, 8887, 8978, 8984, 9015, 9290,
        9420, 9442, 9662, 10018, 10181, 10217, 10411, 10422, 10680, 10850, 10865,
        19881, 10915, 10993, 11150, 11247, 11544, 11597, 11741, 11772, 12009, 12112,
        12153, 12191, 12528, 12530, 12632, 12731, 12740, 12836, 12866, 13053, 13152,
        13250, 13371, 13390, 13488, 13503, 13585, 13756, 13920, 13921, 14378, 14412,
        14414, 14514, 14545, 14546, 14723

Humidity of air and leaf anatomy 6621, 6745, 6936, 7022, 7360, 7544, 7996, 8713,
        10076, 19217, 10242, 10243, 10244, 10245, 11772, 12191, 12257, 12289, 12621,
        12633, 12635, 12836, 13176, 13250, 13554, 13922, 14361, 14414

Humidity of air and other processes than above and below 10459, 10545, 10880, 11270,
     11361, 11431, 11458, 11487, 11623, 11633, 11748, 11767, 11817, 11926, 12111,
     12116, 12267, 12591, 12642, 12723, 12731, 12803, 12836, 13053, 13108, 13242,
     13488, 13503, 13525, 13533, 13542, 13545, 13546, 13547, 13562, 13582, 13600,
     13931, 14144, 14307, 14364, 14416, 14436, 14515, 14686, 14743

Humidity of air and photorespiration 11298, 11855

Humidity of air and respiration 8887, 11671, 11938

Humidity of air and stomata 6889, 6974, 6998, 7092, 7209, 7335, 7343, 7360, 7383,
     7402, 7809, 7996, 8000, 8127, 8744, 8929, 8948, 8951, 9134, 9807, 9913, 9919,
     10037, 10087, 10147, 10424, 10482, 10591, 10651, 10708, 10828, 10866, 10867,
     11018, 11243, 12064, 12078, 12191, 12815, 12817, 12850, 12898, 13488, 13694,
     13778, 14718

Humidity of air and transpiration 6621, 6649, 6726, 6733, 6956, 6998, 7021, 7047,
     7105, 7209, 7253, 7360, 7365, 7431, 7480, 7579, 7620, 7649, 7689, 7739, 7996,
     8198, 8238, 8252, 8262, 8359, 8485, 8534, 8672, 8740, 8801, 8831, 8951, 9044,
     9129, 9134, 9192, 9360, 9397, 9514, 9551, 9583, 9654, 9856, 9920, 9935, 10015,
     10034, 10087, 10107, 10159, 10185, 10205, 10332, 10376, 10401, 10413, 10425,
     10482, 10518, 10590, 10591, 10629, 10818, 10883, 10886, 10890, 10917, 10960,
     11005, 11100, 11239, 11242, 11258, 11342, 11365, 11393, 11394, 11560, 11633,
     11855, 11864, 12161, 12162, 12180, 12191, 12201, 12315, 12401, 12438, 12528,
     12604, 12696, 12699, 12828, 12911, 12969, 13005, 13010, 13176, 13250, 13311,
     13314, 13488, 13533, 13562, 13585, 13628, 13655, 13677, 13721, 13759, 13765,
     13788, 13856, 13910, 13916, 13937, 14070, 14244, 14262, 14335, 14343, 14391,
     14393, 14412, 14478, 14557, 14600, 14610, 14646, 14662, 14703, 14705, 14723,
     14745, 14747, 14765

Humidity of air and water absorption by plant 7021, 9654, 10015, 10518, 10596, 12502,
     13936, 14008

Humidity of air and water status in plant 6726, 7092, 7124, 7131, 7209, 7232, 7348,
     7360, 7365, 7488, 7695, 7735, 7739, 8159, 8213, 8672, 8709, 8712, 8714, 8888,
     9126, 9141, 9297, 9300, 9447, 9838, 10033, 10202, 10239, 10244, 10332, 10482,
     10549, 10791, 10881, 10883, 10973, 11034, 11220, 11268, 11331, 11446, 11468,
     11612, 11685, 11812, 11813, 11855, 11918, 11988, 12191, 12196, 12366, 12415,
     12487, 12621, 12633, 12637, 12808, 12836, 13093, 13094, 13230, 13279, 13314,
     13488, 13503, 13533, 13582, 13598, 13667, 13897, 13910, 13998, 14135, 14414,
     14571, 14587, 14645, 14662, 14686

Humidity of air and water transport in cells 9284, 13094

Humidity of air and water transport in plant 8002, 8159, 11855, 13910,

Humidity of air and wilting 6640, 7352, 7918, 8974, 9141, 9895, 10015, 10883, 12326

Humidity of air, gradients in canopy 6705, 6920, 7764, 7800, 7908, 8315, 8630, 8971,
     9166, 9364, 11276, 11382, 11393, 12505, 12506, 12740, 13010, 13241, 13413,
     13447, 13928, 13929, 14124, 14125, 14271

Humidity of air in leaf intercellular spaces see Transpiration rate, gradients of
     air humidity in leaf intercellular spaces

Humidity of air, methods 8687, 8767, 9148, 9167, 9619, 9906, 10181, 10427, 10868,
     11817, 13653

Hydration level see Water status in plant ...

Hydraulic conductivity see Water transport in plant ...

Hydroactive closure of stomata   see   Water status in plant and stomata

Hydrogen isotopes   see   Deuterium oxide, tritium oxide ...

Hygrometer   see   Humidity of air, methods;   Water potential, methods, dew point
     hygrometer

Hypostomatous leaves   6959, 9306, 9422, 10397, 11466, 12086, 12391, 14718

I

Ideotype   see   Model ...

Infra-red gas analysers   see   Transpiration rate, methods, infra-red gas analysers

Inhibitors   see   Growth substances, hormones, inhibitors etc. ...

Insertion level   see   Leaf insertion level ...

Integrated transpiration   see   Transpiration, integrated

Intercellular spaces   see   Conductance for water vapour and carbon dioxide transfer,
     intercellular spaces;   Transpiration rate, gradients of air humidity in leaf
     intercellular spaces

Irradiance and conductance for water vapour and carbon dioxide transfer 6620, 6647,
     6698, 6726, 6733, 6779, 6840, 6889, 6932, 6934, 6974, 6987, 7008, 7047, 7092,
     7151, 7209, 7225, 7343, 7351, 7358, 7362, 7419, 7449, 7490, 7492, 7540, 7544,
     7545, 7556, 7558, 7578, 7587, 7596, 7621, 7645, 7646, 7654, 7808, 7811, 7877,
     7883, 7939, 7961, 7996, 8010, 8037, 8078, 8127, 8141, 8197, 8238, 8307, 8334,
     8359, 8361, 8417, 8460, 8509, 8576, 8590, 8662, 8673, 8676, 8677, 8729, 8742,
     8744, 8745, 8799, 8801, 8810, 8811, 8838, 8912, 8913, 8914, 8926, 8929, 8950,
     8951, 8952, 8999, 9064, 9066, 9129, 9134, 9141, 9164, 9193, 9372, 9373, 9376,
     9443, 9496, 9497, 9499, 9514, 9550, 9571, 9628, 9644, 9656, 9659, 9660, 9661,
     9693, 9703, 9712, 9728, 9729, 9849, 9850, 9857, 9868, 9951, 9987, 10000,
     10076, 10087, 10117, 10202, 10225, 10313, 10314, 10315, 10332, 10344, 10373,
     10402, 10413, 10424, 10425, 10426, 10437, 10447, 10483, 10484, 10510, 10587,
     10590, 10630, 10642, 10697, 10704, 10729, 10756, 10775, 10779, 10791, 10818,
     10841, 10866, 10889, 11027, 11028, 11031, 11083, 11158, 11180, 11181, 11242,
     11266, 11296, 11332, 11364, 11372, 11414, 11439, 11517, 11582, 11588, 11622,
     11818, 11855, 11881, 11894, 11896, 11943, 11976, 12000, 12060, 12086,
     12167, 12168, 12218, 12237, 12264, 12307, 12321, 12329, 12345, 12346, 12351,
     12382, 12510, 12545, 12608, 12623, 12634, 12699, 12701, 12705, 12719, 12765,
     12850, 13063, 13121, 13123, 13168, 13169, 13173, 13175, 13178, 13202, 13243,
     13344, 13360, 13372, 13382, 13431, 13448, 13530, 13562, 13608, 13616, 13694,
     13699, 13739, 13765, 13777, 13781, 13788, 13789, 13830, 13899, 13949, 13961,
     13978, 14017, 14034, 14048, 14118, 14124, 14171, 14172, 14189, 14206, 14231,
     14232, 14233, 14234, 14236, 14246, 14262, 14447, 14468, 14479, 14522, 14541,
     14542, 14588, 14601, 14615, 14616, 14682

Irradiance and stomata 6662, 6779, 6840, 6888, 6889, 6937, 6974, 7044, 7092, 7093,
     7209, 7210, 7218, 7225, 7335, 7343, 7360, 7362, 7382, 7449, 7545, 7558, 7560,
     7602, 7619, 7645, 7721, 7779, 7782, 7787, 7811, 7939, 7946, 7951, 7968, 7996,
     8066, 8067, 8071, 8115, 8127, 8442, 8654, 8696, 8744, 8758, 8915, 8929, 8950,
     8951, 8952, 9101, 9110, 9128, 9134, 9234, 9231, 9232, 9294, 9295, 9304, 9372,
     9373, 9543, 9571, 9667, 9712, 9771, 9772, 9797, 9904, 9910, 9914, 9919, 9983,
     10029, 10037, 10048, 10087, 10098, 10116, 10131, 10147, 10189, 10273, 10290,
     10380, 10409, 10424, 10486, 10651, 10668, 10669, 10691, 10708, 10731, 10743,
     10775, 10788, 10836, 10866, 11012, 11027, 11158, 11181, 11340, 11494, 11523,
     11573, 11652, 11717, 11798, 11888, 12055, 12056, 12078, 12167, 12208, 12218,
     12310, 12311, 12312, 12329, 12386, 12387, 12545, 12683, 12701, 12710, 12817,
     12850, 12898, 12899, 13147, 13276, 13277, 13359, 13442, 13520, 13589, 13630,
                                                               (continued)

                                                                       (continued)

                                                            (continued)

Irrigation and growth, productivity (continued) 12654, 12663, 12668, 12679, 12688,
       12691, 12696, 12699, 12702, 12724, 12737, 12740, 12744, 12747, 12749, 12751,
       12756, 12761, 12769, 12777, 12789, 12797, 12798, 12800, 12812, 12854, 12875,
       12876, 12877, 12907, 12920, 12921, 12925, 12928, 12929, 12930, 12932, 12954,
       12955, 12958, 12962, 12964, 12975, 12976, 12980, 12984, 13003, 13020, 13039,
       13043, 13050, 13054, 13068, 13074, 13075, 13082, 13083, 13084, 13086, 13134,
       13152, 13179, 13197, 13209, 13212, 13244, 13245, 13246, 13264, 13268, 13291,
       13304, 13310, 13325, 13330, 13341, 13350, 13351, 13355, 13359, 13360, 13362,
       13363, 13368, 13369, 13378, 13380, 13391, 13406, 13419, 13437, 13438, 13443,
       13455, 13459, 13469, 13470, 13483, 13491, 13495, 13507, 13511, 13519, 13521,
       13529, 13550, 13559, 13578, 13579, 13581, 13591, 13615, 13628, 13629, 13631,
       13640, 13665, 13671, 13674, 13732, 13773, 13779, 13794, 13796, 13797, 13806,
       13812, 13861, 13867, 13873, 13875, 13888, 13893, 13894, 13896, 13903, 13950,
       13995, 13999, 14000, 14002, 14015, 14047, 14050, 14088, 14089, 14090, 14108,
       14119, 14138, 14139, 14143, 14165, 14187, 14205, 14225, 14250, 14266, 14276,
       14284, 14291, 14299, 14302, 14308, 14321, 14340, 14341, 14375, 14377, 14397,
       14426, 14431, 14437, 14438, 14440, 14442, 14444, 14448, 14462, 14466, 14469,
       14475, 14479, 14488, 14495, 14496, 14499, 14500, 14503, 14504, 14512, 14518,
       14535, 14545, 14546, 14548, 14550, 14566, 14621, 14634, 14637, 14640, 14642,
       14653, 14654, 14669, 14671, 14672, 14673, 14676, 14703, 14721, 14737, 14755,
       14763

Irrigation and leaf anatomy 6693, 6745, 6874, 7231, 7232, 7241, 7477, 7598, 7690,
       7836, 8005, 8140, 8341, 8399, 8500, 8555, 8717, 8901, 9267, 9409, 10052,
       10246, 10350, 10393, 10431, 10592, 10763, 11258, 11310, 11363, 11447, 11514,
       11796, 11824, 11859, 11865, 11880, 11926, 11934, 11963, 11972, 12620, 12696,
       12761, 12877, 12921, 12958, 13039, 13411, 13438, 13507, 13586, 13959, 13996,
       14090, 14426, 14466, 14680, 14695

Irrigation and other processes than above and below 10324, 11040, 11270, 11344, 11361,
       11432, 11479, 11522, 11664, 11666, 11731, 11764, 11878, 11880, 12020, 12138,
       12157, 12217, 12275, 12290, 12337, 12352, 12474, 12497, 12594, 12636, 12642,
       12645, 12762, 12908, 13017, 13058, 13083, 13179, 13248, 13291, 13309, 13353,
       13355, 13397, 13483, 13510, 13584, 13633, 13888, 13927, 13933, 13943, 14074,
       14089, 14092, 14108, 14135, 14225, 14312, 14381, 14444, 14465, 14499, 14518,
       14546, 14550, 14608, 14671, 14672, 14673, 14686

Irrigation and photorespiration 12316

Irrigation and respiration 7231, 7357, 7655, 11994, 14119

Irrigation and stomata 6829, 7232, 7690, 11796, 11985, 12745, 13359, 14143

Irrigation and transpiration 6608, 7105, 7231, 7318, 7423, 7455, 7598, 7696, 7800,
       7805, 7865, 7992, 8028, 8131, 8173, 8390, 8410, 8760, 8897, 9048, 9267, 9417,
       9508, 9621, 9706, 9722, 9779, 10085, 10107, 10124, 10132, 10325, 10351, 10443,
       10750, 10751, 10810, 10878, 10952, 11005, 11195, 11197, 11287, 11325, 11364,
       11370, 11444, 11492, 11606, 11722, 11731, 11769, 11855, 11914, 11962, 11972,
       12019, 12038, 12165, 12413, 12474, 12654, 12679, 12696, 12699, 12908, 12955,
       12958, 13043, 13212, 13326, 13348, 13395, 13406, 13557, 13566, 13581, 13614,
       13674, 13758, 13957, 14074, 14119, 14143, 14462, 14512, 14545, 14549

Irrigation and water absorption by plant 7232, 7696, 7805, 7963, 8069, 8412, 9508,
       10184, 11137, 11658, 11777, 12468, 12518, 12696, 12876, 13212, 13325, 13566,
       13628, 13856, 14143, 14334

Irrigation and water status in plant 6910, 6931, 7011, 7098, 7105, 7193, 7223, 7231,
       7232, 7241, 7278, 7358, 7582, 7598, 7615, 7690, 7793, 7857, 7997, 8005, 8093,
       8140, 8241, 8412, 8444, 8457, 8658, 8662, 8727, 8962, 9003, 9212, 9267, 9360,
       9401, 9408, 9427, 9428, 9706, 9711, 9759, 9848, 9870, 10124, 10137, 10215,
       10305, 10392, 10426, 10443, 10476, 10559, 10750, 10762, 10763, 10764, 10843,
       10853, 10883, 10941, 11236, 11259, 11363, 11400, 11408, 11430, 11497, 11514,
       11606, 11669, 11670, 11678, 11756, 11764, 11880, 11886, 11932, 11934, 11972,
                                                                              (continued)

Leaf insertion level and water status in plant 6611, 6632, 6769, 6836, 6909, 6975,
     7002, 7163, 7233, 7234, 7331, 7458, 7465, 7679, 7712, 7734, 7745, 7798, 7848,
     7851, 7898, 7900, 8098, 8159, 8364, 8476, 8662, 8709, 8849, 8889, 8905, 8987,
     9040, 9156, 9496, 9622, 9797, 9871, 10030, 10146, 10203, 10482, 10883, 10906,
     10943, 10962, 10972, 10981, 10987, 11118, 11252, 11259, 11343, 11464, 11607,
     11685, 11686, 11716, 11897, 11934, 11938, 12126, 12228, 12246, 12277, 12281,
     12430, 12477, 12586, 12619, 12699, 12736, 12896, 12900, 12922, 13010, 13065,
     13087, 13186, 13203, 13370, 13397, 13428, 13555, 13570, 13700, 13713, 13739,
     13782, 13795, 13985, 14055, 14179, 14295, 14395, 14556, 14602, 14303, 14651,
     14701, 14731

Leaf insertion level and water transport in cells 10958, 12256, 12478, 13713

Leaf insertion level and water transport in plant 10823, 12715, 12755, 13294, 13334,
     13335, 14586

Leaf insertion level and wilting 6682, 6696, 6909, 7233, 7234, 7352, 7898, 7982, 7996,
     10378, 11348, 11607, 13186, 14731

Leaf surface, waxes and trichomes 6669, 6739, 6793, 6794, 6809, 6843, 6968, 7015,
     7028, 7060, 7256, 7276, 7333, 7381, 7422, 7448, 7519, 7520, 7521, 7522, 7656,
     7660, 7748, 7814, 7899, 7996, 8092, 8199, 8288, 8397, 8459, 8471, 8516, 8539,
     8557, 8686, 8747, 8791, 8819, 9035, 9241, 9374, 9575, 9585, 9588, 9809, 9846,
     9849, 10008, 10023, 10234, 10341, 10554, 10562, 10743, 10878, 10998, 11201,
     11219, 11242, 11270, 11276, 11277, 11336, 11388, 11508, 11514, 11582, 11713,
     11810, 11869, 11870, 11957, 11958, 11960, 12233, 12425, 12426, 12489, 12507,
     12555, 12683, 12710, 12717, 12832, 12927, 12955, 12979, 12988, 13048, 13052,
     13067, 13181, 13239, 13265, 13428, 13439, 13599, 13655, 13676, 13866, 13902,
     13905, 14104, 14255, 14277, 14515, 14536, 14614, 14674, 14718, 14719

Leaf surface, waxes and trichomes, seasonal changes 6974, 7028, 7259, 8127, 8824,
     10152, 12537, 12745, 12853, 13905, 14718

Leaf temperature see Transpiration rate and temperature

Leaf wetness 8369, 8370, 11417, 12827,

Light see Irradiance ...

Lipids, fatty acids and transpiration 9243, 9244,

Lipids, fatty acids and water absorption by plant 9121

Lipids, fatty acids and water status in plant 13352

Lipids, fatty acids and water transport in cells 7323

Lipids, fatty acids and wilting 8447, 9894, 10955, 12323

Lysimeters see Evapotranspiration, methods, evaporimeters and lysimeters

M

Mannitol see Osmotically active substances ...

Mass flow see Water transport in plant ...

Mathematical model see Model; Model of canopy

Matric potential in plant tissue (see also Water status in plant ...) 7601, 7712,
     7821, 8989, 9997, 10129, 11052, 12281, 13597, 14036, 14175, 14263

Matric potential in substrate 6687, 6772, 6894, 6997, 7064, 7073, 7130, 7136, 7217,
    7247, 7358, 7369, 7396, 7432, 7598, 7600, 7632, 7696, 7758, 7916, 8120, 8217,
    8237, 8581, 8630, 8751, 8779, 9126, 9166, 9167, 9264, 9440, 9760, 9917, 9980,
    10040, 10237, 10248, 10258, 10272, 10801, 10883, 10977, 11073, 11074, 11111,
    11226, 11259, 11333, 11400, 11430, 11432, 11658, 11708, 11776, 11795, 11833,
    12024, 12047, 12073, 12075, 12294, 12520, 12540, 12549, 12568, 12621, 12625,
    12663, 12699, 12812, 13042, 13125, 13136, 13230, 13326, 13561, 13583, 13594,
    13761, 13841, 13913, 13933, 14070, 14238, 14278, 14279, 14314, 14337, 14378,
    14503

Matric potential, methods 7758, 7821, 9264

Mesophyll conductance (resistance) see Conductance for carbon dioxide transfer,
    mesophyll (intracellular)

Mineral elements and conductance for water vapour and carbon dioxide transfer 6766,
    6889, 7013, 7033, 7281, 7375, 7549, 7550, 7638, 7797, 8104, 8211, 8244, 8423,
    8537, 8540, 8615, 8729, 8744, 8818, 8951, 8952, 8953, 9043, 9066, 9102, 9134,
    9178, 9225, 9226, 9249, 9250, 9337, 9445, 9518, 9577, 9582, 9855, 10156, 10169
    10230, 10269, 10575, 10589, 10659, 10684, 10686, 10718, 10841, 10968, 19969,
    10970, 11026, 11418, 11440, 11644, 11739, 12082, 12303, 12623, 12693, 12694,
    13101, 13131, 13171, 13237, 13382, 13485, 13789, 13831, 13935, 13965, 14093,
    14170, 14196, 14295, 14331, 14520, 14551, 14623

Mineral elements and stomata 6766, 6843, 6889, 6912, 7210, 7218, 7307, 7380, 7388,
    7403, 7666, 7721, 7778, 7782, 8158, 8211, 8744, 8758, 8943, 8951, 8952, 9066,
    9134, 9230, 9295, 9577, 9612, 9613, 9983, 10042, 10116, 10156, 10189, 10273,
    10290, 10424, 10491, 10604, 10632, 10669, 10835, 10927, 11108, 11158, 11660,
    11717, 11798, 12055, 12056, 12057, 12058, 12065, 12078, 12199, 12268, 12310,
    12311, 12312, 12313, 12509, 12704, 12833, 12898, 13085, 13165, 13778, 13877,
    14007, 14077, 14130, 14635, 14641

Mineral elements and transpiration 6619, 6623, 6916, 7271, 7329, 7334, 7374, 7425,
    7464, 7471, 7638, 7762, 7805, 7870, 7924, 7996, 8021, 8051, 8052, 8209, 8244,
    8300, 8372, 8407, 8729, 8992, 9043, 9066, 9102, 9134, 9183, 9337, 9375, 9501,
    9582, 9715, 9855, 9979, 10156, 10574, 10644, 10645, 10659, 10674, 10686, 10716
    10884, 11510, 11528, 11541, 11605, 11613, 11616, 12181, 12303, 12436, 12475,
    12581, 12696, 12884, 13030, 13051, 13176, 13485, 13638, 13789, 13831, 14070,
    14124, 14125, 14140, 14197, 14216, 14331, 14350, 14352, 14524, 14641, 14679,
    14741

Mineral elements and water absorption by plant 6610, 6689, 6993, 7442, 7805, 7928,
    7964, 7966, 8021, 8063, 8108, 8321, 8351, 9043, 9134, 9183, 11257, 11676,
    11899, 11997, 12045, 12146, 12220, 12303, 12473, 12575, 12944, 13051, 13251,
    14066, 14316, 14345, 14362, 14385

Mineral elements and water status in plant 6626, 6630, 6766, 7054, 7233, 7271, 7287,
    7320, 7704, 7735, 7871, 7872, 7971, 7973, 8096, 8151, 8209, 8235, 8349, 8351,
    8372, 8795, 8988, 8992, 9068, 9134, 9178, 9183, 9249, 9250, 9256, 9379, 9416,
    9608, 9710, 9711, 9745, 9873, 9961, 10156, 10215, 10490, 10565, 10570, 10571,
    10574, 10575, 10659, 10684, 10686, 10717, 10718, 10831, 10970, 11254, 11319,
    11510, 11732, 11981, 12045, 12202, 12662, 12696, 12704, 12706, 12857, 13035,
    13060, 13080, 13094, 13131, 13252, 13261, 13274, 13429, 13481, 13789, 13920,
    13925, 13965, 14066, 14070, 14083, 14084, 14140, 14196, 14197, 14295, 14296,
    14311, 14360, 14362, 14452, 14580, 14641, 14741

Mineral elements and water transport in cells 6674, 6843, 6984, 7610, 7698, 7699,
    7966, 9257, 9472, 10283, 13094, 13240

Mineral elements and water transport in plant 8108, 8621, 9043, 10666, 10717, 12884,
    13051, 13319, 14124, 14127, 14197, 14586, 14690

Mutagens, other organic substances and conductance for water vapour and carbon dioxide transfer 8190, 8326, 8360, 8387, 8744, 9158, 9762, 10684, 11357, 11851, 12510, 13463, 13832, 14064, 14131, 14509

Mutagens, other organic substances and stomata 8190, 8387, 8744, 9009, 9359, 9658, 9770, 10097, 10379, 10397, 10904, 11163, 11594, 11939, 12489, 13276, 13277, 14445

Mutagens, other organic substances and transpiration 8168, 8319, 9429, 9648, 10904, 11163, 12671, 13367, 13463, 13577, 13941, 14067, 14699

Mutagens, other organic substances and water absorption by plant 9609, 9819, 10137, 11367, 12223, 13113, 14509

Mutagens, other organic substances and water status in plant 8326, 8360, 8387, 9009, 9134, 9647, 9819, 10137, 10329, 10379, 10389, 10684, 10906, 11163, 11357, 11367, 11631, 11907, 1,1974, 12223, 12671, 12708, 12839, 13367, 13941, 14064, 14067, 14131, 14509, 14699

Mutagens, other organic substances and water transport in plant 13986

Mutagens, other organic substanes and wilting 8350, 8387, 8844, 9134, 9158, 9178, 9647, 11454, 11907, 11974, 11998, 12883, 13367

Mutants and conductance for water vapour and carbon dioxide transfer 11383,

Mutants and stomata 7405, 9983, 11386

Mutants and transpiration 9653, 11383, 12626

Mutants and water status in plant 6941, 7062, 11383, 12369, 13859

N

Nitrogen se Mineral elements ...

Nucleic acids see Proteins, amino acids, nucleic acids ...

O

O_2 see Oxygen ...

O_3 see Pollutants, ozone ...

Ontogeny see age of plant ...

Oscillations see Conductance for water vapour and carbon dioxide transfer, ...; Stomata, ...; Transpiration rate, ...; Water absorption by plant, ...; Water status in plant, ...; Water transport in plant, ...; Wilting, oscillations

Osmotic potential in plant tissue (see also Water status in plant ...) 6609, 6626, 6656, 6660, 6662, 6663, 6681, 6682, 6696, 6726, 6732, 6770, 6782, 6784, 6787, 6896, 6899, 6900, 6910, 6926, 7006, 7087, 7092, 7093, 7110, 7132, 7156, 7158, 7163, 7174, 7197, 7214, 7216, 7232, 7233, 7234, 7235, 7245, 7248, 7255, 7257, 7260, 7278, 7285, 7289, 7290, 7301, 7312, 7343, 7358, 7369, 7378, 7418, 7419, 7427, 7442, 7444, 7458, 7469, 7476, 7494, 7495, 7553, 7576, 7601, 7615, 7637, 7671, 7675, 7676, 7677, 7704, 7712, 7735, 7737, 3355, 7851, 7852, 7871, 7872, 7874, 7888, 7904, 7905, 7906, 7921, 7922, 7941, 7950, 7990, 7995, 7996, 8005, 8014, 8027, 8054, 8055, 8073, 8088, 8126, 8130, 8133, 8151, 8152, 8153, 8188, 8189, 8190, 8219, 8241, 8259, 8282, 8328, 8333, 8349, 8351, 8379, 8398, 8401,
(continued)

Osmotically active substances and stomata 7225, 7782, 8940, 8948, 9612, 9667, 9810,
     9873, 10116, 10424, 10478, 10834, 11116, 11129, 11594, 12054, 12065, 12122

Osmotically active substances and transpiration 7225, 7553, 9911, 10043, 10812,
     11350, 12045, 12527, 12671, 12904, 13765, 14715, 14741, 14747

Osmotically active substances and water absorption by plant 6880, 7294, 7458, 7487,
     7554, 8122, 8274, 9052, 9296, 9643, 9764, 9911, 10043, 10096, 10195, 10196,
     10347, 10874, 11089, 11257, 11475, 11676, 11700, 11773, 11787, 11898, 11899,
     12332, 12388, 12407, 12576, 12592, 12784, 12871, 12926, 13543, 13765, 14079,
     14409, 14411

Osmotically active substances and water status in plant 6770, 6782, 6784, 7005, 7054,
     7152, 7358, 7395, 7554, 7626, 7675, 7676, 7704, 7982, 8080, 8228, 8256, 8258,
     8259, 8282, 8489, 8645, 8750, 8816, 8830, 8875, 8939, 8944, 8970, 9077, 9170,
     9173, 9296, 9317, 9330, 9361, 9561, 9758, 9780, 9873, 10125, 10236, 10271,
     10296, 10547, 10701, 10723, 19872, 10950, 10969, 10974, 11010, 11208, 11284,
     11289, 11621, 11673, 11728, 11760, 11787, 11828, 12176, 12177, 12183, 12332,
     12407, 12527, 12575, 12576, 12663, 12671, 12695, 12696, 12697, 12711, 13364,
     13365, 13481, 13505, 13555, 13809, 13979, 14040, 14041, 14081, 14112, 14419,
     14532, 14715

Osmotically active substances and water transport in cells 6708, 7395, 7714, 8932,
     9538, 10423, 11047, 11049, 11054, 11140, 11142, 11323, 12650, 13514

Osmotically active substances and water transport in plant 9296, 10271, 10874, 13318

Osmotically active substances and wilting 6878, 6980, 7013, 7129, 7358, 7569, 7570,
     7625, 7676, 7823, 7826, 7971, 7982, 8447, 8452, 8453, 8646, 8867, 9205, 9240,
     9317, 9350, 9393, 9520, 9633, 9766, 9911, 10199, 10211, 10236, 10316, 10378,
     10449, 10812, 10892, 10968, 10969, 10970, 11306, 11326, 11348, 11350, 13481,
     13541, 13659, 13678, 13682, 14406, 14528, 14538

Oxygen and conductance for water vapour and carbon dioxide transfer 7319, 7616, 7881,
     8670, 8902, 10343, 11740, 12510, 12956, 13344, 13760, 13781

Oxygen and stomata 7275, 7360, 7881, 8835, 9231, 9910, 9913, 12571

Oxygen and transpiration 6623, 6624, 6887, 7166, 7360, 7374, 7389, 7559, 7561, 8898,
     9043, 9122, 9547, 9910, 10343, 13781

Oxygen and water absorption by plant 6887, 7166, 7881, 8898, 9043, 9547, 11953

Oxygen and water status in plant 6941, 7360, 8902, 9122, 10021, 12213, 12956, 13760

Oxygen and water transport in plant 9043, 12956, 13234

Oxygen and wilting 6887, 8835, 8898, 9043, 9122, 12008

Oxygen isotopes in water 8695, 9464, 9465

Ozone see Pollutants, ozone ...

P

Pathogens and conductance for water vapour and carbon dioxide transfer 6662, 7463,
     8189, 8296, 8387, 8433, 8849, 9096, 9327, 9498, 9639, 9762, 9954, 10111, 10112,
     10635, 10963, 11525, 12100, 12172, 12245, 12516, 13229, 14728

Pathogens and stomata 6662, 7328, 7803, 7899, 8026, 8099, 8185, 8190, 8387, 8459,
     8503, 8744, 10341, 10379, 10907, 10963, 12100, 12771, 13122, 14476, 14718

Pathogens and transpiration 6662, 7095, 7187, 7288, 7463, 7651, 7652, 7653, 7929,
        7934, 8013, 8025, 8175, 8479, 8480, 8646, 8825, 9206, 9327, 9900, 9954,
        10286, 10383, 11527, 11724, 12172, 12884, 13229, 13795, 14114, 14348, 14728

Pathogens and water absorption by plant 6790, 6894, 7396, 7651, 7652, 7653, 8108,
        8120, 9206, 9853, 10193, 12641, 13096

Pathogens and water status in plant 7124, 7125, 7369, 7427, 7518, 8035, 8188, 8189,
        8190, 8387, 8506, 8636, 8744, 8849, 9096, 9143, 9206, 9207, 9352, 9498, 9639,
        9720, 9954, 10111, 10137, 10379, 10524, 10579, 10749, 10859, 10895, 11445,
        11527, 11611, 12024, 12245, 12354, 12516, 12888, 13092, 13136, 13229, 13287,
        13451, 13795, 14114, 14184, 14185, 14226, 14227, 14731

Pathogens and water transport in cells 13805,

Pathogens and water transport in plant 8358, 9358, 9473, 9474, 9485, 9576, 9663,
        11105, 12245, 12587, 12767, 12884, 12888, 12905, 13172, 13229, 13802, 14343

Pathogens and wilting 6730, 6851, 6884, 7125, 7129, 7651, 7652, 7653, 8035, 8190,
        8348, 8387, 8636, 8646, 8744, 9132, 9143, 9206, 9720, 9746, 9962, 10027, 10137
        10270, 10424, 10444, 10524, 10579, 10776, 11526, 11527, 11722, 12641, 12699,
        12967, 14183, 14731

Permanent wilting see Wilting ...

Permeability see Water transpor in cells, permeability

Pesticides, herbicides and conductance for water vapour and carbon dioxide transfer
        7397, 7684, 7809, 11862, 13076

Pesticides, herbicides and stomata 7397, 7684, 7809, 7946, 8092, 10056, 11077, 11212,
        11608, 11829.

Pesticides, herbicides and transpiration 6619, 7397, 7447, 7571, 7684, 7810, 8052,
        8479, 8526, 9254, 11077, 11212, 11608, 11888, 12499, 12940, 13361

Pesticides, herbicides and water absorption by plant 6890, 6948, 9853

Pesticides, herbicides and water status in plant 6846, 7406, 9229, 9352, 9744, 9812,
        10565, 11216, 11457, 11840, 12585, 12779

Pesticides, herbicides and water transport in plant 7447, 7810

Pesticides, herbicides and wilting 6619, 6846, 7447, 8526, 8975

pH and stomata 7300, 7346, 8071, 10074, 10098, 10275, 11743, 12410, 12898, 13085,
        13276, 14718

pH and water absorption by plant 11475

pH and water status in plant 8708, 10114, 10331, 13060,

pH and water transport in cells 9286

Photoperiod and conductance for water vapour and carbon dioxide transfer 9050, 9522,
        11792, 12048, 13472

Photoperiod and stomata 13520, 14344

Photoperiod and transpiration 8171, 8693, 8778, 9551, 11792, 12967, 13588, 14461

Photoperiod and water absorption by plant 8778.

Photoperiod and water status in plant 8693, 19321, 10585, 11491, 11792, 13588,
     14075, 14461

Photoperiod and water transport in plant 14461

Photorespiration see Drought ...; Flooding ...; Humidity of air ...; Irrigation
     ...; Osmotically active substances ...; Salinity ...; Soil moisture ...;
     Water status in plant and photorespiration

Photosynthesis see Carbon dioxide influx ...

Photosynthesis - transpiration ratio see Productivity of transpiration

Phytopathology see Pathogens ...

Plasmolysis see Water transport in cells, plasmolysis

Pollutants, ozone and conductance for water vapour and carbon dioxide transfer 6685,
     6733, 6978, 7058, 7504, 7549, 7550, 7827, 8075, 8077, 8078, 8167, 8249, 8250,
     8290, 8337, 8385, 8411, 8495, 8540, 8546, 8591, 8744, 8823, 8854, 8997, 9007,
     9087, 9109, 9110, 9569, 9570, 9571, 9761, 9776, 9784, 9924, 9953, 9976, 10152,
     10230, 10234, 10368, 10391, 10591, 10622, 11023, 11065, 11066, 11144, 11145,
     11389, 11413, 11422, 11423, 12247, 12248, 12249, 12250, 12385, 12514, 12693,
     12694, 12717, 12983, 13036, 13262, 13307, 13331, 13492, 13699, 13730, 13741,
     13878, 13991, 14022, 14049, 14236, 14560, 14585, 14612

Pollutants, ozone and stomata 6809, 6937, 6979, 7058, 7204, 7300, 7549, 7813, 7814,
     8077, 8249, 8337, 8528, 8557, 8744, 8846, 8854, 8859, 8997, 9109, 9110, 9111,
     9421, 9513, 9570, 9571, 9710, 9784, 10152, 10234, 10318, 10591, 10621, 10622,
     10660, 11066, 11144, 11389, 11592, 11891, 11957, 11958, 12007, 12132, 12249,
     12250, 12385, 12387, 12717, 13599, 13741, 13817, 13878, 13945, 13991, 14052,
     14053, 14585

Pollutants, ozone and transpiration 6685, 6733, 7039, 7271, 7300, 7525, 8076, 8545,
     8546, 8707, 8846, 9072, 9111, 9112, 9113, 9341, 9342, 9489, 9784, 9844, 9908,
     10318, 10391, 10591, 11066, 11144, 11389, 11422, 11613, 11891, 12247, 12248,
     12270, 12292, 12514, 12895, 13098, 13333, 13336, 13698, 13699, 13741, 13819,
     13878, 14236, 14610

Pollutants, ozone and water status in plant 7271, 7525, 8152, 8411, 8707, 8708, 8859,
     9342, 9450, 9776, 9784, 10438, 10622, 11007, 11024, 11066, 11183, 12292,
     12968, 13698, 14251, 14533

Pollutants, ozone and water transport in cells 7769, 9976, 10318, 10967, 11066,
     11119, 11121, 13564, 13698

Pollutants, ozone and wilting 7525, 7549, 8859, 9341, 9844, 10621, 10622, 11007,
     11066, 12229, 12968

Polyethylene glycol see Osmotically active substances ...

Porometer see Stomatal aperture, methods, diffusion porometers; Stomatal aperture,
     methods, viscous flow rate porometers

Potential matric see Matric potential ...

Potential osmotic see Osmotic potential ...

Potential pressure see Pressure potential ...

Potential water see Water potential ...

Potometry 7166, 7308, 7447, 7458, 8840, 8841, 8980, 9296, 10874, 11069, 11603, 11604,
     11690, 12502, 12728, 12926, 13234, 13319, 14441, 14718

Precipitation, dew and canopy architecture 6661, 12353,

Precipitation, dew and carbon dioxide influx 7246, 7386, 9011, 10568, 11809, 14287,

Precipitation, dew and carbon fixation pathways 9624

Precipitation, dew and chlorophyll 8117

Precipitation, dew and conductance for water vapour and carbon dioxide transfer 7228,
     9870, 11809, 13297

Precipitation, dew and growth, productivity 6637, 6653, 6728, 6749, 6750, 6767, 6804,
     6845, 6883, 7010, 7081, 7085, 7086, 7088, 7090, 7303, 7318, 7499, 7500, 7532,
     7533, 7641, 7685, 7686, 7708, 7846, 7984, 7996, 8065, 8103, 8278, 8346, 8373,
     8404, 8427, 8440, 8455, 8477, 8511, 8524, 8552, 8571, 8591, 8604, 8656, 8680,
     8734, 8738, 8769, 8938, 8978, 8979, 9011, 9124, 9319, 9420, 9433, 9493, 9839,
     9870, 9931, 9979, 10058, 10085, 10130, 10145, 10154, 10300, 10361, 10362,
     10447, 10477, 10572, 10573, 10685, 10725, 10777, 10797, 10803, 10804, 10891,
     10901, 10934, 10936, 10946, 11038, 11060, 11135, 11262, 11300, 11426, 11449,
     11479, 11488, 11505, 11506, 11550, 11637, 11650, 11735, 11737, 11794, 11863,
     11915, 11921, 11947, 12038, 12207, 12353, 12392, 12431, 12437, 12465, 12534,
     12562, 12646, 12647, 12652, 12654, 12668, 12696, 12702, 13082, 13184, 13267,
     13268, 13304, 13332, 13355, 13366, 13380, 13384, 13385, 13388, 13391, 13498,
     13529, 13619, 13628, 13654, 13829, 13924, 13956, 14058, 14146, 14147, 14252,
     14274, 14276, 14338, 14346, 14438, 14518, 14526, 14609, 14624, 14634, 14681,
     14690

Precipitation, dew and leaf anatomy 7360, 7426, 8471, 8656, 11859,

Precipitation, dew and other processes than above and below 10787, 11488, 11504,
     11837, 12093, 12646, 12647, 13332, 13398, 13716, 13917, 13987, 14222, 14364,
     14415, 14558, 14692

Precipitation, dew and respiration 7261,

Precipitation, dew and stomata 7228, 7360, 7996

Precipitation, dew and transpiration 7261, 7334, 7360, 7996, 8012, 8236, 8330, 9152,
     9555, 9883, 10085, 11393, 12171, 12219, 13490, 13695, 13977, 14120, 14224,
     14258, 14624, 14733

Precipitation, dew and water absorption by plant 7246, 7998, 8569, 8858, 9480, 11013,
     13628

Precipitation, dew and water status in plant 6707, 7360, 7907, 8629, 9422, 9870,
     9972, 10692, 11230, 11669, 11670, 12213, 12998, 13297, 13936, 13994, 14287,
     14507

Precipitation, dew and water transport in plant 823614558

Precipitation, dew and wilting 6952, 7614, 11230, 14584,,

Precipitation, dew, methods 8166, 8559, 8623, 9619, 10663, 10664, 10946, 12093,
     13423, 14324

Pressure bomb see Water potential, methods, pressure bomb

Pressure potential in plant tissue (see also Water status in plant ...) 6612, 6660,
     6681, 6726, 6746, 6770, 6782, 6784, 6864, 6879, 6900, 6901, 6910, 6912,
     6931, 7005, 7035, 7087, 7093, 7105, 7110, 7163, 7221, 7232, 7233, 7234, 7259,
     7281, 7312, 7350, 7351, 7352, 7419, 7444, 7469, 7476, 7494, 7495, 7576, 7579,
     7601, 7637, 7638, 7677, 7695, 7704, 7712, 7739, 7852, 7873, 7874, 7904, 7905,
                                                                    (continued)

Pressure potential in plant tissue (continued) 7906, 7922, 7940, 7990, 7995, 7996,
    8013, 8073, 8080, 8088, 8123, 8124, 8125, 8126, 8127, 8132, 8133, 8134, 8153,
    8155, 8162, 8188, 8189, 8241, 8256, 8258, 8259, 8301, 8349, 8376, 8377, 8379,
    8386, 8398, 8399, 8401, 8413, 8417, 8448, 8449, 8489, 8490, 8494, 8575, 8612,
    8621, 8666, 8676, 8677, 8706, 8715, 8716, 8743, 8744, 8749, 8750, 8751, 8800,
    8826, 8827, 8857, 8905, 8931, 8932, 8939, 8940, 8956, 8970, 8976, 8981, 8989,
    8998, 9002, 9009, 9031, 9063, 9123, 9134, 9151, 9176, 9177, 9188, 9195, 9261,
    9273, 9330, 9376, 9404, 9434, 9436, 9444, 9456, 9457, 9458, 9499, 9529, 9531,
    9539, 9556, 9558, 9559, 9562, 9564, 9604, 9644, 9646, 9659, 9661, 9675, 9681,
    9701, 9717, 9741, 9754, 9759, 9776, 9780, 9807, 9820, 9827, 9848, 9873, 9892,
    9919, 9926, 9974, 9989, 10000, 10087, 10112, 10129, 10141, 10162, 10200, 10201,
    10202, 10203, 10272, 10305, 10331, 10351, 10357, 10404, 10424, 10436, 10449,
    10469, 10481, 10487, 10488, 10490, 10491, 10520, 10554, 10585, 10586, 10600,
    10630, 10650, 10652, 10653, 10655, 10679, 10712, 10717, 10763, 10764, 10779,
    10791, 10792, 10831, 10853, 10883, 10947, 10948, 10972, 10980, 10981, 10997,
    11000, 11049, 11052, 11053, 11080, 11099, 11123, 11171, 11173, 11190, 11202,
    11239, 11269, 11283, 11319, 11337, 11363, 11382, 11383, 11399, 11476, 11485,
    11516, 11517, 11519, 11554, 11556, 11557, 11574, 11598, 11621, 11706, 11728,
    11754, 11760, 11761, 11762, 11765, 11771, 11790, 11796, 11833, 11854, 11887,
    11897, 11954, 11955, 11962, 11981, 11988, 12018, 12021, 12034, 12041, 12054,
    12066, 12078, 12100, 12126, 12142, 12145, 12155, 12224, 12240, 12241, 12277,
    12281, 12300, 12321, 12350, 12369, 12370, 12378, 12321, 12350, 12369, 12370,
    12378, 12392, 12407, 12417, 12430, 12446, 12459, 12492, 12527, 12547, 12585,
    12586, 12619, 12620, 12621, 12633, 12648, 12649, 12650, 12651, 12655, 12663,
    12671, 12698, 12699, 12711, 12753, 12754, 12763, 12770, 12773, 12805, 12806,
    12808, 12821, 12849, 12856, 12857, 12879, 12894, 12897, 12898, 12922, 12924,
    12954, 12955, 12995, 13038, 13039, 13063, 13065, 13080, 13087, 13092, 13093,
    13094, 13105, 13106, 13120, 13139, 13186, 13190, 13191, 13192, 13194, 13271,
    13287, 13289, 13290, 13298, 13300, 13338, 13340, 13350, 13379, 13418, 13433,
    13437, 13452, 13453, 13462, 13528, 13533, 13534, 13535, 13589, 13597, 13642,
    13673, 13698, 13713, 13717, 13718, 13763, 13771, 13782, 13830, 13835, 13840,
    13901, 13911, 13914, 13916, 13966, 13968, 13969, 13985, 13994, 14014, 14020,
    14031, 14036, 14040, 14041, 14055, 14056, 14075, 14081, 14098, 14102, 14112,
    14116, 14119, 14179, 14196, 14201, 14239, 14242, 14255, 14263, 14275, 14323,
    14359, 14375, 14433, 14434, 14457, 14481, 14508, 14509, 14525, 14541, 14556,
    14557, 14574, 14604, 14606, 14627, 14628, 14697, 14698, 14701, 14731, 14752

Pressure potential, methods 7469, 7906, 8088, 8125, 8126, 8621, 8706, 9434, 9444,
    10652, 11485, 12754, 12879, 14492, 14493, 14574, 14697

Productivity see Growth, productivity ...

Productivity of algae see Deuterium oxide, tritium oxide ...; Osmotically active
    substances ...; Salinity and productivity of algae

Productivity of transpiration 6624, 6631, 6641, 6645, 6726, 6738, 6781, 6823, 6840,
    6875, 6907, 6998, 7027, 7104, 7128, 7162, 7170, 7176, 7177, 7179, 7192, 7195,
    7228, 7295, 7343, 7393, 7398, 7425, 7428, 7578, 7579, 7606, 7617, 7680, 7729,
    7771, 7805, 7820, 7876, 7877, 7878, 7906, 7918, 7943, 7992, 7996, 8010, 8011,
    8015, 8175, 8192, 8203, 8204, 8205, 8238, 8278, 8279, 8311, 8325, 8409, 8465,
    8536, 8576, 8613, 8618, 8764, 8813, 8855, 8857, 8860, 8861, 8909, 8951, 9011,
    9090, 9129, 9134, 9139, 9186, 9217, 9429, 9500, 9510, 9514, 9516, 9567, 9568,
    9606, 9642, 9844, 9891, 9901, 9919, 9920, 9925, 9960, 10020, 10046, 10120,
    10147, 10194, 10339, 10342, 10376, 10383, 10387, 10404, 10413, 10424, 10437,
    10464, 10479, 10482, 10484, 10491, 10530, 10548, 10625, 10626, 10642, 10661,
    10667, 10679, 10687, 10704, 10740, 10775, 10840, 10841, 10890, 11002, 11028,
    11037, 11062, 11076, 11239, 11242, 11278, 11384, 11386, 11414, 11439, 11562,
    11623, 11624, 11661, 11688, 11715, 11757, 11877, 11878, 11955, 12038, 12061,
    12068, 12161, 12162, 12165, 12166, 12254, 12291, 12301, 12303, 12307, 12320,
    12346, 12348, 12379, 12401, 12451, 12470, 12504, 12545, 12557, 12581, 12691,
    12696, 12699, 12751, 12753, 12762, 12834, 12849, 12892, 12909, 12974, 12993,
    13005, 13078, 13082, 13180, 13196, 13202, 13273, 13303, 13311, 13314, 13343,
                                        (continued)

Productivity of transpiration (continued) 13358, 13372, 13444, 13456, 13511, 13562,
          13566, 13590, 13651, 13674, 13720, 13787, 13812, 13819, 13857, 13873, 13937,
          13940, 13954, 13965, 13967, 13988, 14037, 14124, 14258, 14265, 14314, 14350,
          14351, 14354, 14370, 14394, 14405, 14422, 14520, 14637, 14681

Proteins, amino acids, nucleic acids and stomata 9237, 14563

Proteins, amino acids, nucleic acids and transpiration 13958

Proteins, amino acids, nucleic acids and water absorption by plant 6644, 6655, 6898,
          6928, 11839, 12332

Proteins, amino acids, nucleic acids and water status in plant 6741, 6746, 6762, 6787
          7048, 7118, 7233, 7390, 7394, 7395, 7786, 7900, 7907, 8049, 8182, 8245, 8674,
          9134, 9477, 9745, 10077, 10125, 10218, 10232, 10389, 10547, 10723, 11009,
          11732, 11790, 11990, 12019, 12021, 12223, 12546, 12673, 12696, 12846, 12947,
          13522

Proteins, amino acids, nucleic acids and water transport in cells 7395, 9752, 10081,
          11047, 13322

Proteins, amino acids, nucleic acids and water transport in plant 9573, 14586

Proteins, amino acids, nucleic acids and wilting 6622, 6682, 6741, 6746, 6904, 6905,
          6935, 7035, 7117, 7233, 7583, 7907, 7982, 7996, 8049, 8082, 8182, 8245, 9134,
          9463, 9617, 9766, 9894, 10077, 10218, 10232, 10377, 10424, 10557, 10606, 10637
          10953, 10983, 10988, 11070, 11326, 11327, 11990, 12018, 12673, 12947, 13370,
          13523, 14299, 14406, 14407

Psychrometry see Water potential, methods, psychrometry

P/T ratio see Productivity of transpiration

R

Radiation see Irradiance ...

Rain see Precipitation, dew ...

Reactivity of stomata see Stomatal reaction rate; Stomatal reactivity during
          ontogeny

Rehydration 6611, 6622, 6648, 6649, 6688, 6689, 6714, 6786, 6853, 6896, 6900, 6970,
          7006, 7011, 7035, 7193, 7234, 7288, 8351, 7432, 7449, 7554, 7599, 7610, 7719,
          7723, 7835, 7985, 7996, 8035, 8070, 8073, 8113, 8131, 8156, 8182, 8239, 8245,
          8271, 8294, 8296, 8345, 8367, 8413, 8431, 8448, 8488, 8515, 8522, 8589, 8641,
          8665, 8744, 8751, 8772, 8782, 8818, 8821, 8843, 8861, 8874, 8899, 8907, 8911,
          8932, 8951, 8954, 8959, 8994, 9015, 9063, 9108, 9134, 9159, 9176, 9177, 9278,
          9317, 9350, 9353, 9354, 9366, 9390, 9403, 9404, 9506, 9520, 9563, 9602, 9647,
          9659, 9710, 9713, 9734, 9765, 9766, 9788, 9825, 9878, 9949, 10011, 10092,
          10122, 10149, 10158, 10162, 10201, 10248, 10250, 10260, 10278, 10280, 10296,
          10310, 10378, 10404, 10424, 10430, 10441, 19472, 10522, 10563, 10569, 10593,
          10594, 10595, 10653, 10670, 10681, 10732, 10805, 10809, 10812, 10816, 10846,
          10871, 10955, 10980, 10981, 10985, 11014, 11047, 11178, 11235, 11249, 11271,
          11318, 11326, 11327, 11345, 11359, 11446, 11454, 11473, 11490, 11499, 11516,
          11554, 11586, 11745, 11752, 11786, 11834, 11856, 11984, 12018, 12021, 12209,
          12245, 12258, 12273, 12285, 12321, 12334, 12357, 12378, 12430, 12466, 12470,
          12583, 12487, 12492, 12500, 12501, 12606, 12619, 12663, 12696, 12697, 12699,
          12776, 12807, 12834, 12841, 12856, 12857, 12860, 12866, 12897, 12903, 12914,
          12935, 12943, 12947, 12961, 13000, 13001, 13037, 13077, 13078, 13111, 13186,
          13259, 13284, 13350, 13370, 13393, 13403, 13412, 13431, 13534, 13590, 13713,
(continued)

Rehydration (continued) 13727, 13728, 13809, 13822, 13869, 13887, 13914, 13968,
        13974, 14044, 14045, 14046, 14055, 14069, 14086, 14142, 14183, 14242, 14268,
        14285, 14314, 14355, 14367, 14389, 14392, 14394, 14395, 14434, 14466, 14480,
        14538, 14542, 14570, 14596, 14628, 14647, 14667

Relative water content see Water saturation deficit

Resistance see Conductance ...

Respiration see Deuterium oxide, tritium oxide ...; Drought ...; Flooding ...;
        Humidity of air ...; Irrigation ...; Osmotically active substances ...;
        Precipitation, dew ...; Root, underground part ...; Salinity ...; Soil
        moisture ...; Water status in plant and respiration

Root pressure, exudation 6775, 6803, 6872, 6890, 6928, 7103, 7161, 7378, 7418, 7469,
        7872, 8122, 8292, 8436, 8437, 8504, 8577, 8778, 9052, 9182, 9184, 9436, 9636,
        9670, 9717, 9799, 9806, 10045, 10424, 10491, 10507, 10518, 10586, 10717, 10814,
        10854, 10884, 11069, 11113, 11271, 11328, 11367, 11484, 11620, 11692, 11773,
        11890, 11953, 11973, 11997, 12094, 12132, 12220, 12332, 12576, 12648, 12682,
        12956, 13176, 13232, 13570, 13639, 13705, 13772, 13843, 13901, 13907, 14305,
        14461, 14586, 14606, 14626

Root pressure, exudation, methods 8122, 13985, 14305

Root removal see Defoliation, decapitation, ear, root removal ...

Root system, dimensions, volume etc. and water absorption by plant 6890, 6913, 7041,
        7183, 7197, 7459, 7947, 7996, 8472, 8527, 9153, 9917, 10088, 10184, 10456,
        10757, 10767, 11029, 11030, 11043, 11218, 11257, 11655, 11690, 12045, 12074,
        12075, 12575, 12608, 12663, 12695, 12699, 12728, 12734, 12753, 13187, 13301,
        13457, 13593, 14385

Root, underground part and carbon dioxide influx 11424

Root, underground part and conductance for water vapour and carbon dioxide transfer
        6911, 9695, 9818, 11424, 12705, 13187

Root, underground part and growth, productivity 8292, 8354, 8367, 8381, 8414, 8548,
        8565, 8796, 8972, 8983, 9140, 9271, 9297, 9391, 9525, 9618, 9623, 9695, 9883,
        9918, 10363, 10456, 10516, 10556, 10671, 10712, 10865, 10883, 11090, 11093,
        11109, 11189, 11197, 11210, 11252, 11332, 11359, 11364, 11424, 11428, 11430,
        11533, 11550, 11564, 11655, 11662, 11722, 11756, 11795, 11805, 11864, 11927,
        12022, 12034, 12073, 12074, 12092, 12105, 12114, 12254, 12286, 12295, 12576,
        12598, 12699, 12751, 12752, 12764, 12802, 12889, 12914, 13014, 13050, 13076,
        13187, 13188, 13231, 13252, 13253, 13366, 13390, 13412, 13457, 13570, 13593,
        13750, 13758, 13794, 13815, 13831, 13845, 13916, 14180, 14185, 14218, 14293,
        14340, 14384, 14388, 14426, 14446, 14479, 14533, 14609, 14630, 14702, 14704.

Root, underground part and stomata 10292, 10986

Root, underground part and transpiration 6891, 9818, 9916, 10987, 11343, 11754,
        12022, 13187, 13451

Root, underground part and water status in plant 6772, 6911, 6931, 7131, 7236, 7996,
        9818, 11085, 11252, 11290, 11367, 11662, 12022, 12073, 12074, 12155, 12576,
        12695, 12696, 12699, 12705, 12889, 13187, 14293, 14409, 14411

Root, underground part and water transport in plant 7947, 14557

Root, underground part and wilting 6730, 7608, 8772, 9134, 10556, 10986, 11033,
        11218, 11252, 11290, 11533, 11955, 12696, 12831, 12837, 13079, 14411

S

Saccharides and conductance for water vapour and carbon dioxide transfer 7808, 11740

Saccharides and stomata 7968, 8405, 10835, 12313, 14563

Saccharides and transpiration 7684, 8480, 11089, 11740

Saccharides and water transport in plant 12332

Saccharides and water status in plant 6682, 6787, 6926, 7233, 7871, 7872, 7873, 7996
        8118, 8073, 8152, 8182, 8482, 8636, 8708, 8772, 9134, 10077, 10125, 10351,
        11202, 11430, 12223, 12546, 12575, 12699, 13035, 13352, 13542, 13710, 14574

Saccharides and water transport in cells 7314, 8808, 11047

Saccharides and water transport in plant 11680, 12512, 14124

Saccharides and wilting 6682, 6829, 7213, 7233, 7452, 7823, 7965, 7996, 8018, 8073,
        8429, 8482, 8522, 8622, 8636, 9617, 10077, 10404, 10983, 13370, 13642, 13710,
        14407

Saline water see Irrigation water quality

Salinity and canopy architecture 7026

Salinity and carbon dioxide influx 6618, 6693, 6752, 6885, 6888, 7113, 7175, 7237,
        7266, 7292, 7409, 7482, 7541, 7576, 7628, 7694, 8023, 8048, 8130, 8228, 8264,
        8342, 8415, 8587, 8657, 8660, 8673, 8674, 8750, 8810, 8811, 8815, 8829, 8886,
        9604, 9684, 9750, 9761, 9781, 9911, 9952, 9996, 10102, 10103, 10364, 10589,
        10625, 10681, 10730, 10968, 11099, 11196, 11213, 11214, 11353, 11397, 11531,
        11539, 11547, 11684, 11747, 11768, 11873, 11875, 11985, 12033, 12204, 12284,
        12418, 12531, 12538, 12557, 12757, 12890, 12924, 13005, 13006, 13072, 13424,
        13425, 13484, 13576, 13658, 13683, 13717, 13790, 13825, 14001, 14006, 14031,
        14032, 14110, 14123, 14181, 14228, 14246, 14412, 14413, 14603

Salinity and carbon fixation pathways 6694, 7026, 7175, 7237, 7541, 7671, 8048, 8080
        8164, 8673, 8674, 9107, 9596, 9597, 9598, 9604, 10036, 10103, 10142, 10190,
        10210, 10690, 10713, 11543, 11643, 11768, 11876, 11884, 12182, 12225, 12538,
        12873, 13658, 14110, 14408, 14412, 14674

Salinity and carotenoids 7026, 7964, 8657, 9242, 9602, 10183, 11214, 12033, 12912,
        14195

Salinity and chlorophyll 6618, 6710, 6939, 7026, 7237, 7266, 7401, 7409, 7541, 7628,
        7964, 8102, 8144, 8145, 8164, 8342, 8351, 8380, 8646, 8657, 9242, 9322, 9596,
        9598, 9680, 9684, 9749, 9750, 9915, 10083, 10103, 10183, 10298, 11213, 11214,
        11308, 11496, 11547, 11556, 11557, 11601, 11684, 11687, 11883, 12033, 12179,
        12252, 12284, 12418, 12511, 12538, 12566, 12609, 12707, 12910, 12912, 12972,
        13007, 13055, 13404, 13424, 13425, 13426, 13504, 13658, 13825, 14031, 14095,
        14110, 14111, 14195, 14408, 14455, 14603

Salinity and chloroplasts 6940, 7026, 7401, 8023, 9692, 10183, 11147, 11531, 11755,
        11876, 11905, 12446, 12538, 12758, 12942, 13007, 13349, 13504

Salinity and conductance for water vapour and carbon dioxide transfer 6889, 7026,
        7113, 7266, 7544, 7704, 7996, 8130, 8449, 8673, 8810, 8811, 9604, 10364, 1058
        10730, 10812, 10968, 11067, 11099, 11747, 11768, 11961, 12193, 12446, 12538,
        12557, 12805, 13005, 13006, 13171, 13252, 13582, 13717, 13718, 13744, 13760,
        13825, 14031, 14123, 14246, 14319, 14362, 14674

Soil moisture and carbon dioxide influx (continued) 13273, 13303, 13311, 13406,
        13407, 13519, 13560, 13605, 13641, 13711, 13731, 13940, 14016, 14043, 14086,
        14124, 14244, 14332, 14340, 14350, 14447, 14454, 14516, 14542, 14592

Soil moisture and carbon fixation pathways 6732, 9642, 10142, 10806, 11001, 11084,
        11643, 11963, 12405, 12406

Soil moisture and carotenoids 7830, 8053, 10155, 11799, 14516, 14595

Soil moisture and chlorophyll 6686, 6732, 7146, 7264, 7517, 7733, 7750, 7830, 9381,
        10079, 10317, 10715, 10806, 11083, 11725, 11799, 12202, 12677, 12781, 13238,
        13489, 13522, 13733, 14095, 14516, 14542, 14595, 14636

Soil moisture and chloroplasts 7751, 10877, 12456, 12678, 12725, 13028, 14569

Soil moisture and conductance for water vapour and carbon dioxide transfer 6726,
        7002, 7104, 7130, 7170, 7442, 7449, 7544, 7621, 7626, 7739, 7996, 8160, 8197,
        8198, 8221, 8378, 8383, 8406, 8411, 8413, 8457, 8515, 8549, 8719, 8744, 8751,
        8768, 8831, 8843, 8999, 9001, 9074, 9129, 9141, 9321, 9329, 9397, 9422, 9506,
        9528, 9529, 9536, 9659, 9741, 9830, 9870, 9902, 10061, 10076, 10120, 10339,
        10363, 10388, 10404, 10447, 10500, 10501, 10589, 10756, 10841, 10883, 10985,
        11026, 11083, 11222, 11249, 11264, 11429, 11518, 11548, 11722, 11847, 11855,
        11856, 11858, 11923, 11925, 12019, 12021, 12152, 12191, 12226, 12246, 12273,
        12405, 12516, 12572, 12696, 12699, 12701, 12717, 12725, 12751, 12955, 13089,
        13121, 13169, 13230, 13238, 13252, 13284, 13395, 13560, 13605, 13626, 13794,
        13977, 14020, 14043, 14086, 14093, 14177, 14286, 14295, 14332, 14362, 14447,
        14479, 14508, 14542

Soil moisture and electron transport chain 7750, 7860, 8031, 12456, 14762

Soil moisture and growth, productivity 6608, 6614, 6625, 6631, 6642, 6653, 6659,
        6663, 6690, 6700, 6704, 6722, 6726, 6732, 6742, 6749, 6750, 6755, 6766, 6770,
        6777, 6807, 6816, 6824, 6829, 6834, 6847, 6851, 6860, 6862, 6883, 6945, 6955,
        6956, 6973, 6982, 6992, 7000, 7009, 7014, 7018, 7032, 7041, 7043, 7045, 7056,
        7064, 7068, 7069, 7083, 7084, 7091, 7098, 7102, 7109, 7112, 7117, 7134, 7135,
        7136, 7137, 7146, 7154, 7165, 7178, 7181, 7196, 7203, 7221, 7223, 7235, 7236,
        7238, 7239, 7251, 7264, 7265, 7280, 7312, 7313, 7318, 7334, 7345, 7348, 7353,
        7356, 7359, 7373, 7386, 7390, 7400, 7410, 7412, 7430, 7442, 7449, 7450, 7452,
        7457, 7471, 7472, 7480, 7493, 7496, 7498, 7503, 7508, 7532, 7534, 7539, 7542,
        7568, 7575, 7580, 7598, 7621, 7628, 7633, 7638, 7644, 7668, 7696, 7706, 7708,
        7709, 7727, 7730, 7765, 7766, 7799, 7801, 7804, 7820, 7833, 7846, 7849, 7851,
        7852, 7853, 7863, 7884, 7886, 7887, 7896, 7913, 7915, 7925, 7930, 7942, 7949,
        7951, 7959, 7960, 7962, 7969, 7970, 7971, 7975, 7985, 7994, 7996, 8013, 8019,
        8029, 8030, 8036, 8041, 8042, 8045, 8046, 8053, 8058, 8074, 8084, 8085, 8103,
        8110, 8111, 8118, 8119, 8143, 8146, 8147, 8151, 8154, 8181, 8188, 8192, 8195,
        8202, 8206, 8222, 8278, 8287, 8293, 8314, 8355, 8363, 8367, 8378, 8395, 8399,
        8402, 8408, 8409, 8411, 8428, 8432, 8435, 8439, 8440, 8446, 8499, 8511, 8548,
        8549, 8551, 8556, 8560, 8565, 8583, 8594, 8607, 8616, 8617, 8618, 8653, 8671,
        8685, 8704, 8796, 8828, 8834, 8843, 8864, 8870, 8885, 8893, 8900, 8903, 8908,
        8917, 8936, 8945, 8946, 8959, 8962, 8966, 8983, 9001, 9006, 9017, 9028, 9029,
        9056, 9073, 9074, 9131, 9133, 9134, 9138, 9140, 9145, 9147, 9155, 9170,
        9171, 9183, 9199, 9200, 9201, 9204, 9214, 9215, 9217, 9218, 9220, 9222, 9247,
        9262, 9269, 9271, 9275, 9283, 9292, 9297, 9298, 9301, 9303, 9313, 9315, 9320,
        9333, 9377, 9384, 9392, 9398, 9399, 9400, 9410, 9413, 9414, 9415, 9420, 9430,
        9432, 9433, 9453, 9467, 9482, 9483, 9488, 9506, 9507, 9519, 9524, 9529, 9545,
        9558, 9567, 9578, 9582, 9583, 9586, 9591, 9608, 9618, 9634, 9669, 9678, 9690,
        9693, 9697, 9720, 9728, 9732, 9734, 9745, 9748, 9753, 9760, 9779, 9791, 9794,
        9802, 9814, 9820, 9830, 9842, 9845, 9861, 9870, 9872, 9883, 9896, 9903, 9917,
        9941, 9950, 9955, 9958, 9984, 10017, 10024, 10026, 10040, 10044, 10053, 10058,
        10059, 10069, 10082, 10101, 10108, 10132, 10133, 10155, 10160, 10167, 10168,
        10206, 10216, 10226, 10227, 10229, 10237, 10266, 10267, 10268, 10272, 10281,
        10289, 10291, 10300, 10317, 10323, 10326, 10335, 10338, 10358, 10359, 10363,
                                                                            (continued)

                                                              (continued)

Soil moisture and other processes than above and below (continued) 13619, 13766,
        13853, 13861, 13943, 13967, 13970, 14054, 14092, 14107, 14160, 14223, 14240,
        14248, 14259, 14260, 14274, 14346, 14363, 14370, 14459, 14503, 14507, 14508,
        14549, 14558, 14571, 14622, 14632, 14691, 14729, 14746

Soil moisture and photorespiration 8719

Soil moisture and respiration 6726, 7134, 7386, 7421, 7655, 8719, 9090, 9884, 10338,
        10903, 11856, 12152, 12918, 13761, 14332, 14340, 14454, 14516, 14542

Soil moisture and stomata 6912, 7259, 7360, 7449, 7864, 9902, 9919, 10498, 10501,
        11018, 12021, 12191, 13147, 13238, 13940, 14011

Soil moisture and transpiration 6621, 6642, 6643, 6726, 6808, 6836, 6956, 7014, 7170,
        7230, 7247, 7360, 7377, 7386, 7399, 7410, 7434, 7435, 7457, 7471, 7598, 7600,
        7634, 7665, 7723, 7743, 7865, 7911, 7932, 7948, 7992, 7996, 8012, 8013, 8160,
        8198, 8207, 8237, 8252, 8263, 8267, 8352, 8383, 8390, 8625, 8629, 8672, 8751,
        8799, 8857, 8959, 8981, 8999, 9001, 9044, 9066, 9090, 9129, 9139, 9152, 9278,
        9283, 9326, 9364, 9397, 9483, 9506, 9507, 9516, 9536, 9615, 9642, 9648, 9654,
        9696, 9697, 9779, 9818, 9824, 9920, 9984, 10019, 10067, 10107, 10124, 10240,
        10319, 10334, 10337, 10339, 10342, 10388, 10404, 10447, 10498, 10500, 10542,
        10543, 10561, 10563, 10645, 10751, 10773, 10883, 10985, 11072, 11076, 11191,
        11195, 11235, 11264, 11279, 11288, 11325, 11342, 11444, 11541, 11606, 11658,
        11661, 11662, 11722, 11731, 11847, 11855, 11901, 12019, 12021, 12085, 12108,
        12152, 12181, 12191, 12377, 12405, 12409, 12412, 12414, 12423, 12436, 12438,
        12470, 12471, 12474, 12475, 12520, 12577, 12663, 12696, 12699, 12725, 12828,
        12876, 12955, 12978, 12979, 12981, 13030, 13043, 13047, 13077, 13078, 13097,
        13155, 13196, 13225, 13238, 13263, 13275, 13284, 13303, 13395, 13406, 13444,
        13490, 13557, 13560, 13566, 13581, 13605, 13647, 13707, 13711, 13727, 13756,
        13856, 13928, 13940, 13947, 13964, 13971, 13972, 13989, 14020, 14070, 14124,
        14238, 14244, 14268, 14281, 14293, 14296, 14332, 14337, 14362, 14396, 14404,
        14479, 14508, 14723, 14765

Soil moisture and water absorption by plant 7928, 8218, 8392, 8417, 9262, 9654,
        9735, 10088, 10184, 10206, 11402, 11403, 11700, 11802, 12220, 13566, 13724,
        13916, 13995, 14069, 14107, 14385, 14509

Soil moisture and water status in plant 6649, 6705, 6712, 6726, 6732, 6770, 6836,
        6900, 7002, 7076, 7183, 7247, 7313, 7348, 7350, 7360, 7399, 7457, 7495, 7503,
        7579, 7588, 7626, 7665, 7739, 7857, 7905, 7990, 7996, 8160, 8280, 8304, 8406,
        8444, 8457, 8549, 8629, 8658, 8672, 8701, 8709, 8714, 8715, 8857, 8874, 8888,
        8962, 8999, 9001, 9126, 9141, 9170, 9195, 9212, 9283, 9300, 9321, 9326, 9422,
        9519, 9528, 9529, 9555, 9586, 9601, 9688, 9741, 9745, 9825, 9845, 10061, 10067,
        10079, 10112, 10124, 10126, 10215, 10219, 10239, 10244, 10332, 10363, 10388,
        10476, 10500, 10501, 10502, 10516, 10524, 10563, 10692, 10738, 10756, 10805,
        10843, 10871, 10883, 10985, 11083, 11107, 11191, 11222, 11254, 11263, 11268,
        11428, 11429, 11472, 11548, 11606, 11656, 11685, 11722, 11745, 11812, 11813,
        11833, 11847, 11855, 11858, 11918, 11923, 11995, 12019, 12073, 12074, 12085,
        12090, 12105, 12191, 12202, 12226, 12234, 12246, 12273, 12295, 12440, 12516,
        12572, 12585, 12606, 12619, 12633, 12640, 12663, 12696, 12698, 12699, 12722,
        12725, 12808, 12828, 12867, 12915, 12955, 13038, 13089, 13155, 13169, 13223,
        13279, 13284, 13338, 13395, 13437, 13451, 13519, 13555, 13607, 13626, 13693,
        13700, 13707, 13710, 13727, 13768, 13794, 13897, 13901, 13915, 13396, 13998,
        14020, 14069, 14075, 14093, 14143, 14177, 14180, 14286, 14288, 14293, 14295,
        14314, 14362, 14370, 14375, 14376, 14414, 14418, 14451, 14507, 14508, 14568,
        14580, 14628, 14636, 14637, 14751, 14756

Soil moisture and water transport in plant 6836, 7740, 8267, 9326, 100010, 19883,
        11107, 11191, 11252, 11472, 12699, 12807, 13129, 13435, 13607, 13700

Soil moisture and wilting 6935, 6952, 7236, 7918, 7996, 8406, 8701, 8751, 8874, 8893,
        9141, 9262, 9336, 10003, 10067, 10883, 11675, 11843, 12021, 13489, 13583,
        13707, 14086

Soil moisture control, methods 6692, 6789, 9276, 9333, 9431, 13723, 14356

Soil moisture, methods 6692, 6735, 6814, 6837, 6902, 7122, 7202, 7211, 7372, 7589,
    7707, 7731, 7773, 7774, 776, 7777, 7866, 7909, 7930, 7980, 7998, 8017, 8039,
    8234, 8581, 8785, 9062, 9290, 9444, 10264, 10400, 10564, 11659, 11701, 11806,
    11808, 11953, 11959, 12214, 12658, 12713, 12787, 13342, 13413, 13449, 13482,
    13513, 13517, 13775, 13643, 13645, 13675, 13687, 13775, 13824, 13981, 14278,
    14279, 14294, 14373

Soil water potential see Water potential in substrate

Solar radiation see Irradiance

Stomata see Age of plant ...; Altitude, pressure ...; Anaerobic atmosphere ...;
    Anatomical structure ...; Antibiotics ...; Carbon dioxide ...; Conductance
    for water vapour and carbon dioxide transfer ...; Cultivars ,,.; Defoliation
    decapitation, ear, root removal ...; Drought ...; Ecotypes, geographical
    types ...; Enzyme inhibitors ...; Enzymes ...; Farming practices ...;
    Flooding ...; Gases organic ...; Genetics ...; Growth substances, hormones,
    inhibitors etc. ...; Humidity of air ...; Irradiance ...; Irrigation ...;
    Leaf insertion level ...; Mineral elements ...; Mutagens, other organic
    substances ...; Mutants ...; Osmotically active substances ...; Oxygen ...;
    Pathogens ...; Pesticides, herbicides ...; pH ...; Photoperiod ...; Pollu-
    tants, ozone ...; Precipitaion, dew ...; Proteins, amino acids, nucleic
    acids ...; Root, underground part ...; Saccharides ...; Salinity ...; Soil
    moisture ...; Taxons ...; Temperature ...; Water status in plant ...; Wind
    and stomata

Stomata and conductance for water vapour and carbon dioxide transfer 6877, 6907,
    6912, 7092, 7093, 7210, 7335, 7397, 7468, 7476, 7619, 7684, 7721, 7918, 7939,
    8000, 8066, 8158, 8413, 8503, 8557, 8719, 8744, 8745, 8929, 8952, 9543, 9548,
    9613, 9650, 10087, 11334, 11717, 12065, 12815, 13067, 13616, 13852, 14319,
    14718

Stomata and epidermis, heterogeneity of single leaf blade 6959, 7405, 7522, 7662,
    8948, 9016, 9809, 9897, 10140, 11018, 11176, 11424, 11494, 11975, 12249,
    12250, 12305, 13312, 13660, 13769, 14007, 14275

Stomata and photosynthetic rate 6623, 6717, 6738, 6756, 6848, 6998, 7031, 7070, 7104,
    7228, 7343, 7382, 7491, 7492, 7561, 7576, 7577, 7617, 7794, 7876, 7955, 7957,
    7996, 8011, 8015, 8224, 8238, 8239, 8261, 8306, 8470, 8515, 8525, 8740, 8742,
    8744, 8745, 8857, 8861, 8891, 8894, 8913, 8953, 8991, 9041, 9134, 9360, 9551,
    9582, 9729, 9761, 9763, 9784, 9793, 9801, 9848, 9901, 9904, 9925, 9946, 10013,
    10037, 10051, 10087, 10158, 10165, 10170, 10212, 10344, 10363, 10373, 10376,
    10476, 10482, 10529, 10540, 10554, 10625, 10641, 10682, 10712, 10729, 10841,
    10866, 10886, 10968, 10969, 11033, 11082, 11185, 11332, 11379, 11384, 11414,
    11466, 11470, 11482, 11483, 11519, 11610, 11628, 11624, 11663, 11790, 11796,
    11809, 11827, 11856, 11864, 11906, 11932, 11955, 11962, 11963, 11964, 11975,
    11976, 11988, 12021, 12086, 12090, 12126, 12132, 12168, 12191, 12234, 12260,
    12356, 12528, 12529, 12572, 12696, 12699, 12730, 12763, 12834, 12835, 12849,
    12850, 12911, 12919, 13056, 13168, 13173, 13344, 13358, 13446, 13448, 13530,
    13562, 13580, 13609, 13677, 13715, 13720, 13820, 13825, 13830, 13833, 13852,
    13868, 13937, 13965, 13978, 14017, 14124, 14126, 14243, 14335, 14394, 14422,
    14436, 14539, 14552, 14687, 14736, 14765

Stomata and transpiration rate 6623, 6660, 6739, 6976, 6998, 7066, 7104, 7318, 7470,
    7558, 7561, 7574, 7617, 7753, 7877, 8000, 8015, 8171, 8239, 8590, 8591, 8729,
    8740, 8744, 8745, 8800, 8857, 8953, 8991, 8999, 9066, 9074, 9134, 9141, 9243,
    9429, 9443, 9510, 9523, 9551, 9901, 9925, 10087, 10202, 10308, 10424, 10682,
    10866, 10883, 10907, 11031, 11035, 11242, 11265, 11458, 11511, 11519, 11662,
    11663, 11756, 11760, 11790, 11818, 11864, 11927, 11954, 11955, 11975, 12021,
    12086, 12090, 12141, 12301, 12528, 12634, 12663, 12696, 12699, 12802, 12849,
(continued)

                                                                     (continued)

Stomata, mechanisms of movement (continued) 13293, 13344, 13445, 13494, 13553, 13630, 13778, 13830, 13849, 13850, 13877, 13891, 13904, 14076, 14124, 14342, 14357, 14464, 14473, 14474, 14561, 14562, 14635, 14696, 14718, 14748, 14759

Stomata on other organs than leaf 6868, 7692, 7967, 8180, 8620, 9069, 9185, 9255, 9302, 9497, 9668, 11008, 11443, 11538, 12096, 12391, 12752, 13801

Stomata, physiology of movements 6624, 7031, 7104, 7151, 7209, 7210, 7310, 7335, 7343, 7380, 7403, 7502, 7558, 7779, 7781, 7782, 7783, 7937, 8038, 8060, 8071, 8115, 8116, 8158, 8533, 8654, 8696, 8697, 8744, 8758, 8929, 8943, 8947, 8952, 8953, 8991, 9005, 9038, 9049, 9128, 9134, 9294, 9312, 9348, 9372, 9513, 9534, 9543, 9640, 9810, 9868, 9904, 9951, 10037, 10042, 10048, 10189, 10235, 10274, 10424, 10481, 10493, 10631, 10653, 10668, 10731, 10828, 10835, 10867, 11012, 11057, 11129, 11130, 11181, 11182, 11243, 11398, 11410, 11476, 11652, 11660, 11691, 11717, 11790, 11975, 12056, 12057, 12064, 12065, 12078, 12086, 12168, 12185, 12218, 12268, 12269, 12310, 12385, 12386, 12655, 12696, 12763, 12833, 12850, 12898, 12899, 13048, 13059, 13195, 13276, 13277, 13288, 13293, 13344, 13553, 13778, 13817, 13830, 13839, 13849, 13850, 13877, 13891, 14005, 14077, 14124, 14393, 14473, 14474, 14562, 14563, 14635, 14696, 14718, 14759.

Stomata, role of subsidiary cells in stomatal movement 6733, 7813, 8696, 11476, 11660, 12078, 12086, 12269, 12383, 12630, 12696, 13293, 13878, 14464, 14635, 14718, 14748

Stomata size 6618, 6673, 6691, 6720, 6737, 6802, 6806, 6855, 6947, 7027, 7031, 7189, 7232, 7256, 7275, 7304, 7335, 7338, 7404, 7405, 7422, 7524, 7562, 7563, 7585, 7648, 7692, 7697, 7753, 7813, 7814, 7879, 7918, 7944, 8066, 8067, 8160, 8187, 8254, 8323, 8451, 8463, 8467, 8645, 8791, 8797, 8855, 8860, 9016, 9023, 9069, 9074, 9219, 9262, 9329, 9374, 9422, 9503, 9543, 9575, 9577, 9609, 9626, 9660, 9712, 9719, 9863, 9864, 9866, 9897, 9902, 10100, 10159, 10212, 10608, 10686, 10756, 10798, 10878, 10907, 10986, 10992, 11018, 11201, 11291, 11336, 11368, 11371, 11443, 11466, 11494, 11550, 11573, 11580, 11622, 11705, 11798, 11866, 11943, 12004, 12086, 12096, 12109, 12199, 12251, 12363, 12389, 12537, 12554, 12663, 12686, 12710, 12717, 12771, 13047, 13056, 13067, 13141, 13181, 13239, 13312, 13408, 13431, 13442, 13562, 13570, 13630, 13715, 13749, 13774, 13912, 14011, 14104, 14154, 14253, 14277, 14323, 14419, 14483, 14536, 14718

Stomatal aperture (see also Growth substances, hormones, inhibitors etc. and stomata) 6877, 6907, 6912, 7055, 7092, 7093, 7210, 7335. 7346, 7397, 7468, 7476, 7557, 7619, 7659, 7684, 7721, 7918, 7937, 8000, 8066, 8158, 8413, 8503, 8557, 8719, 8745, 8929, 8952, 9543, 9548, 9613, 9650, 10087, 11398, 12040, 12086, 12311

Stomatal aperture, diurnal course 6848, 6889, 6912, 6974, 70092, 7093, 7259, 7297, 7397, 7398, 7581, 7619, 7684, 7885, 7939, 7951, 7996, 8127, 8729, 8790, 8899, 8950, 8951, 9009, 9134, 9597, 9629, 9806, 9919, 10121, 10152, 10363, 10498, 10658, 10731, 10828, 10966, 11003, 11796, 12437, 12745, 13057, 13147, 13520, 13694, 13940, 14005, 14292, 14319, 14718

Stomatal aperture, methods diffusion porometers 6716, 6974, 6996, 7156, 7293, 7595, 8231, 8233, 8340, 8460, 8603, 8662, 8740, 8744, 8762, 8801, 8990, 9444, 9505, 10151, 10428, 10842, 11346, 11420, 11665, 11846, 11855, 11894, 12109, 12499, 12755, 12872, 13221, 13471, 13823

Stomatal aperture, methods, direct observations 8696, 10428, 11855, 11975, 12251, 12872, 14053

Stomatal aperture, methods, epidermal strips 6946, 7218, 7362, 7782, 8161, 8405, 8533, 8744, 8929, 8990, 9312, 9613, 9626, 9904, 9983, 10397, 11340, 11798, 11855, 12208, 12310, 12630, 12898, 14718

Stomatal aperture, methods, infiltration 7093, 7156, 8744, 8990, 9797, 10428, 11855, 12745

T

Taxons and transpiration 6643, 6648, 6739, 7088, 7283, 7311, 7593, 7945, 8999, 9001,
    9395, 9516, 9708, 10067, 10072, 10105, 10159, 10387, 10447, 10478, 10846,
    11283, 11370, 11530, 11891, 12166, 12203, 12270, 12301, 12409, 12504, 12547,
    12895, 12946, 13005, 13098, 13107, 13119, 13229, 13252, 13258, 13314, 13433,
    13654, 14026, 14030, 14078, 14166, 14168, 14201, 14349, 14354, 14394, 14422,
    14508, 14600, 14679

Taxons and water absorption by plant 7378, 11236, 12347, 12547

Taxons and water absorption by plant 6648, 6739, 6893, 7076, 7184, 7712, 7793, 7840,
    7945, 7997, 8140, 8422, 8850, 8999, 9123, 9322, 9617, 10067, 10370, 10392,
    10438, 10478, 10520, 10541, 10559, 11067, 11236, 11263, 11283, 11686, 11771,
    12155, 12203, 12277, 12301, 12492, 12547, 12675, 12801, 12808, 12857, 12895,
    12943, 12946, 13025, 13119, 13120, 13167, 13178, 13215, 13229, 13252, 13257,
    13314, 13373, 13433, 13462, 13592, 13595, 13626, 13771, 13796, 13828, 13830,
    13939, 14014, 14089, 14098, 14116, 14143, 14166, 14167, 14168, 14177, 14201,
    14339, 14354, 14376, 14394, 14433, 14434, 14435, 14460, 14508, 14568, 14597

Taxons and water transport in plant 11686, 12956, 13229, 13984, 14304

Taxons and wilting 6648, 6660, 8681, 10846, 11771, 12547, 12857

Temperature and conductance for water vapour and carbon dioxide transfer 6620, 6641,
    6647, 6658, 6717, 6726, 6842, 6934, 6998, 7008, 7092, 7104, 7170, 7209, 7269,
    7361, 7351, 7371, 7419, 7490, 7501, 7502, 7526, 7544, 7556, 7572, 7578, 7645,
    7739, 7955, 7996, 8037, 8127, 8130, 8252, 8260, 8307, 8417, 8460, 8509, 8576,
    8677, 8744, 8745, 8787, 8799, 8801, 8810, 8811, 8860, 8894, 8895, 8913, 8929,
    8960, 9007, 9032, 9064, 9085, 9166, 9460, 9494, 9514, 9625, 9628, 9703, 9724,
    9849, 9960, 9991, 10000, 10005, 10051, 10087, 10117, 10153, 10202, 10314,
    10373, 10376, 10381, 10401, 10413, 10479, 10510, 10589, 10590, 10718, 10729,
    10775, 10779, 10818, 10841, 10866, 10883, 10889, 11003, 11026, 11028, 11035,
    11118, 11152, 11239, 11314, 11429, 11517, 11574, 11575, 11610, 11809, 11855,
    11943, 11992, 12034, 12037, 12086, 12159, 12218, 12349, 12528, 12623, 12674,
    12705, 12719, 12825, 12844, 12986, 13101, 13173, 13175, 13178, 13202, 13288,
    13344, 13530, 13720, 13721, 13765, 13826, 13885, 14017, 14018, 14118, 14246,
    14273, 14288, 14359, 14447, 14460, 14463, 14467, 14557

Temperature and stomata 6717, 6792, 6998, 7092, 7093, 7209, 7360, 7545, 7645, 7721,
    7809, 7939, 7946, 7996, 8127, 8699, 8744, 8758, 8929, 9285, 9588, 9865, 9914,
    9919, 10005, 10087, 10147, 10424, 10478, 10554, 10614, 10668, 10718, 10743,
    10775, 10782, 10788, 10866, 11035, 11574, 12034, 12122, 12392, 12630, 12853,
    12898, 13530, 13877, 13945, 14113, 14473, 14474, 14483, 14718

Temperature and transpiration 6641, 6712, 6726, 6825, 6887, 6956, 6998, 7008, 7021,
    7105, 7158, 7170, 7192, 7198, 7349, 7351, 7360, 7371, 7389, 7480, 7526, 7544,
    7545, 7567, 7578, 7620, 7649, 7696, 7710, 7877, 7918, 7932, 7996, 8051, 8056,
    8171, 8252, 8270, 8315, 8393, 8485, 8576, 8591, 8625, 8764, 8801, 8836, 8909,
    8935, 8960, 9033, 9044, 9122, 9290, 9395, 9429, 9514, 9625, 9696, 9724, 9856,
    9865, 9920, 10019, 10020, 10087, 10106, 10107, 10124, 10159, 10185, 10205,
    10241, 10308, 10319, 10376, 10401, 10403, 10413, 10420, 10458, 10478, 10590,
    10773, 10818, 10890, 10917, 10960, 11005, 11028, 11035, 11139, 11239, 11271,
    11324, 11616, 11712, 11749, 11948, 12034, 12036, 12201, 12261, 12315, 12317,
    12528, 12561, 12696, 12828, 12844, 12869, 13016, 13049, 13126, 13174, 13185,
    13207, 13384, 13432, 13677, 13721, 13759, 13765, 13826, 13855, 13916, 14152,
    14244, 14273, 14359, 14391, 14402, 14410, 14592, 14646, 14745, 14748

Temperature and water absorption by plant 6706, 6872, 6887, 7349, 7594, 8691, 7874,
    8881, 8910, 9052, 9553, 9636, 9683, 9951, 10550, 10596, 11035, 11453, 11722,
    11787, 12063, 12392, 12699, 12739, 12784, 12853, 13113, 13147, 13461, 13964,
    14008, 14079

                                                            (continued)

Transpiration coefficient (continued) 11478, 11518, 11519, 11541, 11544, 11555,
    11584, 11597, 11606, 11618, 11630, 11644, 11648, 11650, 11666, 11702, 11703,
    11714, 11717, 11719, 11730, 11731, 11754, 11756, 11775, 11776, 11785, 11819,
    11845, 11864, 11901, 11914, 11920, 11929, 12030, 12053, 12085, 12140, 12144,
    12174, 12175, 12187, 12272, 12296, 12347, 12348, 12364, 12408, 12412, 12413,
    12420, 12429, 12435, 12436, 12471, 12496, 12570, 12577, 12590, 12594, 12595,
    12597, 12623, 12640, 12654, 12691, 12696, 12699, 12751, 12753, 12761, 12762,
    12819, 12867, 12873, 12901, 12920, 12928, 12951, 12969, 12979, 12985, 12993,
    13010, 13014, 13043, 13045, 13046, 13047, 13067, 13097, 13157, 13158, 13185,
    13196, 13225, 13324, 13343, 13347, 13366, 13374, 13384, 13385, 13414, 13446,
    13447, 13456, 13459, 13495, 13500, 13581, 13601, 13602, 13603, 13605, 13615,
    13640, 13642, 13643, 13652, 13758, 13768, 13777, 13860, 13875, 13880, 13881,
    13938, 13947, 13967, 13972, 13974, 13975, 14026, 14067, 14114, 14121, 14124,
    14125, 14126, 14129, 14132, 14143, 14169, 14172, 14208, 14232, 14236, 14268,
    14269, 14287, 14295, 14296, 14323, 14332, 14377, 14378, 14388, 14404, 14405,
    14421, 14436, 14442, 14444, 14479, 14487, 14495, 14503, 14512, 14522, 14524,
    14545, 14546, 14547, 14549, 14622, 14623, 14634, 14639, 14669, 14673, 14722

Transpiration curves 6660, 6891, 6924, 7399, 7574, 7009, 7618, 7657, 8040, 8061,
    8772, 9134, 9888, 10156, 10159, 10179, 10404, 10600, 10724, 10907, 11035,
    11081, 12178, 12725, 12960, 13239, 13728, 13941, 14285, 14350, 14537, 14568,
    14685, 14756

Transpiration cuticular 6793, 7399, 7525, 7557, 7657, 8040, 8516, 9241, 9243, 9244,
    9254, 9365, 10180, 10404, 11163, 11550, 11720, 11891, 13239, 13278, 13655,
    14393

Transpiration integrated 6932, 6956, 7049, 7138, 7228, 7620, 7936, 8013, 8014, 8043,
    8051, 8693, 8729, 8840, 8944, 9074, 9262, 9337, 9516, 9583, 9654, 9732, 10039,
    10106, 10229, 10630, 10751, 10778, 10878, 10937, 10984, 11239, 11292, 11301,
    11350, 11511, 11550, 11644, 11662, 11683, 11722, 11792, 11891, 12034, 12072,
    12315, 12317, 12520, 12569, 12665, 12671, 12699, 12802, 12813, 12869, 12901,
    12928, 12946, 13119, 13196, 13347, 13488, 13557, 13581, 13654, 13692, 13739,
    13841, 13861, 14070, 14124, 14271, 14309, 14362, 14405, 14595, 14624, 14681,
    14741

Transpiration rate and antitranspirants see Antitranspirants ...

Transpiration rate and fine structure 7571, 14617

Transpiration rate and leaf temperature 7343, 8591, 9033, 9046, 9268, 10087, 11342,
    11721, 11814, 11855, 12699, 12952, 13721

Transpiration rate and stomata see Stomata and transpiration rate

Transpiration rate, comparison of plants with different types of carbon metabolism
    7389, 8168, 10629, 11002, 11291, 12166, 12228, 12920, 13947, 14020, 14121,
    14436, 14714

Transpiration rate, diurnal changes 6647, 6658, 6712, 6726, 6738, 7002, 7047, 7088,
    7089, 7224, 7259, 7302, 7351, 7352, 7424, 7425, 7457, 7474, 7526, 7545, 7565,
    7620, 7634, 7649, 7677, 7683, 7711, 7723, 7771, 7772, 7911, 7932, 7936, 7978,
    7979, 8028, 8051, 8079, 8080, 8198, 8207, 8216m 8218, 8219, 8238, 8252, 8262,
    8330, 8359, 8435, 8538, 8682, 8688, 8709, 8729, 8731, 8740, 8743, 8764, 8790,
    8801, 8838, 8841, 8857, 8860, 8899, 8913, 8919, 8921, 8925, 8965, 8999, 9009,
    9035, 9036, 9043, 9044, 9090, 9152, 9253, 9439, 9452, 9514, 9515, 9628, 9642,
    9648, 9654, 9660, 9703, 9740, 9806, 9841, 9876, 9901, 9919, 9920, 9954, 9960,
    10004, 10015, 10030, 10039, 10043, 10061, 10087, 10090, 10122, 10126, 10159,
    10185, 10186, 10240, 10288, 10309, 10319, 10387, 10404, 10408, 10420, 10424,
    10425, 10467, 10498, 10518, 10561, 10563, 10590, 10629, 10634, 10665, 10673,
    10686, 10724, 10750, 10751, 10773, 10778, 10817, 10841, 10842, 10884, 10886,
    10960, 11002, 11003, 11031, 11037, 11113, 11138, 11141, 11172, 11191, 11223,
                                                                        (continued)

                                                                          (continued)

Transpiration rate in natural conditions (continued) 9013, 9035, 9044, 9066, 9072,
        9082, 9090, 9166, 9195, 9216, 9243, 9267, 9283, 9297, 9327, 9328, 9440, 9483,
        9491, 9499, 9501, 9512, 9514, 9515, 9579, 9583, 9627, 9634, 9644, 9654, 9671,
        9703, 9706, 9708, 9710, 9724, 9736, 9740, 9748, 9755, 9778, 9841, 9855, 9876,
        9901, 9914, 9932, 9937, 9984, 10010, 10014, 10034, 10046, 10061, 10090, 10110,
        10124, 10159, 10186, 10240, 10242, 10288, 10339, 10387, 10425, 10443, 10448,
        10498, 10591, 10600, 10629, 10634, 10636, 10638, 10645, 10655, 10724, 10728,
        10778, 10786, 10791, 10821, 10833, 10886, 10905, 10943, 10944, 10985, 11003,
        11031, 11037, 11100, 11102, 11138, 11155, 11172, 11191, 11223, 11252, 11258,
        11279, 11363, 11364, 11365, 11366, 11389, 11408, 11414, 11419, 11444, 11458,
        11478, 11492, 11511, 11512, 11514, 11549, 11550, 11584, 11588, 11616, 11652,
        11662, 11686, 11691, 11730, 11745, 11754, 11832, 11861, 11892, 11893, 11943,
        11948, 11960, 11962, 11972, 11988, 12004, 12022, 12072, 12120, 12126, 12141,
        12201, 12203, 12224, 12228, 12244, 12292, 12309, 12314, 12353, 12357, 12363,
        12438, 12462, 12471, 12475, 12499, 12520, 12523, 12537, 12547, 12565, 12574,
        12577, 12665, 12696, 12699, 12725, 12751, 12755, 12813, 13869, 12892, 12908,
        12911, 12913, 12928, 12940, 12946, 12954, 12958, 12970, 12979, 12992, 12995,
        13010, 13030, 13049, 13087, 13185, 13241, 13263, 13275, 13284, 13295, 13325,
        13384, 13415, 13433, 13446, 13488, 13497, 13560, 13562, 13566, 13570, 13580,
        13632, 13650, 13661, 13674, 13692, 13693, 13699, 13734, 13744, 13747, 13748,
        13759, 13784, 13786, 13789, 13795, 13827, 13830, 13857, 13929, 13940, 13964,
        13985, 14004, 14005, 14020, 14037, 14048, 14067, 14074, 14119, 14134, 14148,
        14166, 14168, 14189, 14201, 14243, 14275, 14281, 14287, 14288, 14315, 14323,
        14331, 14343, 14354, 14387, 14391, 14394, 14410, 14412, 14419, 14421, 14497,
        14507, 14508, 14509, 14539, 14569, 14592, 14623, 14630, 14646, 14681, 14687,
        14699, 14702, 14705, 14714, 14715, 14765

Transpiration rate, methods, gasometric systems, conditioning of air 8270, 10587,
        10811, 10842, 13653, 13677

Transpiration rate, methods, gasometric systems, generally 8270, 9569, 9858, 10151,
        10811, 11592, 11855, 12014, 13688

Transpiration rate, methods, gasometric systems, open 6623, 6685, 6729, 6818, 6886,
        7341, 7556, 7807, 8076, 8473, 8913, 8921, 8923, 9858, 9940, 10151, 10484,
        10704, 10811, 10819, 10842, 11260, 11346, 11441, 11605, 11623, 11697, 11739,
        11892, 11984, 12098, 12601, 12859, 13677, 14427

Transpiration rate, methods, gasometric systems, semiclosed and closed 7275, 8051,
        8470, 8603, 10587, 11346, 12262, 13299, 13554, 13690

Transpiration rate, methods, gravimetric 7744, 7747, 8510, 9047, 9888, 10159, 10179,
        10587

Transpiration rate, methods, infra-red gas analysers 6818, 7341, 8913, 8921, 10587,
        10811, 10842, 11260, 11605, 12859, 13299, 13320, 14038, 14133

Transpiration rate, methods, other hygrometers 6734, 8965, 10159, 10587, 11602,
        11604, 11697, 13260, 13677

Transpiration rate, oscillations 7170, 7225, 7600, 7860, 8860, 9129, 9565, 9916,
        10159, 10308, 10750, 11372, 12289, 12561, 13078, 13721, 12908, 14129

Transpiration rate, seasonal changes 6647, 6711, 6726, 2853, 7002, 7080, 7158, 7162,
        7261, 7302, 7311, 7316, 7318, 7379, 7416, 7424, 7435, 7464, 7486, 7536, 7543,
        7593, 7649, 7657, 7677, 7711, 7720, 7759, 7934, 7943, 7979, 7996, 8218, 8263,
        8311, 8330, 8390, 8456, 8493, 8494, 8562, 8629, 8659, 8765, 8825, 8851, 8857,
        8925, 9011, 9012, 9035, 9041, 9061, 9072, 9082, 9152, 9164, 9191, 9195, 9328,
        9446, 9452, 9453, 9478, 9508, 9514, 9555, 9606, 9607, 9654, 9841, 9943, 9960,
        10105, 10107, 10126, 10184, 10240, 10408, 10419, 10425, 10448, 10463, 10500,
        10543, 10561, 10563, 10665, 10724, 10869, 10913, 10942, 10960, 11100, 11127,
        11193, 11224, 11270, 11325, 11366, 11370, 11389, 11419, 11511, 11512, 11514,
        11528, 11532, 11541, 11572, 11584, 11605, 11616, 11722, 11775, 11815, 11932,
                                                                        (continued)

Transpiration rate, seasonal changes (continued) 11965, 11988, 11989, 12151, 11224, 12364, 12412, 12452, 12565, 12597, 12608, 12679, 12699, 12823, 12864, 12954, 12955, 13049, 13058, 13098, 13119, 13174, 13343, 13347, 13355, 13406, 13414, 13434, 13444, 13490, 13495, 13581, 13628, 13632, 13636, 13675, 13695, 13734, 13739, 13748, 13759, 13784, 13916, 13964, 14005, 14024, 14166, 14168, 14201, 14202, 14209, 14220, 14258, 14274, 14287, 14288, 14303, 14359, 14387, 14391, 14394, 14403, 14404, 14453, 14498, 14403, 14524, 14550, 14557, 14619, 14646, 14681, 14702, 14705

Transpiration rate, theoretical background 6726, 7114, 7115, 7474, 7497, 7515, 7649, 7650, 7681, 7720, 7747, 7749, 7764, 7791, 7801, 7880, 7894, 7932, 7979, 8203, 8204, 8262, 8315, 8316, 8494, 8517, 8591, 8630, 8638, 8731, 8732, 8740, 8857, 8891, 8925, 9033, 9036, 9047, 9048, 9061, 9064, 9075, 9089, 9167, 9192, 9288, 9438, 9439, 9446, 9452, 9565, 9627, 9644, 9893, 10087, 10090, 10185, 10241, 10263, 10277, 10408, 10638, 10665, 10778, 10883, 11017, 11141, 11277, 11278, 11287, 11342, 11372, 11393, 11477, 11593, 11749, 11830, 11855, 12014, 12038, 12120, 12140, 12357, 12505, 12515, 12608, 12634, 12696, 12699, 12803, 12826, 12828, 12913, 12952, 12963, 13127, 13148, 13247, 13405, 13446, 13449, 13490, 13495, 13721, 13739, 13786, 13928, 14952, 14214, 14387, 14416, 14417, 14497, 14505, 14572, 14658, 14733, 14748

Transpiration stomatal 7095, 7525, 7557, 7657, 8040, 9243, 9244, 9254, 10180, 10404, 11550, 12813, 14393, 14765

Transport of water see Water transport ...

Trichomes see Leaf surface, waxes and trichomes

Turgor pressure see Pressure potential ...

V

Vapour pressure deficit see Humidity of air ...

Virus diseases see Pathogens ...

W

Water absorption by leaf discs 8004, 11682

Water absorption by part of plant 6657, 6924, 7740, 7767, 7768, 7966, 8021, 9357, 9540, 10962, 11029, 11030, 11080, 11089, 11257, 11269, 11311, 11475, 11647, 11682, 12140, 12209, 12388, 12604, 12726, 12727, 12728, 13163, 14039, 14100, 14392

Water absorption by plant see Age of plant ...; Anatomical structure ...; Carbon dioxide ...; Cultivars ...; Defoliation, decapitation, ear, root removal ...; Deuterium oxide, tritium oxide ...; Drought ...; Enzyme inhibitors ...; Enzymes ...; Farming practices ...; Flooding ...; Genetics ...; Growth substances, hormones, inhibitors etc. ...; Humidity of air ...; Irradiance ...; Irrigation ...; Leaf insertion level ...; Lipids, fatty acids ...; Mineral elements ...; Mutagens, other organic substances ...; Osmotically active substances ...; Oxygen ...; Pathogens ...; Pesticides, herbicides ...; pH ...; Precipitation, dew ...; Protein, amino acids, nucleic acids ...; Root system, dimensions, volume etc. ...; Saccharides ...; Salinity ...; Soil moisture ...; Taxons ...; Temperature ...; Water status in plant ...; Wind and water absorption by plant

Water absorption by plant and ion uptake 6674, 6766, 6863, 6951, 6993, 7197, 7217, 7220, 7280, 7287, 7442, 7611, 8063, 8069, 8321, 8354, 8421, 8586, 8868, 9005, 9037, 9134, 9183, 9266, 9305, 9436, 9531, 9541, 9600, 9817, 10069, 10088,
                                                                                    (continued)

Water absorption by plant and ion uptake (continued) 10109, 10144, 10399, 10452,
    10491, 10851, 10883, 11014, 11105, 11174, 11257, 11285, 11364, 11514, 11524,
    11604, 11676, 11677, 11690, 11706, 11831, 11898, 11899, 12045, 12146, 12157,
    12282, 12298, 12519, 12575, 12608, 12648, 12699, 12739, 12770, 12991, 13983,
    13091, 13146, 13268, 13301, 13317, 13318, 13385, 13400, 13526, 13843, 13942,
    14066, 14070, 14107, 14124, 14127, 14304, 14305, 14352, 14362, 14366, 14385,
    14399, 14400, 14555, 14603

Water absorption by plant, diurnal changes 7376, 8840, 8857, 9043, 9654, 9717, 10015
    10518, 11069, 11689, 12094, 13147, 13216, 13765

Water absorption by plant, effect of anaerobic conditions 10364, 11286

Water absorption by plant, oscillations 11069

Water absorption by plant, seasonal changes 6661, 6700, 6701, 6711, 6773, 7126, 7138,
    7423, 7971, 8093, 8825, 8857, 9211, 9557, 9568, 9654, 10022, 10067, 10160,
    10240, 10406, 10596, 12163, 12455, 12468, 12469, 12556, 12645, 12728, 13090,
    13310, 13375, 13376, 13414, 13444, 13628, 13662, 13787, 13916, 13936,
    14202, 14288, 14334, 14388

Water absorption by seeds 6622, 6644, 6655, 6677, 6688, 6689, 6778, 6785, 6813, 6816,
    6861, 6880, 6898, 6906, 6971, 7142, 7143, 7147, 7152, 7200, 7294, 7337, 7340,
    7373, 7439, 7481, 7624, 7725, 7727, 7785, 7788, 7824, 7725, 7727, 7785, 7788,
    7824, 7826, 7838, 7847, 7850, 7874, 7916, 7923, 8006, 8008, 8070, 8095, 8183,
    8274, 8312, 8392, 8402, 8469, 8498, 8501, 8550, 8610, 8847, 8881, 8910, 8937,
    9070, 9076, 9120, 9121, 9138, 9203, 9314, 9363, 9390, 9391, 9553, 9554, 9643,
    9651, 9683, 9687, 9764, 9773, 9792, 9838, 9840, 9853, 9860, 9975, 10096,
    10187, 10188, 10195, 10196, 10347, 10353, 10424, 10441, 10550, 10558, 10883,
    10902, 10931, 10965, 10974, 11043, 11045, 11094, 11126, 11286, 11297, 11332,
    11402, 11403, 11412, 11453, 11474, 11566, 11570, 11571, 11578, 11698, 11754,
    11787, 11791, 11839, 11865, 12010, 12063, 12089, 12113, 12223, 12308, 12407,
    12466, 12483, 12500, 12526, 12537, 12540, 12542, 12592, 12665, 12696, 12784,
    12813, 12822, 12870, 12871, 13110, 13161, 13177, 13182, 13183, 13251, 13255,
    13390, 13402, 13417, 13461, 13487, 13491, 13518, 13527, 13551, 13573, 13708,
    13714, 13716, 13869, 13924, 13947, 13976, 13993, 14079, 14100, 14212, 14267,
    14298, 14345, 14348, 14429, 14430, 14480, 14527, 14581, 14645

Water absorption from atmosphere 6706, 7020, 7042, 7246, 7487, 7641, 7656, 7679,
    7748, 8339, 8672, 9060, 9876, 10796, 10961, 11220, 11461, 11700, 11820, 12734,
    13716, 14307,

Water absorption from soil 6610, 6695, 6766, 6836, 6913, 7020, 7021, 7041, 7066,
    7162, 7166, 7183, 7236, 7242, 7247, 7275, 7340, 7376, 7424, 7433, 7457, 7475,
    7479, 7641, 7845, 7881, 7947, 7948, 7964, 7996, 8044, 8143, 8267, 8292, 8339,
    8390, 8416, 8417, 8472, 8527, 8569, 8630, 8651, 8653, 8677, 8727, 8732, 8743,
    8857, 8925, 9037, 9123, 9140, 9266, 9297, 9396, 9509, 9541, 9582, 9627, 9654,
    9735, 9914, 10022, 10206, 10229, 10285, 10294, 10299, 10352, 10393, 10404,
    10424, 10491, 10500, 10613, 10671, 10814, 10943, 10944, 11029, 11030, 11035,
    11072, 11102, 11113, 11137, 11165, 11226, 11236, 11264, 11325, 11333, 11364,
    11387, 11433, 11444, 11472, 11514, 11602, 11603, 11604, 11630, 11700, 11701,
    11722, 11767, 11777, 11795, 11845, 11901, 11914, 11919, 11959, 12091, 12108,
    12130, 12141, 12157, 12163, 12195, 12296, 12347, 12357, 12392, 12460, 12468,
    12469, 12518, 12519, 12530, 12547, 12608, 12695, 12696, 12699, 12729, 12734,
    12753, 12764, 12768, 12802, 12853, 12876, 12985, 13090, 13096, 13125, 13129,
    13147, 13187, 13212, 13216, 13325, 13375, 13376, 13449, 13457, 13459, 13550,
    13581, 13593, 13614, 13631, 13638, 13661, 13675, 13693, 13724, 13739, 13765,
    13786, 13793, 13796, 13830, 13907, 13913, 13916, 13954, 13995, 14008, 14066,
    14069, 14143, 14202, 14203, 14208, 14288, 14334, 14337, 14366, 14376, 14385,
    14387, 14509, 14639, 14673, 14674

                                                                          (continued)

Water content in plant tissues (continued) 7322, 7325, 7331, 7332, 7340, 7341, 7347,
        7358, 7399, 7406, 7438, 7444, 7446, 7450, 7478, 7505, 7507, 7525, 7536, 7597,
        7603, 7622, 7628, 7630, 7632, 7647, 7674, 7724, 7725, 7726, 7738, 7741, 7745,
        7747, 7760, 7826, 7835, 7848, 7861, 7862, 7868, 7869, 7922, 7935, 7937, 7953,
        7973, 7982, 7989, 7996, 8003, 8035, 8086, 8094, 8095, 8096, 8144, 8169, 8174,
        8184, 8200, 8209, 8213, 8220, 8253, 8289, 8298, 8310, 8328, 8353, 8362, 8364,
        8377, 8379, 8382, 8389, 8391, 8412, 8444, 8475, 8478, 8481, 8496, 8535, 8547,
        8561, 8573, 8590, 8591, 8596, 8597, 8598, 8609, 8630, 8634, 8646, 8648, 8649,
        8663, 8672, 8690, 8693, 8699, 8707, 8708, 8730, 8739, 8741, 8778, 8784, 8806,
        8817, 8822, 8871, 8889, 8890, 8911, 8915, 8920, 8924, 8934, 8937, 8987, 8988,
        9035, 9040, 9068, 9088, 9156, 9159, 9173, 9177, 9180, 9189, 9194, 9195, 9207,
        9209, 9213, 9227, 9287, 9308, 9314, 9322, 9336, 9353, 9363, 9381, 9409, 9411,
        9412, 9423, 9424, 9425, 9426, 9454, 9477, 9490, 9527, 9530, 9532, 9535, 9560,
        9561, 9586, 9587, 9588, 9602, 9631, 9643, 9649, 9650, 9666, 9680, 9682, 9709,
        9727, 9731, 9742, 9744, 9765, 9785, 9798, 9817, 9834, 9838, 9840, 9845, 9860,
        9862, 9865, 9876, 9880, 9886, 9887, 9888, 9890, 9894, 9895, 9919, 9923, 9929,
        9987, 9993, 9994, 9999, 10000, 10006, 10016, 10018, 10019, 10021, 10029,
        10030, 10041, 10060, 10079, 10089, 10123, 10131, 10141, 10179, 10199, 10213,
        10221, 10265, 10271, 10292, 10321, 10322, 10330, 10345, 10348, 10365, 10366,
        10370, 10375, 10381, 10396, 10410, 10421, 10428, 10438, 10443, 10447, 10491,
        10497, 10522, 10523, 10532, 10538, 10539, 10541, 10559, 10567, 10600, 10612,
        10620, 10655, 10670, 10689, 10724, 10790, 10792, 10796, 10803, 10817, 10827,
        10831, 10849, 10850, 10857, 10859, 10881, 10883, 10897, 10908, 10922, 10971,
        10987, 10988, 11010, 11021, 11025, 11032, 11034, 11051, 11071, 11079, 11096,
        11106, 11110, 11115, 11127, 11161, 11162, 11163, 11183, 11204, 11205, 11207,
        11208, 11220, 11230, 11232, 11247, 11249, 11258, 11259, 11283, 11304, 11305,
        11319, 11345, 11366, 11402, 11412, 11438, 11445, 11446, 11451, 11453, 11462,
        11464, 11468, 11483, 11486, 11490, 11491, 11553, 11581, 11590, 11591, 11612,
        11631, 11638, 11653, 11694, 11704, 11705, 11709, 11714, 11716, 11723, 11752,
        11762, 11778, 11787, 11832, 11840, 11855, 11879, 11889, 11909, 11910, 11935,
        11938, 11941, 11953, 11954, 11972, 11977, 11984, 11995, 12010, 12023, 12027,
        12043, 12045, 12060, 12066, 12084, 12085, 12089, 12112, 12113, 12131, 12173,
        12178, 12186, 12196, 12197, 12200, 12205, 12221, 12222, 12253, 12266, 12271,
        12293, 12299, 12317, 12319, 12339, 12354, 12357, 12359, 12360, 12362, 12363,
        12365, 12366, 12380, 12415, 12417, 12418, 12433, 12443, 12444, 12445, 12446,
        12477, 12530, 12537, 12538, 12540, 12542, 12564, 12583, 12604, 12605, 12612,
        12661, 12665, 12666, 12669, 12706, 12708, 12722, 12725, 12736, 12741, 12748,
        12784, 12795, 12801, 12810, 12814, 12839, 12845, 12846, 12848, 12856, 12860,
        12895, 12900, 12912, 12924, 12948, 12960, 12989, 12991, 13004, 13019, 13025,
        13027, 13040, 13044, 13069, 13103, 13136, 13140, 13149, 13159, 13164, 13167,
        13170, 13182, 13185, 13193, 13194, 13200, 13201, 13203, 13228, 13249, 13252,
        13257, 13259, 13261, 13266, 13272, 13284, 13285, 13305, 13315, 13316, 13367,
        13372, 13396, 13409, 13411, 13428, 13429, 13436, 13461, 13465, 13474, 13487,
        13493, 13502, 13503, 13536, 13540, 13555, 13570, 13589, 13594, 13598, 13663,
        13667, 13669, 13670, 13690, 13700, 13704, 13728, 13733, 13734, 13735, 13739,
        13740, 13744, 13753, 13770, 13791, 13798, 13799, 13814, 13816, 13827, 13828,
        13839, 13858, 13859, 13868, 13876, 13879, 13887, 13890, 13921, 13993, 14004,
        14005, 14031, 14043, 14054, 14070, 14083, 14084, 14087, 14100, 14109, 14115,
        14140, 14149, 14150, 14157, 14168, 14184, 14194, 14201, 14211, 14213, 14219,
        14221, 14251, 14264, 14285, 14339, 14347, 14368, 14369, 14391, 14409, 14410,
        14418, 14427, 14429, 14451, 14456, 14459, 14461, 14477, 14511, 14513, 14516,
        14517, 14527, 14537, 14565, 14568, 14569, 14575, 14576, 14580, 14581, 14584,
        14597, 14645, 14651, 14662, 14663, 14667, 14674, 14681, 14711, 14714, 14739,
        14756, 14757

Water cycle in biosphere 6997, 7614, 8316, 8339, 8494, 9191, 9694, 10003, 10004,
        10175, 10258, 10480, 10883, 11230, 11237, 11279, 11282, 11342, 11815, 12353,
        12452, 12515, 12886, 12945, 12950, 12978, 13434, 13449, 13537, 13563, 13841,
        13952, 13963, 13977, 13983, 14202, 14209, 14387, 14450, 14690

Water heavy see Deuterium oxide, tritium oxide ...

Water loss curves see Transpiration curves

Water potential in plant (see also Water status in plant ...) 6609, 6611, 6612,
        6620, 6625, 6632, 6648, 6649, 6654, 6656, 6663, 6664, 6667, 6681, 6682, 6684,
        6696, 6698, 6705, 6707, 6712, 6721, 6726, 6732, 6736, 6737, 6739, 6746, 6756,
        6769, 6770, 6774, 6780, 6782, 6784, 6836, 6846, 6851, 6891, 6893, 6899, 6900,
        6901, 6909, 6910, 6911, 6920, 6923, 6934, 6946, 6958, 6987, 7002, 7006, 7011,
        7015, 7029, 7035, 7047, 7076, 7092, 7093, 7098, 7105, 7106, 7110, 7113, 7116,
        7118, 7119, 7120, 7131, 7143, 7156, 7163, 7183, 7184, 7185, 7193, 7209, 7222,
        7226, 7227, 7231, 7232, 7233, 7234, 7241, 7245, 7247, 7248, 7278, 7281, 7288,
        7289, 7304, 7310, 7312, 7313, 7327, 7343, 7348, 7358, 7365, 7377, 7383, 7384,
        7399, 7415, 7419, 7432, 7442, 7444, 7457, 7458, 7459, 7469, 7476, 7479, 7483,
        7488, 7494, 7495, 7501, 7502, 7503, 7506, 7518, 7535, 7554, 7576, 7579, 7581,
        7582, 7588, 7601, 7610, 7612, 7615, 7621, 7625, 7626, 7637, 7661, 7677, 7679,
        7689, 7690, 7704, 7712, 7723, 7734, 7735, 7737, 7739, 7740, 7741, 7742, 7775,
        7793, 7794, 7797, 7798, 7802, 7834, 7837, 7840, 7851, 7852, 7857, 7865, 7871,
        7873, 7874, 7898, 7900, 7904, 7905, 7906, 7907, 7918, 7922, 7932, 7935, 7939,
        7947, 7949, 7951, 7954, 7965, 7983, 7990, 7995, 7996, 7997, 7999, 8000, 8001,
        8005, 8007, 8018, 8035, 8049, 8058, 8061, 8072, 8087, 8088, 8093, 8098, 8106,
        8109, 8126, 8131, 8132, 8133, 8134, 8136, 8137, 8140, 8141, 8153, 8159, 8160,
        8182, 8188, 8189, 8192, 8203, 8204, 8205, 8216, 8221, 8224, 8228, 8239, 8241,
        8245, 8252, 8265, 8266, 8267, 8281, 8286, 8289, 8295, 8299, 8306, 8311, 8315,
        8326, 8333, 8349, 8360, 8365, 8374, 8377, 8379, 8381, 8382, 8383, 8387, 8398,
        8400, 8401, 8406, 8411, 8412, 8413, 8417, 8422, 8435, 8443, 8444, 8445, 8448,
        8449, 8457, 8468, 8470, 8474, 8475, 8482, 8494, 8496, 8506, 8511, 8515, 8522,
        8524, 8525, 8530, 8532, 8543, 8544, 8549, 8553, 8575, 8589, 8591, 8593, 8627,
        8629, 8636, 8645, 8651, 8653, 8658, 8662, 8664, 8665, 8666, 8667, 8670, 8671,
        8672, 8674, 8676, 8677, 8679, 8684, 8693, 8701, 8707, 8709, 8712, 8714, 8715,
        8716, 8723, 8727, 8729, 8741, 8743, 8744, 8745, 8749, 8750, 8751, 8772, 8799,
        8800, 8810, 8811, 8821, 8824, 8826, 8827, 8831, 8837, 8840, 8849, 8850, 8851,
        8855, 8857, 8860, 8861, 8877, 8888, 8889, 8899, 8902, 8907, 8911, 8920, 8924,
        8925, 8927, 8939, 8949, 8951, 8955, 8956, 8960, 8962, 8970, 8973, 8976, 8985,
        8989, 8998, 8999, 9001, 9002, 9003, 9005, 9009, 9022, 9030, 9031, 9035, 9039,
        9041, 9042, 9054, 9057, 9063, 9066, 9077, 9078, 9094, 9096, 9099, 9108, 9110,
        9123, 9126, 9134, 9141, 9149, 9151, 9166, 9170, 9173, 9174, 9177, 9178, 9194,
        9195, 9197, 9209, 9211, 9212, 9213, 9225, 9226, 9228, 9229, 9249, 9250, 9261,
        9273, 9278, 9280, 9290, 9296, 9297, 9300, 9316, 9321, 9326, 9329, 9352, 9360,
        9366, 9376, 9378, 9401, 9403, 9404, 9408, 9418, 9422, 9427, 9428, 9429, 9436,
        9444, 9447, 9450, 9453, 9475, 9494, 9496, 9497, 9498, 9499, 9515, 9519, 9528,
        9529, 9530, 9531, 9555, 9556, 9558, 9560, 9563, 9564, 9601, 9604, 9617, 9622,
        9623, 9627, 9628, 9631, 9639, 9644, 9657, 9660, 9661, 9676, 9681, 9688, 9695,
        9698, 9701, 9703, 9710, 9711, 9720, 9732, 9741, 9745, 9754, 9759, 9765, 9776,
        9783, 9784, 9799, 9801, 9806, 9807, 9818, 9819, 9820, 9825, 9826, 9841, 9848,
        9849, 9868, 9870, 9871, 9879, 9902, 9916, 9917, 9919, 9922, 9924, 9925, 9950,
        9954, 9961, 9969, 9989, 10000, 10001, 10010, 10028, 10033, 10039, 10061,
        10067, 10077, 10087, 10106, 10112, 10121, 10124, 10126, 10129, 10132, 10137,
        10141, 10142, 10148, 10149, 10158, 10162, 10163, 10199, 10200, 10201, 10202,
        10203, 10204, 10205, 10211, 10215, 10218, 10219, 10232, 10239, 10244, 10247,
        10248, 10249, 10258, 10272, 10280, 10288, 10302, 10303, 10305, 10310, 10314,
        10319, 10327, 10329, 10332, 10351, 10357, 10373, 10379, 10382, 10386, 10388,
        10390, 10392, 10404, 10419, 10424, 10425, 10426, 10428, 10436, 10445, 10447,
        10456, 10460, 10461, 10465, 10467, 10476, 10478, 10482, 10487, 10490, 10491,
        10500, 10501, 10502, 10505, 10508, 10510, 10516, 10520, 10524, 10536, 10549,
        10563, 10565, 10570, 10571, 10574, 10575, 10579, 10585, 10586, 10601, 10605,
        10610, 10623, 10627, 10628, 10629, 10630, 10636, 10637, 10638, 10639, 10650,
        10653, 10655, 10659, 10681, 10684, 10686, 10692, 10697, 10700, 10701, 10717,
        10718, 10738, 10750, 10756, 10757, 10762, 10763, 10764, 10779, 10786, 10791,
        10792, 10804, 10805, 10841, 10843, 10846, 10853, 10855, 10867, 10871, 10874,
        10875, 10883, 10889, 10941, 10942, 10943, 10960, 10969, 10970, 10972, 10973,
        10980, 10981, 10984, 10985, 10987, 10991, 10997, 11000, 11003, 11019, 11023,
        11024, 11044, 11051, 11052, 11053, 11062, 11067, 11080, 11083, 11085, 11095,
        11099, 11113, 11118, 11144, 11155, 11162, 11163, 11168, 11171, 11173, 11187,
        11190, 11191, 11202, 11217, 11222, 11236, 11241, 11242, 11246, 11249, 11252,
        11254, 11258, 11259, 11263, 11264, 11268, 11269, 11277, 11278, 11280, 11289,
                                                                    (continued)

                                                                    (continued)

Water potential in substrate (continued) 7730, 7743, 7752, 7757, 7758, 7788, 7789,
        7790, 7866, 7878, 7917, 7918, 7923, 7962, 7996, 8007, 8013, 8122, 8137, 8143,
        8147, 8151, 8160, 8192, 8228, 8237, 8239, 8252, 8262, 8267, 8283, 8291, 8306,
        8352, 8383, 8411, 8413, 8417, 8418, 8422, 8443, 8445, 8472, 8493, 8494, 8501,
        8581, 8630, 8659, 8672, 8677, 8744, 8751, 8777, 8780, 8781, 8782, 8796, 8831,
        8851, 8864, 8875, 8888, 8925, 8931, 8939, 8944, 8946, 8985, 9011, 9056, 9129,
        9138, 9141, 9190, 9195, 9214, 9240, 9258, 9259, 9290, 9317, 9321, 9326, 9422,
        9430, 9431, 9444, 9453, 9459, 9486, 9487, 9519, 9555, 9608, 9639, 9679, 9690,
        9702, 9711, 9720, 9735, 9756, 9766, 9779, 9801, 9815, 9825, 9826, 9879, 9884,
        9944, 9958, 9980, 10033, 10039, 10112, 10142, 10163, 10184, 10193, 10196,
        10227, 10236, 10258, 10272, 10288, 10296, 10310, 10331, 10332, 10353, 10433,
        10456, 10466, 10476, 10480, 10520, 10650, 10750, 10760, 10791, 10792, 10804,
        10872, 10883, 10943, 10944, 11010, 11024, 11111, 11122, 11137, 11191, 11216,
        11229, 11249, 11256, 11259, 11263, 11264, 11278, 11312, 11326, 11341, 11362,
        11402, 11411, 11472, 11474, 11508, 11549, 11567, 11572, 11584, 11596, 11621,
        11629, 11657, 11658, 11686, 11708, 11728, 11753, 11773, 11779, 11796, 11827,
        11828, 11856, 11864, 11865, 11875, 11893, 11902, 11925, 11929, 11932, 11933,
        11948, 11988, 12035, 12039, 12075, 12106, 12144, 12176, 12177, 12224, 12277,
        12295, 12361, 12364, 12434, 12437, 12449, 12474, 12519, 12520, 12526, 12527,
        12539, 12549, 12550, 12556, 12575, 12580, 12625, 12670, 12671, 12695, 12696,
        12699, 12702, 12713, 12729, 12779, 12808, 12866, 12889, 12913, 12015, 12918,
        12924, 12026, 12936, 12955, 12981, 12994, 13018, 13030, 13041, 13071, 13077,
        13078, 13079, 13086, 13089, 13129, 13155, 13173, 13177, 13196, 13199, 13256,
        13318, 13338, 13441, 13464, 13550, 13555, 13591, 13592, 13625, 13626, 13697,
        13700, 13711, 13718, 13724, 13761, 13793, 13824, 13909, 13913, 13914, 13916,
        13924, 13936, 13968, 13979, 14012, 14029, 14040, 14070, 14082, 14099, 14124,
        14188, 14245, 14257, 14278, 14279, 14333, 14373, 14394, 14441, 14450, 14479,
        14486, 14507, 14519, 14571, 14573, 14619, 14693, 14725, 14741

Water potential, methods, dew point hygrometer 7432, 7588, 8136, 8299, 8687, 9444,
        9556, 9826, 10467, 10628, 10825, 19825, 11260, 11395, 12381, 12461, 12513,
        13093, 13102, 13260, 14327, 14328, 14329, 14374

Water potential, methods, other than above and below 7076, 8468, 8712, 10719, 19757

Water potential methods, pressure bomb 6632, 6697, 7105, 7226, 7527, 7600, 7821,
        7857, 7997, 8054, 8295, 8299, 8398, 8709, 9039, 9174, 9273, 9387, 9444, 9556,
        9564, 9922, 10201, 10424, 10467, 10487, 10505, 10628, 10757, 11051, 11053,
        11399, 11457, 11561, 11897, 11995, 12278, 12381, 12461, 12513, 12753, 12806,
        12841, 12915, 13211, 13260, 13595, 13639, 13728, 13886, 13920, 14326, 14329,
        14492, 14493, 14574, 14602, 14697, 14764

Water potential, methods, psychrometry 6788, 8956, 9273, 9444, 9556, 9631, 9944,
        10424, 10428, 10505, 10826, 10922, 11399, 12753, 12806, 12915, 13920, 14326,
        14327, 14677, 14707

Water potential, methods, psychrometry in situ 7588, 7908, 10424, 12461, 12513,
        13102, 13908, 14325, 14326, 14327, 14328, 14329, 14602

Water saturation deficit 6609, 6634, 6635, 6664, 6666, 6682, 6721, 6777, 6893, 6896,
        6899, 6901, 6910, 7006, 7092, 7093, 7108, 7134, 7135, 7156, 7183, 7212, 7234,
        7245, 7247, 7282, 7293, 7312, 7334, 7384, 7413, 7431, 7479, 7494, 7495, 7619,
        7621, 7628, 7665, 7677, 7679, 7712, 7741, 7749, 7789, 7858, 7873, 7895, 7898,
        7905, 7911, 7935, 7996, 7997, 8001, 8003, 8005, 8040, 8073, 8241, 8247, 8271,
        8289, 8304, 8360, 8379, 8398, 8406, 8475, 8476, 8544, 8666, 8677, 8701, 8709,
        8715, 8716, 8743, 8750, 8751, 8770, 8771, 8772, 8787, 8821, 8831, 8843, 8857,
        8874, 8877, 8927, 8951, 8954, 8979, 8989, 8992, 9005, 9057, 9074, 9095, 9119,
        9123, 9134, 9151, 9159, 9207, 9249, 9250, 9261, 9273, 9283, 9292, 9293, 9316,
        9317, 9329, 9342, 9392, 9393, 9403, 9404, 9499, 9520, 9556, 9564, 9617, 9627,
        9659, 9688, 9706, 9731, 9925, 9950, 9954, 10001, 10007, 10061, 10075, 10077,
        10106, 10156, 10201, 10272, 10363, 10379, 10388, 10404, 10424, 10447, 10482,
        10491, 10496, 10503, 10520, 10522, 10532, 10536, 10544, 10570, 10586, 10600,
                                                                    (continued)

Water saturation deficit (continued) 10650, 10679, 10692, 10701, 10723, 10724,
      10748, 10832, 10871, 10883, 10896, 10962, 10987, 10988, 11010, 11035, 11052,
      11053, 11110, 11163, 11187, 11190, 11223, 11239, 11249, 11271, 11284, 11315,
      11332, 11364, 11366, 11379, 11399, 11438, 11514, 11550, 11553, 11563, 11577,
      11588, 11607, 11612, 11688, 11711, 11721, 11722, 11762, 11771, 11786, 11790,
      11832, 11834, 11855, 11864, 11907, 11916, 11917, 11927, 11932, 11938, 11948,
      11954, 11962, 11970, 11972, 12060, 12074, 12126, 12133, 12141, 12172, 12176,
      12177, 12191, 12202, 12203, 12223, 12224, 12228, 12234, 12245, 12246, 12277,
      12292, 12295, 12354, 12369, 12378, 12485, 12486, 12487, 12492, 12493, 12519,
      12528, 12537, 12538, 12547, 12556, 12557, 12585, 12586, 12589, 12612, 12619,
      12663, 12667, 12675, 12696, 12699, 12704, 12711, 12844, 12856, 12857, 12878,
      12905, 12922, 12946, 12960, 12998, 12999, 13000, 13001, 13002, 13009, 13010,
      13038, 13047, 13061, 13094, 13120, 13155, 13181, 13229, 13253, 13271, 13282,
      13300, 13340, 13367, 13379, 13393, 13403, 13404, 13418, 13431, 13433, 13452,
      13474, 13533, 13534, 13610, 13612, 13638, 13642, 13654, 13655, 13693, 13713,
      13717, 13727, 13728, 13757, 13760, 13768, 13771, 13782, 13798, 13815, 13830,
      13879, 13883, 13911, 13965, 13966, 13968, 13969, 13994, 14014, 14020, 14025,
      14055, 14056, 14064, 14071, 14081, 14098, 14109, 14116, 14131, 13152, 14166,
      14169, 14196, 14242, 14243, 14275, 14350, 14355, 14392, 14410, 14419, 14433,
      14434, 14452, 14461, 14481, 14516, 14541, 14556, 14557, 14569, 14580, 14596,
      14597, 14605, 14628, 14637, 14662, 14663, 14737

Water saturation deficit, methods 6957, 7413, 8004, 8715, 9832, 9942, 10328, 10360,
      10428, 10735, 11577, 12486, 12753, 12896, 13112

Water status in plant see also Age of plant ...; Altitude, pressure ...; Anatomi-
      cal structure ...; Antibiotics ...; Carbon dioxide ...; Cultivars ...;
      Defoliation, decapitation, ear, root removal ...; Deuterium oxide, tritium
      oxide ...; Drought ...; Ecotypes, geographical types ...; Enzyme inhibi-
      tors ...; Enzymes ...; Farming practices ...; Flooding ...; Genetics ...;
      Growth substances, hormones, inhibitors etc. ...; Humidity of air ...; Ir-
      radiance ...; Irrigation ...; Leaf insertion level ...; Lipids, fatty
      acids ...; Mineral elements ...; Mutagens, other organic substances ...;
      Mutants ...; Osmotically active substances ...; Oxygen ...; Pathogens ...;
      Pesticides, herbicides ...; pH ...; Photoperiod ...; Pollutants, ozone ...;
      Precipitation, dew ...; Proteins, amino acids, nucleic acids ...; Root,
      underground part ...; Saccharides ...; Salinity ...; Soil moisture ...;
      Taxons ...; Temperature ...; Wind and water status in plant

Water status in plant and canopy architecture 6697, 7361, 8474, 8732, 9122, 9848,
      12025, 12621, 12699, 13111, 13350, 13387

Water status in plant and carbon dioxide influx 6698, 6732, 6769, 6871, 6892, 6925,
      6968, 6970, 6998, 7001, 7003, 7008, 7013, 7035, 7070, 7120, 7170, 7246, 7248,
      7269, 7297, 7341, 7343, 7347, 7361, 7384, 7484, 7486, 7540, 7555, 7576, 7577,
      7578, 7579, 7719, 7723, 7738, 7793, 7794, 7918, 7939, 7951, 7956, 7995, 7996,
      8001, 8023, 8051, 8072, 8132, 8133, 8189, 8205, 8216, 8224, 8228, 8239, 8242,
      8306, 8352, 8445, 8470, 8596, 8597, 8598, 8619, 8677, 8683, 8857, 8861, 8890,
      8892, 8902, 8909, 8981, 8985, 9099, 9108, 9122, 9134, 9253, 9278, 9330, 9353,
      9354, 9360, 9395, 9412, 9424, 9426, 9534, 9558, 9599, 9605, 9615, 9639, 9699,
      9720, 9761, 9765, 9781, 9807, 9822, 9946, 9961, 9993, 10005, 10006, 10011,
      10119, 10120, 10124, 10158, 10164, 10204, 10205, 10245, 10247, 10321, 10332,
      10339, 10363, 10373, 10386, 10388, 10404, 10415, 10447, 19476, 10491, 10521,
      10522, 10567, 10576, 10581, 10589, 10653, 10681, 10729, 10791, 10796, 10841,
      10855, 10871, 10883, 10984, 11019, 11062, 11082, 11118, 11144, 11167, 11172,
      11202, 11249, 11252, 11263, 11272, 11318, 11324, 11332, 11364, 11377, 11379,
      11380, 11421, 11442, 11483, 11489, 11499, 11519, 11545, 11550, 11560, 11581,
      11583, 11586, 11610, 11633, 11662, 11704, 11705, 11711, 11722, 11756, 11827,
      11855, 11861, 11864, 11873, 11910, 11911, 11927, 11929, 11954, 11955, 11962,
      11964, 11984, 11985, 11990, 11991, 12000, 12026, 12050, 12051, 12069, 12084,
      12090, 12101, 12102, 12175, 12192, 12234, 12279, 12286, 12299, 12306, 12316,
      12346, 12351, 12356, 12357, 12378, 12392, 12401, 12404, 12430, 12485, 12501,
(continued)

Water status in plant and carbon dioxide influx (continued) 12557, 12572, 12576,
        12614, 12634, 12696, 12699, 12719, 12722, 12753, 12775, 12802, 12810, 12845,
        12846, 12859, 12860, 12873, 12900, 12911, 13010, 13062, 13078, 13079, 13099,
        13118, 13123, 13169, 13180, 13185, 13196, 13200, 13201, 13214, 13270, 13286,
        13298, 13311, 13472, 13519, 13590, 13605, 13607, 13704, 13750, 13752, 13762,
        13763, 13792, 13795, 13798, 13821, 13822, 13830, 13914, 13961, 13965, 13973,
        14016, 14033, 14074, 14086, 14096, 14124, 14178, 14181, 14287, 14314, 14323,
        14350, 14351, 14355, 14370, 14390, 14395, 14427, 14463, 14553, 14567, 14568,
        14571, 14617, 14628, 14662, 14762, 14754

Water status in plant and carbon fixation pathways 6892, 7180, 7297, 7996, 8673,
        8865, 8902, 9122, 9534, 10104, 10162, 10370, 10493, 10681, 10855, 11084,
        11318, 11586, 11711, 11722, 11990, 12031, 12192, 12699, 12775, 12792, 12845,
        12846, 12851, 13472, 13706, 13751, 13830, 14457

Water status in plant and carotenoids 7830, 7856, 8874, 10345, 10429, 14595, 14596

Water status in plant and chlorophyll 7035, 7107, 7108, 7325, 7334, 7355, 7387,
        7443, 7573, 7750, 7830, 7856, 7933, 7996, 8156, 8247, 8360, 8430, 8589, 8646,
        8699, 8786, 8874, 8969, 9134, 9188, 9224, 9353, 9354, 9769, 9782, 9783, 9804,
        10060, 10207, 10208, 10218, 10346, 10415, 10429, 10461, 10511, 10581, 10617,
        10657, 10698, 10983, 11083, 11175, 11451, 11554, 11556, 11557, 11632, 11711,
        11763, 11970, 12238, 12501, 12704, 12781, 12960, 13061, 13062, 13392, 13489,
        13571, 13689, 13733, 13830, 13837, 13838, 13890, 14021, 14068, 14178, 14571,
        14595, 14596, 14636

Water status in plant and chloroplasts 6802, 6961, 7014, 7107, 7108, 7330, 7719,
        7745, 7798, 7862, 8023, 8345, 8861, 8865, 8874, 8953, 8969, 9122, 9134, 9188,
        9309, 9758, 9804, 9875, 10207, 10209, 10522, 10679, 10752, 10883, 11070,
        11083, 11318, 11378, 11379, 11716, 12324, 12495, 12660, 12775, 12862, 12938,
        13010, 13352, 13830, 14176, 14318, 14628

Water status in plant and conductance for water vapour and carbon dioxide transfer
        6611, 6612, 6625, 6648, 6660, 6698, 6871, 6893, 6931, 6946, 6974, 6998, 7003,
        7006, 7008, 7013, 7113, 7134, 7135, 7151, 7156, 7170, 7209, 7227, 7231, 7232,
        7269, 7343, 7350, 7351, 7356, 7365, 7377, 7383, 7384, 7431, 7476, 7485, 7501,
        7535, 7578, 7579, 7582, 7587, 7612, 7621, 7645, 7661, 7691, 7739, 7883, 7939,
        7995, 7996, 8061, 8127, 8131, 8133, 8134, 8189, 8190, 8221, 8224, 8239, 8252,
        8306, 8360, 8383, 8387, 8406, 8413, 8417, 8515, 8525, 8549, 8597, 8629, 8645,
        8662, 8667, 8676, 8677, 8727, 8744, 8745, 8750, 8751, 8771, 8800, 8821, 8861,
        8902, 8907, 8927, 8929, 8944, 8951, 8999, 9005, 9074, 9094, 9134, 9141, 9224,
        9225, 9226, 9250, 9261, 9273, 9278, 9297, 9306, 9321, 9329, 9360, 9404, 9422,
        9426, 9494, 9498, 9499, 9558, 9628, 9639, 9661, 9681, 9732, 9741, 9806, 9848,
        9868, 9902, 9950, 10000, 10039, 10042, 10051, 10061, 10087, 10112, 10120,
        10121, 10122, 10149, 10158, 10162, 10200, 10202, 10219, 10245, 10247, 10248,
        10280, 10314, 10315, 10332, 10339, 10363, 10373, 10382, 10388, 10392, 10404,
        10424, 10425, 10447, 10465, 10476, 10510, 10586, 10589, 10600, 10653, 10697,
        10718, 10729, 10756, 10764, 10779, 10791, 10804, 10805, 10841, 10871, 10883,
        10889, 10972, 10984, 11000, 11003, 11019, 11024, 11028, 11031, 11062, 11072,
        11118, 11144, 11173, 11202, 11211, 11222, 11242, 11249, 11264, 11277, 11332,
        11349, 11363, 11364, 11400, 11442, 11514, 11516, 11574, 11610, 11622, 11633,
        11677, 11686, 11691, 11701, 11704, 11722, 11756, 11760, 11761, 11765, 11771,
        11790, 11834, 11848, 11855, 11864, 11866, 11886, 11923, 11925, 11927, 11932,
        11943, 11954, 11955, 11960, 11962, 11964, 11976, 11990, 12000, 12019, 12043,
        12060, 12086, 12090, 12100, 12165, 12209, 12245, 12246, 12277, 12279, 12306,
        12321, 12329, 12345, 12346, 12351, 12392, 12430, 12485, 12486, 12488, 12501,
        12557, 12572, 12595, 12606, 12608, 12619, 12621, 12622, 12623, 12634, 12655,
        12663, 12696, 12699, 12705, 12751, 12753, 12775, 12802, 12805, 12828, 12873,
        12878, 12911, 12919, 12955, 12965, 13010, 13038, 13062, 13063, 13093, 13123,
        13155, 13169, 13175, 13178, 13180, 13214, 13230, 13259, 13267, 13284, 13297,
        13298, 13311, 13346, 13466, 13472, 13485, 13488, 13508, 13530, 13535, 13590,
        13592, 13605, 13692, 13713, 13739, 13771, 13782, 13783, 13833, 13838, 13846,
        13900, 13961, 13965, 13978, 13994, 14069, 14071, 14086, 14121, 14135, 14143,
                                                                    (continued)

Water status in plant and conductance for water vapour and carbon dioxide trahsfer
(continued) 14170, 14171, 14172, 14189, 14196, 14242, 14268, 14285, 14286,
14288, 14314, 14370, 14386, 14433, 14434, 14463, 14467, 14468, 14479, 14541,
14556, 14600, 14601, 14602, 14628, 14714, 14736, 14754

Water status in plant and electron transport chain 6732, 6961, 7538, 7573, 7697,
7750, 7933, 8023, 8109, 8247, 8519, 8589, 8960, 9765, 9782, 9783, 9804, 9828,
9981, 10060, 10158, 10162, 10207, 10209, 10346, 10415, 10461, 10581, 10855,
11083, 11378, 11379, 11381, 11545, 11557, 11970, 11990, 11991, 12102, 12404,
12544, 13007, 13034, 13062, 13704, 13830, 13838, 13900, 14174, 14175, 14178,
14355, 14371, 14762

Water status in plant and growth, productivity 6625, 6656, 6681, 6682, 6713, 6769,
6770, 6799, 6812, 6829, 6864, 6910, 6945, 6973, 7003, 7013, 7075, 7100, 7134,
7154, 7160, 7227, 7230, 7327, 7343, 7361, 7415, 7437, 7440, 7441, 7509, 7610,
7621, 7635, 7661, 7691, 7696, 7723, 7834, 7837, 7864, 7918, 7991, 7995, 7996,
8005, 8041, 8072, 8082, 8126, 8330, 8376, 8377, 8399, 8433, 8457, 8511, 8524,
8553, 8614, 8619, 9624, 8633, 8677, 8700, 8712, 8768, 8907, 8924, 8970, 8998,
9006, 9022, 9088, 9108, 9122, 9126, 9211, 9271, 9292, 9316, 9360, 9391, 9427,
9428, 9444, 9453, 9477, 9510, 9519, 9558, 9617, 9720, 9765, 9766, 9807, 9823,
9840, 9892, 9950, 9964, 10004, 10024, 10053, 10054, 10114, 10129, 10132,
10148, 10219, 10233, 10351, 10404, 10415, 10424, 10476, 10490, 10491, 10496,
10553, 10601, 10627, 10639, 10737, 10757, 10767, 10790, 10875, 10883, 10980,
10981, 10997, 11032, 11118, 11133, 11169, 11241, 11263, 11305, 11332, 11349,
11364, 11377, 11380, 11400, 11428, 11430, 11486, 11500, 11585, 11587, 11651,
11664, 11701, 11722, 11731, 11754, 11790, 11855, 11864, 11865, 11868, 11932,
11954, 11862, 11964, 12041, 12042, 12073, 12123, 12133, 12350, 12392, 12540,
12542, 12562, 12572, 12616, 12620, 12639, 12675, 12696, 12699, 12704, 12814,
12837, 12965, 13040, 13061, 13067, 13162, 13191, 13197, 13253, 13286, 13290,
13346, 13350, 13389, 13403, 13488, 13489, 13501, 13573, 13624, 13750, 13784,
13785, 13799, 13800, 13869, 13895, 13916, 13966, 13967, 13969, 14057, 14072,
14124, 14128, 14181, 14227, 14261, 14311, 14347, 14360, 14375, 14424, 14525,
14568, 14575, 14576, 14637, 14753

Water status in plant and leaf anatomy 6682, 6696, 6901, 6910, 7014, 7019, 7108,
7229, 7265, 7334, 7360, 7485, 7544, 7621, 7834, 7898, 7905, 7991, 7996, 8005,
8018, 8101, 8135, 8280, 8671, 8786, 8802, 8821, 8844, 8873, 8874, 8907, 9074,
9122, 9278, 9558, 10008, 10395, 10476, 10522, 10585, 10589, 10653, 10737,
10763, 10883, 10980, 10981, 11080, 11168, 11169, 11318, 11332, 11350, 11379,
11400, 11499, 11517, 11556, 11585, 11865, 11964, 12285, 12378, 12556, 12621,
12633, 12696, 12699, 12748, 12753, 12781, 12793, 12805, 12965, 13001, 13055,
13220, 13286, 13290, 13297, 13346, 13350, 13403, 13489, 13506, 13605, 13713,
13914, 14055, 14057, 14069, 14086, 14458

Water status in plant and other processes than above and below 10511, 10539, 10570,
11203, 11283, 11545, 11554, 11581, 11612, 11633, 11748, 11764, 11810, 11929,
12165, 12166, 12441, 12780, 13022, 13024, 13118, 13177, 13180, 13200, 13549,
13590, 13771, 14191, 14264, 14314, 14463, 14506, 14593, 14753

Water status in plant and photorespiration 8902, 9122, 9424, 9639, 10162, 10415,
10536, 10681, 11133, 11855, 11990, 12316, 13123, 13752, 13965

Water status in plant and respiration 6786, 7035, 7261, 7269, 7341, 7347, 7484, 7869,
7996, 8072, 8216, 8228, 8298, 8517, 8598, 8633, 8831, 8890, 9122, 9353, 9354,
9366, 9412, 9424, 9426, 9639, 9709, 9765, 9807, 10006, 10162, 10218, 10345,
10386, 10415, 10522, 10883, 11252, 11489, 11545, 11694, 11865, 11911, 11984,
12051, 12234, 12351, 12357, 12485, 12614, 12696, 12699, 12859, 13078, 13099,
13316, 13751, 13752, 13792, 13821, 13914, 14463, 14568, 14620, 14645, 14754

Water status in plant and stomata 6792, 6802, 6829, 6871, 6891, 6912, 6946, 6974,
6998, 7135, 7209, 7227, 7259, 7343, 7360, 7377, 7383, 7476, 7495, 7581, 7587,
7619, 7645, 7864, 7939, 7995, 7996, 8000, 8127, 8190, 8413, 8696, 8744, 8857,
8909, 8929, 8951, 8953, 9116, 9134, 9159, 9422, 9534, 9641, 9650, 9681, 9807,
(continued)

Water status in plant and stomata (continued) 9902, 9914, 9919, 10037, 10042,
        10087, 10121, 10147, 10180, 10363, 10424, 10476, 10493, 10554, 10653, 10668,
        10718, 10731, 10867, 10907, 11018, 11057, 11116, 11122, 11129, 11476, 11516,
        11760, 11955, 12064, 12203, 12329, 12404, 12655, 12696, 12704, 12745, 13077,
        13079, 13194, 13713, 14458, 14601, 14718, 14736

Water status in plant and transpiration 6648, 6649, 6660, 6698, 6808, 6871, 6891,
        6916, 6998, 7008, 7014, 7047, 7113, 7151, 7170, 7209, 7230, 7259, 7261, 7360,
        7365, 7399, 7418, 7431, 7475, 7476, 7579, 7621, 7723, 7739, 7740, 7801, 7865,
        7918, 7930, 7936, 7996, 8061, 8131, 8171, 8352, 8359, 8383, 8417, 8435, 8517,
        8662, 8667, 8799, 8831, 8857, 8861, 8909, 8944, 8951, 8981, 8985, 9122, 9134,
        9166, 9192, 9253, 9278, 9297, 9326, 9498, 9510, 9627, 9710, 9740, 9916, 9950,
        10042, 10087, 10122, 10126, 10205, 10240, 10247, 10332, 10339, 10404, 10415,
        10424, 10425, 10467, 10563, 10600, 10750, 10786, 10841, 10883, 10984, 11003,
        11031, 11044, 11113, 11242, 11252, 11332, 11350, 11363, 11364, 11379, 11419,
        11511, 11519, 11550, 11560, 11626, 11686, 11721, 11722, 11731, 11855, 11861,
        11864, 11892, 11893, 11948, 11954, 11962, 12019, 12021, 12279, 12286, 12317,
        12346, 12519, 12528, 12556, 12557, 12561, 12606, 12608, 12633, 12663, 12696,
        12699, 12755, 12775, 12828, 12908, 12911, 12946, 12955, 12965, 13010, 13049,
        13077, 13078, 13079, 13119, 13196, 13284, 13311, 13432, 13466, 13485, 13488,
        13590, 13605, 13692, 13727, 13795, 13839, 13909, 13920, 13961, 13973, 13984,
        14020, 14070, 14102, 14124, 14189, 14261, 14343, 14354, 14386, 14600, 14705,
        14714, 14723, 14741, 14765

Water status in plant and water absorption by plant 10874, 11898

Water status in plant and water transport in cells 7035, 7234, 7245, 7439, 7996,
        8376, 8377, 9134, 9188, 9192, 9356, 10948, 11142, 13192, 13564, 13713

Water status in plant and water transport in plant 6836, 7621, 8159, 8376, 8743,
        9166, 9740, 10007, 10846, 10883, 11472, 13909

Water status in plant, comparison of plants with different types of carbon metabolism
        7576, 8402, 9617, 13887, 14020

Water status in plant, diurnal changes 6609, 6632, 6635, 6681, 6712, 6726, 6769,
        6836, 6910, 7002, 7047, 7092, 7093, 7120, 7184, 7185, 7241, 7247, 7248, 7259,
        7293, 7352, 7377, 7501, 7502, 7576, 7581, 7582, 7588, 7612, 7621, 7677, 7695,
        7723, 7735, 7739, 7740, 7747, 7840, 7857, 7905, 7918, 7935, 7937, 7939, 7949,
        7954, 7990, 7996, 7997, 7999, 8093, 8098, 8133, 8140, 8141, 8159, 8204, 8216,
        8220, 8221, 8262, 8295, 8299, 8365, 8374, 8399, 8417, 8435, 8448, 8470, 8525,
        8543, 8629, 8649, 8662, 8672, 8674, 8677, 8688, 8709, 8714, 8715, 8729, 8741,
        8743, 8810, 8837, 8839, 8849, 8851, 8855, 8857, 8860, 8907, 8999, 9001, 9009,
        9035, 9041, 9077, 9094, 9141, 9149, 9166, 9194, 9280, 9297, 9300, 9326, 9329,
        9376, 9427, 9434, 9447, 9496, 9497, 9498, 9499, 9515, 9558, 9622, 9623, 9627,
        9628, 9660, 9661, 9703, 9717, 9732, 9841, 9848, 9849, 9889, 9902, 9917, 9954,
        10061, 10122, 10126, 10132, 10149, 10156, 10162, 10203, 10215, 10239, 10244,
        10271, 10288, 10302, 10319, 10357, 10363, 10366, 10373, 10392, 10425, 10426,
        10456, 10460, 10467, 10469, 10544, 10549, 10563, 10586, 10629, 10686, 10724,
        10756, 10762, 10763, 10764, 10791, 10804, 10841, 10883, 10919, 10941, 10943,
        10973, 10980, 10981, 11034, 11044, 11067, 11095, 11113, 11141, 11190, 11191,
        11223, 11258, 11278, 11363, 11383, 11408, 11419, 11429, 11458, 11472, 11514,
        11549, 11550, 11574, 11577, 11582, 11587, 11588, 11634, 11641, 11677, 11678,
        11686, 11691, 11706, 11756, 11760, 11771, 11813, 11855, 11861, 11864, 11865,
        11866, 11887, 11892, 11893, 11918, 11927, 11934, 11936, 11948, 11980, 11988,
        11995, 12072, 12105, 12126, 12165, 12173, 12224, 12234, 12246, 12321, 12334,
        12346, 12369, 12401, 12415, 12430, 12487, 12488, 12537, 12547, 12561, 12593,
        12608, 12619, 12634, 12642, 12663, 12696, 12699, 12705, 12735, 12753, 12754,
        12755, 12807, 12808, 12874, 12887, 12908, 12913, 12914, 12943, 12946, 12999,
        13010, 13062, 13063, 13087, 13105, 13114, 13119, 13128, 13140, 13149, 13155,
        13175, 13178, 13212, 13215, 13271, 13274, 13279, 13284, 13297, 13373, 13432,
        13433, 13451, 13466, 13534, 13542, 13582, 13592, 13626, 13650, 13661, 13667,
        13692, 13693, 13700, 13724, 13725, 13727, 13835, 13840, 13846, 13901, 13915,
                                                                              (continued)

Water status in plant, diurnal changes (continued) 13916, 13964, 13994, 13996,
    13998, 14004, 14005, 14014, 14043, 14074, 14089, 14124, 14143, 14148, 14167,
    14177, 14179, 14197, 14203, 14207, 14226, 14228, 14263, 14273, 14308, 14315,
    14354, 14376, 14386, 14391, 14395, 14435, 14460, 14509, 14539, 14556, 14460,
    14509, 14539, 14556, 14568, 14580, 14584, 14587, 14653, 14681, 14705, 14740

Water status in plant, heterogeneity of single leaf blade 7734, 8970, 9002, 9531,
    10502, 10962, 11171, 11762, 13506, 13728, 14603, 14701

Water status in plant, oscillations 7554, 9916, 10310, 13908, 14102

Water status in plant, seasonal changes 6625, 6626, 6697, 6705, 6726, 6767, 6893,
    6911, 6926, 7093, 7156, 7222, 7241, 7247, 7267, 7301, 7320, 7325, 7415, 7446,
    7503, 7612, 7621, 7647, 7677, 7695, 7712, 7739, 7741, 7743, 7775, 7793, 7794,
    7840, 7895, 7932, 7935, 7945, 7990, 8005, 8054, 8093, 8106, 8143, 8219, 8239,
    8253, 8289, 8311, 8349, 8365, 8422, 8445, 8629, 8659, 8676, 8677, 8684, 8707,
    8708, 8727, 8784, 8824, 8857, 8899, 8979, 9011, 9041, 9059, 9094, 9095, 9099,
    9126, 9149, 9194, 9195, 9209, 9211, 9228, 9321, 9360, 9427, 9490, 9497, 9498,
    9555, 9622, 9666, 9690, 9732, 9785, 9826, 9889, 9902, 9924, 9972, 9973, 10010,
    10067, 10126, 10184, 10215, 10218, 10239, 10348, 10363, 10419, 10425, 10426,
    10447, 10520, 10549, 10563, 10586, 10605, 10650, 10724, 10779, 10801, 10805,
    10883, 10896, 10942, 11003, 11035, 11067, 11230, 11332, 11366, 11419, 11428,
    11429, 11474, 11486, 11550, 11563, 11582, 11587, 11686, 11764, 11804, 11847,
    11858, 11886, 11912, 11918, 11954, 11976, 11980, 11988, 12018, 12031, 12142,
    12178, 12224, 12253, 12277, 12327, 12365, 12434, 12437, 12537, 12539, 12550,
    12604, 12608, 12812, 12821, 12887, 12888, 12900, 12946, 12954, 12955, 13000,
    13063, 13119, 13120, 13175, 13178, 13215, 13254, 13261, 13271, 13297, 13338,
    13351, 13452, 13462, 13469, 13470, 13592, 13626, 13645, 13663, 13667, 13669,
    13704, 13725, 13734, 13735, 13782, 13827, 13915, 13916, 13936, 13964, 13994,
    14005, 14014, 14020, 14035, 14043, 14055, 14056, 14116, 14135, 14143, 14166,
    14168, 14180, 14201, 14245, 14255, 14262, 14263, 14282, 14286, 14287, 14288,
    14308, 14347, 14359, 14375, 14391, 14394, 14427, 14460, 14537, 14556, 14557,
    14653, 14667, 14705, 14739

Water status in plant, theoretical background 7312, 7364, 7996, 8148, 8281, 9134,
    9714, 9812, 9832, 9847, 10129, 10175, 10404, 10414, 10492, 10495, 10735,
    11084, 11855, 11953, 12050, 12101, 13454, 13901, 13019

Water stress development see Wilting ...

Water stress in plant see Wilting ...

Water transport in cells (see also Age of plant ...; Carbon dioxide ...; Deu-
    terium oxide, tritium oxide ...; Drought ...; Enzyme inhibitors ...;
    Flooding ...; Growth substances, hormones, inhibitors etc. ...; Humidity
    of air ...; Irradiance ...; Leaf insertion level ...; Lipids, fatty
    acids ...; Mineral elements ...; Osmotically active substances ...;
    Pathogens ...; pH ...; Pollutants, ozone ...; Proteins, amino acids,
    nucleic acids ...; Salinity ...; Temperature ...; Water status in plant
    and water transport in cells) 6708, 6709, 7046, 7132, 7249, 7255, 7284,
    7292, 7314, 7592, 7791, 7831, 8126, 8257, 8258, 8302, 8376, 8425, 8426, 8574,
    8602, 8606, 8612, 8742, 8780, 8794, 8808, 8930, 9030, 9142, 9187, 9258, 9260,
    9286, 9312, 9436, 9526, 9548, 9562, 9676, 9752, 9780, 9827, 10517, 10653,
    10808, 10838, 11049, 11054, 11337, 11706, 11760, 11829, 11855, 12001, 12064,
    12280, 12476, 12540, 12611, 12629, 12648, 12649, 12650, 12692, 12785, 12898,
    13085, 13106, 13165, 13204, 13233, 13234, 13235, 13418, 14036, 14039, 14040,
    14041, 14493, 14604, 14698

Water transport in cells, cell wall structure, modulus of elasticity 6864, 6896, 6899,
    6910, 7234, 7245, 7495, 7506, 7673, 7769, 7864, 7906, 7940, 7996, 8016, 8126,
    8301, 8302, 8376, 8377, 8508, 8621, 8706, 8716, 8749, 9030, 9108, 9123, 9151,
    9258, 9436, 9456, 9457, 9458, 9539, 9562, 9627, 9675, 9827, 9925, 9938, 9951,
    10481, 10487, 10517, 10520, 10650, 10652, 10838, 10947, 10948, 11049, 11080,
                                                                        (continued)

Water transport in cells, cell wall structure, modulus of elasticity (continued)
        11104, 11123, 11190, 11373, 11485, 11790, 11854, 11954, 12054, 12183, 12224,
        12228, 12476, 12492, 12586, 12620, 12629, 12648, 12649, 12651, 12655, 12734,
        12773, 12856, 12879, 13080, 13120, 13190, 13192, 13240, 13271, 13283, 13289,
        13340, 13418, 13441, 13713, 13830, 14036, 14098, 14116, 14255, 14433, 14457,
        14481, 14493, 14604, 14627

Water transport in cells, heterogeneity of single leaf blade 12006

Water transport in cells, membrane structure 7323, 7418, 7531, 7591, 7630, 7673,
        7714, 7769, 7950, 7996, 8123, 8808, 9257, 9455, 9538, 9740, 9752, 10468,
        10838, 10967, 11104, 11323, 11954, 12001, 12256, 12470, 12040, 14749, 14750

Water transport in cells, methods 7592, 9457, 9539, 9676, 11049, 11123, 12648,
        12650, 14040

Water transport in cells, permeability 6831, 6864, 7067, 7174, 7255, 7363, 7439,
        7445, 7592, 7819, 7864, 7906, 7937, 8301, 8302, 8376, 8437, 8577, 8706, 9030,
        9037, 9142, 9184, 9187, 9345, 9356, 9436, 9456, 9457, 9458, 9539, 9562, 9574,
        9675, 9676, 9827, 9959, 10398, 10423, 10491, 10517, 10745, 10808, 10838,
        10929, 10947, 10948, 11047, 11049, 11051, 11123, 11485, 11706, 11854, 11953,
        12001, 12006, 12276, 12319, 12476, 12629, 12648, 12649, 12650, 12651, 12692,
        12770, 12840, 12879, 13008, 13094, 13191, 13204, 13283, 13289, 13315, 13487,
        13569, 13780, 14101, 14215, 14493, 14604, 14626, 14647, 14697

Water transport in cells, plasmolysis 6672, 7057, 7067, 7093, 7245, 7257, 7358, 7591,
        7698, 7699, 7786, 7950, 8124, 8155, 8442, 8502, 8584, 8753, 8832, 8862, 8906,
        8932, 10403, 10405, 10423, 10528, 10588, 10799, 10808, 11140, 11257, 12015,
        12280, 12318, 12583, 12862, 12917, 13302, 13322, 13418, 13564, 13805, 13934,
        14036, 14458, 14482

Water transport in cells, vacuole development 8063, 8110, 13514

Water transport in plant see also Age of plant ...; Cultivars ...; Deuterium
        oxide, tritium oxide ...; Drought ...; Ecotypes, geographical types ...;
        Enzyme inhibitors ...; Flooding ...; Growth substances, hormones, inhi-
        bitors etc. ...; Humidity of air ...; Irradiance ...; Irrigation ...;
        Leaf insertion level ...; Mineral elements ...; Mutagens, other organic
        substances ...; Osmotically active substances ...; Oxygen ...; Pathogens
        ...; Pesticides, herbicides ...; Photoperiod ...; Precipitation, dew ...;
        Proteins, amino acids, nucleic acids ...; Root, underground part ...;
        Saccharides ...; Salinity ...; Soil moisture ...; Taxons ...; Tempera-
        ture ...; Water status in plant and water transport in plant

Water transport in plant, capacities 8743, 9030, 9037, 9166, 9627, 9854, 10007,
        10655, 10731, 11625, 11626, 11855, 11892, 11893, 11933, 11953, 12228, 12807,
        12905, 13155, 13433, 13441, 13692, 13693, 13964, 13986, 14457

Water transport in plant, comparison of plants with different types of carbon metabo-
        lism 12228

Water transport in plant, conductances 6836, 6887, 6891, 7447, 7458, 7734, 7740,
        7742, 8126, 8267, 8302, 8330, 8375, 8621, 8743, 8860, 9030, 9037, 9116, 9166,
        9296, 9326, 9495, 9627, 9636, 9818, 9932, 10007, 10288, 10332, 10655, 10717,
        10731, 10791, 10814, 10867, 11044, 11080, 11155, 11191, 11252, 11271, 11383,
        11472, 11549, 11625, 11626, 11686, 11692, 11855, 11892, 11893, 11933, 11953,
        12228, 12245, 12357, 12461, 12674, 12699, 12767, 12807, 12905, 12913, 12956,
        13063, 13090, 13108, 13119, 13155, 13172, 13229, 13294, 13319, 13334, 13335,
        13420, 13433, 13441, 13639, 13654, 13662, 13692, 13693, 13700, 13718, 13724,
        13739, 13793, 13830, 13885, 13907, 13909, 13984, 14101, 14102, 14197, 14215,
        14304, 14305, 14314, 14343, 14461, 14705

Water transport in plant, diurnal changes 6836, 7475, 7749, 7936, 8098, 8159, 8359,
    8709, 8741, 8851, 8899, 9043, 9636, 9854, 10250, 10823, 11383, 11472, 11892,
    11893, 12044, 12315, 12807, 12887, 12913, 13119, 13140, 13295, 14606, 14646

Water transport in plant, evaporation sites and comparison of liquid and gaseous
    phases 10846, 11485, 11953

Water transport in plant, methods 6813, 8359, 9358, 10007, 10063, 12991, 13802,
    13864, 13865

Water transport in plant, oscillations 11892, 11893

Water transport in plant, radial transport in tree stems 8857, 8743, 9037, 9663,
    9754, 10007, 10424, 10655, 10883, 11387, 11485, 11625, 11855, 11892, 11893,
    11953, 11954, 12276, 12696, 12828, 12887, 13090, 13190, 13420

Water transport in plant, seasonal changes 8857, 9854, 11387, 11892, 12315, 13090,
    13119, 13323, 13327, 13964, 14646

Water transport in plant, theoretical background 8302, 8386, 8401, 8621, 9030, 9288,
    9436, 9541, 9670, 9799, 10517, 11626, 11953, 12383, 12905, 13441, 13786,
    13841

Water transport in plant, transport in leaf 6908, 7458, 7734, 8000, 8098, 8125,
    8621, 9037, 9539, 9695, 9754, 10235, 10491, 10509, 10655, 10731, 10867,
    10883, 11049, 11055, 11080, 11252, 11373, 11625, 11626, 11691, 11655, 11953,
    11954, 11993, 12064, 12228, 12383, 12461, 12696, 12699, 12715, 12734, 12751,
    12753, 12887, 13094, 13420, 14215, 14393, 14414, 14557, 14586

Water transport in plant, transport in other organes than above and below 7208,
    7734, 8098, 8955, 9037, 9117, 9265, 9346, 9995, 10271, 10950, 10978, 11042,
    11240, 11680, 11855, 12080, 12887, 12888, 12992, 13137, 13177, 13864, 13865

Water transport in plant, transport in root 6872, 6891, 7236, 7309, 7310, 7417,
    7447, 7458, 7459, 7600, 7734, 7740, 7837, 7947, 8136, 8267, 8375, 8401, 8514,
    8527, 8855, 8860, 8868, 8889, 9030, 9037, 9116, 9134, 9139, 9153, 9182, 9184,
    9266, 9271, 9272, 9296, 9436, 9495, 9531, 9541, 9636, 9663, 9670, 9695, 9754,
    9799, 9806, 9818, 9932, 9951, 9995, 10229, 10424, 10491, 10655, 10814, 10874,
    10883, 10943, 11059, 11105, 11113, 11252, 11271, 11401, 11516, 11625, 11626,
    11692, 11854, 11855, 11924, 11933, 11953, 11954, 12228, 12291, 12383, 12502,
    12648, 12674, 12696, 12699, 12705, 12729, 12751, 12753, 12770, 12828, 12841,
    12956, 12959, 12991, 13108, 13172, 13187, 13229, 13234, 13318, 13420, 13639,
    13654, 13662, 13724, 13830, 13907, 13909, 13942, 14101, 14124, 14127, 14197,
    14304, 14305, 14461, 14557, 14626

Water transport in plant, transport in xylem, methods 7747, 7749, 8043, 8098, 8331,
    8468, 9491, 9854, 10063, 10706, 10749, 10814, 10823, 12376, 12774, 13233,
    13235, 13294, 13295, 13317, 14606

Water transport in plant, transport in xylem of herbaceous stem 6957, 7310, 7604,
    7749, 8002, 8527, 8793, 9037, 9573, 9576, 9754, 10424, 10491, 10655, 10765,
    10823, 10883, 11031, 11069, 11187, 11191, 11192, 11252, 11625, 11626, 11855,
    11892, 11893, 11954, 12029, 12228, 12314, 12383, 12696, 12699, 12751, 12753,
    12807, 12828, 13155, 13163, 13187, 13420, 13432, 13435, 13601, 13638, 13645,
    13693, 13910, 14215, 14701

Water transport in plant, transport in xylem of woody stem 6836, 6837, 7475, 7740,
    7936, 8098, 8107, 8159, 8330, 8331, 8358, 8359, 8527, 8709, 8741, 8743, 8802,
    8824, 8855, 8857, 8899, 9037, 9100, 9139, 9358, 9491, 9627, 9663, 9695, 9754,
    9854, 10007, 10062, 10424, 10491, 10610, 10611, 10655, 10666, 10883, 11187,
    11387, 11472, 11625, 11855, 11953, 11954, 12296, 12276, 12315, 12383, 12512,
                                                                    (continued)

Water transport in plant, transport in xylem of woody stem (continued) 12753,
        12755, 12774, 12828, 12884, 12905, 13090, 13140, 13172, 13294, 13295, 13323,
        13327, 13334, 13335, 13607, 13645, 13786, 13986, 14557, 14606, 14646

Water transport in plant, transport soil - root 6605, 6819, 7247, 7281, 7310, 7418,
        7600, 7739, 7740, 7947, 8044, 8136, 8292, 8401, 8443, 8651, 9030, 9037, 9134,
        9153, 9266, 9436, 9695, 9754, 9951, 9995, 10010, 10424, 10655, 10717, 10883,
        10944, 11102, 11401, 11549, 11625, 11855, 11924, 11933, 11953, 11954, 12271,
        12696, 12729, 12828, 13129, 13155, 13319, 13550, 13662, 13724, 13830, 13841,
        13913, 14557, 14625

Water transport in plant, vascular bundle structure 6681, 6991, 7123, 7309, 7639,
        7749, 8328, 8375, 8524, 8802, 8855, 9037, 9139, 9265, 9271, 9272, 9473, 9474,
        9576, 9672, 9899, 9939, 10254, 10776, 10814, 11055, 11058, 11147, 11155,
        11186, 11187, 11373, 11387, 11680, 11685, 11686, 11754, 11855, 11933, 11966,
        12164, 12209, 12277, 12308, 12383, 12548, 12696, 12836, 12905, 12906, 12987,
        13090, 13140, 13153, 13259, 13334, 13335, 13420, 14306, 14414, 14659, 14708

Water transport in soil 6610, 6695, 6701, 6819, 6865, 6982, 7211, 7217, 7236, 7243,
        7281, 7424, 7539, 7600, 7658,7664, 7740, 7815, 7981, 7996, 8017, 8044, 8236,
        8237, 8267, 8268, 8269, 8472, 8494, 8651, 8692, 8777, 9030, 9037, 9055, 9134,
        9153, 9166, 9169, 9191, 9264, 9266, 9313, 9321, 9389, 9487, 9722, 9748, 9814,
        9815, 9944, 10110, 10299, 10352, 10491, 10655, 10807, 10883, 10943, 10944,
        11013, 11072, 11102, 11124, 11191, 11411, 11550, 11591, 11658, 11715, 11802,
        11855, 11924, 11948, 11953, 12047, 12075, 12288, 12419, 13460, 12696, 12699,
        12713, 12729, 12762, 12768, 12828, 12843, 12913, 13032, 12071, 13129, 13145,
        13177, 13325, 13386, 13449, 13550, 13594, 13596, 13664, 13691, 13873, 13913,
        13970, 14164, 14182, 14238, 14256, 14290, 14558, 14666, 14690

Waxes see Leaf surface, waxes and trichomes

Wetness of leaf see Leaf wetness

Wilting see also Age of plant ...; Anatomical structure ...; Carbon dioxide ...;
        Cultivars ...; Defoliation, decapitation, ear, root removal ...; Deuterium
        oxide, tritium oxide ...; Drought ...; Ecotypes, geographical types ...;
        Enzyme inhibitors ...; Enzymes ...; Farming practices ...; Growth sub-
        stances, hormones, inhibitors etc. ...; Humidity of air ...; Irradiance
        ...; Irrigation ...; Leaf insertion level ...; Lipids, fatty acids ...;
        Mineral elements ...; Mutagens, other organic substances ...; Osmotically
        active substances ...; Oxygen ...; Pathogens ...; Pesticides, herbicides
        ...; Pollutants, ozone ...; Precipitation, dew ...; Proteins, amino acids,
        nucleic acids ...; Root, underground part ...; Saccharides ...; Salinity
        ...; Soil moisture ...; Taxons ...; Temperature ...; Wind and wilting

Wilting and ion absorption, transport 6617, 7016, 7233, 7665, 7925, 7977, 8068,
        8202, 8343, 8351, 8416, 8489, 8490, 8705, 8755, 8864, 8879, 8968, 9004, 9019,
        9077, 9238, 9330, 9398, 9521, 9633, 9704, 9710, 9807, 9892, 9926, 10162,
        10255, 10306, 10333, 10336, 10574, 11117, 11258, 11301, 11415, 11600, 11621,
        11728, 11746, 11898, 11899, 11940, 12002, 12003, 12066, 12462, 12519, 12563,
        12632, 12777, 12805, 12857, 12943, 13017, 13274, 13403, 13610, 13925, 13944,
        14073, 14105, 14108, 14296, 14441, 14628, 14741, 14747

Wilting and other processes than above and below 6607, 6611, 6634, 6654, 6667, 6672,
        6678, 6710, 6721, 6723, 6724, 6758, 6759, 6761, 6782, 6850, 6856, 6857, 6879,
        6897, 6929, 6941, 6980, 6983, 7013, 7024, 7035, 7050, 7052. 7054, 7108, 7116,
        7175, 7233, 7273, 7290, 7299, 7384, 7414, 7415, 7439, 7440, 7445, 7477, 7483,
        7564, 7569, 7570, 7582, 7625, 7626, 7661, 7676, 7732, 7756, 7798, 7825, 7874,
        7904, 7977, 8003, 8030, 8064, 8068, 8089, 8146, 8155, 8163, 8178, 8179, 8212,
        8243, 8248, 8272, 8345, 8431, 8434, 8447, 8458, 8481, 8489, 8498, 8522, 8556,
        8631, 8634, 8664, 8677, 8705, 8720, 8756, 8761, 8780, 8781, 8782, 8787, 8817,
                                                                              (continued)